Active and Intelligent Food Packaging Polymers

Active and Intelligent Food Packaging Polymers

Editors

Victor G. L. Souza
Lorenzo M. Pastrana
Ana Luisa Fernando

Basel • Beijing • Wuhan • Barcelona • Belgrade • Novi Sad • Cluj • Manchester

Editors

Victor G. L. Souza
Food Processing and
Nutrition
International Iberian
Nanotechnology Laboratory
Braga
Portugal

Lorenzo M. Pastrana
Food Processing and
Nutrition
International Iberian
Nanotechnology Laboratory
Braga
Portugal

Ana Luisa Fernando
MEtRICs, CubicB,
Departamento de Química
NOVA School of Science and
Technology—FCT NOVA,
Universidade Nova de Lisboa
Lisboa
Portugal

Editorial Office
MDPI
St. Alban-Anlage 66
4052 Basel, Switzerland

This is a reprint of articles from the Special Issue published online in the open access journal *Polymers* (ISSN 2073-4360) (available at: www.mdpi.com/journal/polymers/special_issues/Act_Intell_Food_Packag_Polym).

For citation purposes, cite each article independently as indicated on the article page online and as indicated below:

Lastname, A.A.; Lastname, B.B. Article Title. *Journal Name* **Year**, *Volume Number*, Page Range.

ISBN 978-3-0365-9735-5 (Hbk)
ISBN 978-3-0365-9734-8 (PDF)
doi.org/10.3390/books978-3-0365-9734-8

© 2023 by the authors. Articles in this book are Open Access and distributed under the Creative Commons Attribution (CC BY) license. The book as a whole is distributed by MDPI under the terms and conditions of the Creative Commons Attribution-NonCommercial-NoDerivs (CC BY-NC-ND) license.

Contents

About the Editors vii

Preface ix

Ana G. Azevedo, Carolina Barros, Sónia Miranda, Ana Vera Machado, Olga Castro, Bruno Silva, et al.
Active Flexible Films for Food Packaging: A Review
Reprinted from: *Polymers* 2022, 14, 2442, doi:10.3390/polym14122442 1

Angel Jr Basbasan, Bongkot Hararak, Charinee Winotapun, Wanwitoo Wanmolee, Wannee Chinsirikul, Pattarin Leelaphiwat, et al.
Lignin Nanoparticles for Enhancing Physicochemical and Antimicrobial Properties of Polybutylene Succinate/Thymol Composite Film for Active Packaging
Reprinted from: *Polymers* 2023, 15, , doi:10.3390/polym15040989 33

Pedro V. Rodrigues, Dalila M. Vieira, Paola Chaves Martins, Vilásia Guimarães Martins, M. Cidália R. Castro and Ana V. Machado
Evaluation of Active LDPE Films for Packaging of Fresh Orange Juice
Reprinted from: *Polymers* 2023, 15, 50, doi:10.3390/polym15010050 57

Carol López-de-Dicastillo, Gracia López-Carballo, Pedro Vázquez, Florian Schwager, Alejandro Aragón-Gutiérrez, José M. Alonso, et al.
Designing an Oxygen Scavenger Multilayer System Including Volatile Organic Compound (VOC) Adsorbents for Potential Use in Food Packaging
Reprinted from: *Polymers* 2023, 15, 3899, doi:10.3390/polym15193899 71

Alba Maldonado, Paulina Cheuquepan, Sofía Gutiérrez, Nayareth Gallegos, Makarena Donoso, Carolin Hauser, et al.
Study of Ethylene-Removing Materials Based on Eco-Friendly Composites with Nano-TiO$_2$
Reprinted from: *Polymers* 2023, 15, 3369, doi:10.3390/polym15163369 89

Alejandro Aragón-Gutiérrez, Raquel Heras-Mozos, Antonio Montesinos, Miriam Gallur, Daniel López, Rafael Gavara, et al.
Pilot-Scale Processing and Functional Properties of Antifungal EVOH-Based Films Containing Methyl Anthranilate Intended for Food Packaging Applications
Reprinted from: *Polymers* 2022, 14, 3405, doi:10.3390/polym14163405 105

Sonja Jamnicki Hanzer, Rahela Kulčar, Marina Vukoje and Ana Marošević Dolovski
Assessment of Thermochromic Packaging Prints' Resistance to UV Radiation and Various Chemical Agents
Reprinted from: *Polymers* 2023, 15, 1208, doi:10.3390/polym15051208 125

Nur Nabilah Hasanah, Ezzat Mohamad Azman, Ashari Rozzamri, Nur Hanani Zainal Abedin and Mohammad Rashedi Ismail-Fitry
A Systematic Review of Butterfly Pea Flower (*Clitoria ternatea* L.): Extraction and Application as a Food Freshness pH-Indicator for Polymer-Based Intelligent Packaging
Reprinted from: *Polymers* 2023, 15, 2541, doi:10.3390/polym15112541 141

Seongyoung Kwon and Seonghyuk Ko
Colorimetric Freshness Indicator Based on Cellulose Nanocrystal–Silver Nanoparticle Composite for Intelligent Food Packaging
Reprinted from: *Polymers* 2022, 14, 3695, doi:10.3390/polym14173695 161

Tomy J. Gutiérrez, Ignacio E. León, Alejandra G. Ponce and Vera A. Alvarez
Active and pH-Sensitive Nanopackaging Based on Polymeric Anthocyanin/Natural or Organo-Modified Montmorillonite Blends: Characterization and Assessment of Cytotoxicity
Reprinted from: *Polymers* **2022**, *14*, 4881, doi:10.3390/polym14224881 . . . . **173**

João Ricardo Afonso Pires, Victor Gomes Lauriano Souza, Pablo Fuciños, Lorenzo Pastrana and Ana Luísa Fernando
Methodologies to Assess the Biodegradability of Bio-Based Polymers—Current Knowledge and Existing Gaps
Reprinted from: *Polymers* **2022**, *14*, 1359, doi:10.3390/polym14071359 . . . . **191**

About the Editors

Victor G. L. Souza

Dr. Victor Souza is a researcher at the International Iberian Nanotechnology Laboratory in Portugal. He holds a Bachelor's degree in Food Engineering and a PhD in Food Quality. His main interests are active packaging; intelligent packaging; smart packaging; bio-based polymers; nanotechnology; natural additives; biosensors; shelf life extension; food preservation; and food safety.

Lorenzo M. Pastrana

Prof. Dr. Lorenzo Pastrana is currently Chair of the Research Office (Scientific Directorate) and Group Leader of the Food Processing Group at the International Iberian Nanotechnology Laboratory (INL). He joined the INL as Head of the Life Sciences Department in 2015. He is also a Professor of Food Science at the University of Vigo and was a Visiting Professor at Università Cattolica del Sacro Cuore (Italy) and the Universidades Federal Rural de Pernambuco and Federal de Santa Catarina (Brazil). He was the Director of the Centre of Research Transference and Innovation (CITI) and Head of the Knowledge Transfer Office (2009-2010) at the University of Vigo. He founded the Galician Agri-Food Technology Platform (2006). Currently, he takes part in the scientific board of the Portugal Foods innovation cluster and is President of the European Sustainable Nanotechnology Solutions Association. His research is marked-oriented with a multidisciplinary approach integrating methods and concepts of biotechnology and nanotechnology with applications in food micro and nanostructures, encapsulation technologies for food personalization and active and intelligent food packaging.

Ana Luisa Fernando

Prof. Dr. Ana Luisa Fernando is an Associate Professor at NOVA School of Science and Technology, Universidade Nova de Lisboa, Portugal. She is a researcher at MEtRiCS, Mechanical Engineering and Resource Sustainability Center, and her research interests are related to the development of innovative materials for food packaging and innovative technologies for food preservation and to the sustainable production and valorization of industrial crops. She has been working in the field of Food Technology and Safety, testing natural compounds extracted from plants and agro-food wastes into biopolymers for food packaging or as additives for food preservation. Also, she has been working with energy crops for more than 25 years, with special interest in studies related to the sustainability of energy crops production (use of marginal land; efficient use of water and mineral resources; environmental impact assessment studies to detect options for systems improvement).

Preface

In the dynamic landscape of modern food technology, packaging plays a pivotal role that extends beyond conventional containment. While safeguarding industrialized food against external contamination remains crucial, the ever-evolving demands of both the industry and consumers have ushered in a new era of packaging innovation. This paradigm shift is characterized by the emergence of active, intelligent, and smart food packaging materials designed not merely to encase food but to interact intentionally with it.

In contrast to traditional packaging, this advanced class of materials deliberately engages with the packaged food, offering additional features. These groundbreaking packaging materials have the capability to monitor and extend the shelf life of food products.

This Special Issue aimed to compile significant contributions to the field, providing an in-depth overview on the development and application of active and intelligent films for food packaging.

The scope of this Special Issue is extensive, covering various facets of active and intelligent food packaging, among others, such as the development and characterization of these innovative materials, strategies for extending shelf life, the application of nanotechnology in packaging, the creation of novel scavenger systems for food packaging, colorimetric freshness indicators, antimicrobial and antioxidant packaging solutions, the use of bio-based polymers, and a comprehensive review of biodegradation methodologies for bio-based polymers.

As we embark on this exploration of the forefront of food packaging technology, we anticipate that the contributions gathered within this Special Issue will not only reflect the current state of the field but also act as a catalyst for future advancements. Together, we aim to unravel the intricacies of active and intelligent food packaging, paving the way for a safer, more sustainable, and technologically enriched future in the realm of food preservation.

Victor G. L. Souza, Lorenzo M. Pastrana, and Ana Luisa Fernando
Editors

Review

Active Flexible Films for Food Packaging: A Review

Ana G. Azevedo [1], Carolina Barros [2], Sónia Miranda [3], Ana Vera Machado [2], Olga Castro [4], Bruno Silva [3], Margarida Saraiva [5], Ana Sanches Silva [6], Lorenzo Pastrana [1], Olga Sousa Carneiro [2] and Miguel A. Cerqueira [1,*]

[1] International Iberian Nanotechnology Laboratory, Av. Mestre José Veiga s/n, 4715-330 Braga, Portugal; ana.azevedo@inl.int (A.G.A.); lorenzo.pastrana@inl.int (L.P.)
[2] IPC—Institute for Polymers and Composites, University of Minho, Campus de Azurém, 4800-058 Guimarães, Portugal; b8453@dep.uminho.pt (C.B.); avm@dep.uminho.pt (A.V.M.); olgasc@dep.uminho.pt (O.S.C.)
[3] PIEP—Centre for Innovation in Polymer Engineering, University of Minho, Campus de Azurém, 4800-058 Guimarães, Portugal; sonia.miranda@piep.pt (S.M.); bruno.silva@piep.pt (B.S.)
[4] Vizelpas—Flexible Films, S.A., Rua da Fundição, 8, Vilarinho, 4795-791 Santo Tirso, Portugal; olgacastro@vizelpas.pt
[5] INSA—National Institute of Health Doutor Ricardo Jorge, Rua Alexandre Herculano, 321, 4000-055 Porto, Portugal; margarida.saraiva@insa.min-saude.pt
[6] National Institute for Agricultural and Veterinary Research I.P., Portugal and CECA-Center for Study in Animal Science, ICETA, University of Porto, Vairão, 4099-002 Vila do Conde, Portugal; ana.silva@iniav.pt
* Correspondence: miguel.cerqueira@inl.int

Abstract: Active food packaging is a dynamic area where the scientific community and industry have been trying to find new strategies to produce innovative packaging that is economically viable and compatible with conventional production processes. The materials used to develop active packaging can be organized into scavenging and emitting materials, and based on organic and inorganic materials. However, the incorporation of these materials in polymer-based flexible packaging is not always straightforward. The challenges to be faced are mainly related to active agents' sensitivity to high temperatures or difficulties in dispersing them in the high viscosity polymer matrix. This review provides an overview of methodologies and processes used in the production of active packaging, particularly for the production of active flexible films at the industrial level. The direct incorporation of active agents in polymer films is presented, focusing on the processing conditions and their effect on the active agent, and final application of the packaging material. Moreover, the incorporation of active agents by coating technologies and supercritical impregnation are presented. Finally, the use of carriers to help the incorporation of active agents and several methodologies is discussed. This review aims to guide academic and industrial researchers in the development of active flexible packaging, namely in the selection of the materials, methodologies, and process conditions.

Keywords: antimicrobial film; antioxidant film; food packaging; active packaging

1. Introduction

Food packaging is used to enclose food products and presents as main functions to: contain and protect foods from the environment, namely gases, ultraviolet radiation, and chemical and microbiological contamination. Therefore, it has a crucial role in guaranteeing the quality and safety of food products, and their shelf-life. Food packaging can be in the form of a bag, bottle, can, box, wrapped pouch, or other type of containers and can also be used to communicate with the consumer and be used as a utility. The food packaging industry has suffered significant changes over the years, and new materials and technologies have been developed to reach the industry and consumers' demands.

Currently, various approaches, such as modified atmosphere packaging (MAP) and active packaging, prevent or reduce food product damage and spoilage, and extend their shelf-life. Therefore, these approaches allow to save energy, decrease the production costs, protect the sensorial and nutritional quality of food, and protect the consumer's health [1,2]. Moreover, intelligent or smart packaging has been presented as a unique technology to help industry and consumers to monitor the quality and safety of food products.

Active packaging is one of the most recent approaches used in the food packaging area. According to the European Commission (EC) Regulation No 450/2009, active packaging are systems projected "to extend the shelf-life or to maintain or improve the condition of packaged food. They are designed to deliberately incorporate components that would release or absorb substances into or from the packaged food or the environment surrounding the food". Therefore, this type of packaging is developed to increase the shelf-life of foods while maintaining their nutritional quality and ensuring safety. Currently, active packaging continues to be explored and there is an increasing interest in its application in the food area. The active packaging interacts with the packaging environment (i.e., headspace) or directly with the food product. So, the active agent can be a scavenger, which absorbs the residual oxygen of headspace, moisture or water, and ethylene resultant from food maturation. Alternatively, the active agent can be released/emitted over time, in a controlled way, from packaging to headspace, or to the food, inhibiting the development of bacterial microorganisms, for example [1,3]. The active agent can be an individual substance or a combination of substances (EC Regulation No 450/2009). It can be a synthetic material, metal, inorganic material, salt, and enzyme, and is used according to the intended activity [3]. The interest in natural or organic materials, such as plant extracts, biopolymers, and essential oils (EOs) has increased in the past decade. These are used to substitute some synthetic materials, such as antioxidants and antimicrobials, as reported in several studies [4–6]. Another approach is the use of nanoparticles or carriers loaded with active agents. The nanomaterial concept is defined by the European Union (EU) in the Recommendation No 2011/696/EU, 2011. Currently, and according to EU Regulation No 10/2011 and its amendments, the use of some nanostructures is allowed in the manufacture of plastic materials for food contact.

Plastic films are one of the most popular products used for food packaging. The main materials used in the production of these films are synthetic polymers. This is due to their unique properties, such as ease of processing, transparency, flexibility, lightweight, and low cost. The most often used polymers to produce flexible and rigid food packaging are low-density polyethylene (PE-LD), high-density polyethylene (PE-HD), polypropylene (PP), poly(ethylene terephthalate) (PET), poly(vinyl chloride) (PVC), ethylene vinyl alcohol (EVOH), and polystyrene (PS), among others. Among all these materials, PE-HD and PE-LD are extensively used in film packaging [7,8]. These materials present low cost, low water vapor permeability values, high resistance to tear by presenting outstanding elongation at break values, good thermal stability and, at the same time, low heat seal temperature [9,10]. Currently, food packaging based on biopolymers obtained through the synthesis of bio-derived monomers (such as polylactide acid (PLA)), or produced by microorganisms (such as polyhydroxyalkanoates, PHAs), are being produced. However, these solutions are still not common, because they are expensive when compared with the synthetic alternatives [11,12]. There is also the possibility of producing flexible films using the biopolymers extracted from biomass, such as polysaccharides and proteins [13]. These materials have been extensively investigated for application in food packaging and some of them can already be processed in the existing conventional production lines, as, for example, the film extrusion process. However, these materials still present some drawbacks when compared with their synthetic counterparts [14]. In recent years, cellulose-based materials have been another option to substitute plastics in flexible films. This is a consequence of their low price, low weight, extensive availability, printability and good mechanical properties [12,15]. However, and since there are several reviews focused on these materials [16–19], they will not be included in the present one.

Processes such as extrusion, injection molding and thermoforming are used to produce polymer-based food packaging. The extrusion process is the major polymer processing technology in which the polymeric material is melted and shaped into a constant cross-section continuous product. Films, sheets, pipes, and profiles are some examples of extruded products. In addition, this process allows producing structures with a single layer (monolayer), or two or more layers (multilayer) using a co-extrusion process [10]. Another technique used to produce multilayer structures in food packaging is lamination. This technique is generally combined with polymer extrusion/co-extrusion and is used to produce multilayer flexible films using different materials (different types of polymers or different types of materials, such as polymers, paper and aluminum (Al)). The coating process is another technique used to produce multilayer films. Usually, the coating is used to provide the films' aesthetic and physical properties derived from the coating material. The coating is usually applied in-line (e.g., in an extrusion line), or off-line, by spray, rolls, or dipping, to produce a thin layer on films' surface. All these processing technologies have been used to study the addition of active agents to flexible films. It is worth mentioning that there are already some active packaging incorporating active agents in mono and multilayer flexible films produced by conventional industrial processes, as mentioned in active flexible packaging section.

Recently, two new technologies have been reported for the addition of the active agent: the impregnation of the active agents by super critical carbon dioxide (SC-CO_2), and the loading of the active agents in carrier materials by encapsulation technologies, absorption and integration processes. These technologies proved to be efficient alternatives for preventing the volatilization or degradation of the active agent when subjected to high temperatures during extrusion. However, these strategies are still very difficult to use at the industrial scale due to the lack of devoted industrial equipment [4,20].

This review reports the recent studies about technologies used to produce active flexible films with monolayer and multilayers structures for food packaging purposes. Despite the availability of several review articles on active packaging, reporting the active agents and materials used [1,3,21], there are no reviews exploring and discussing the methodologies and processes used in the production of active packaging. This review aims to fill this gap and to discuss the methods and processes that the industry can use to produce active packaging, focusing on flexible films.

2. Active Flexible Packaging

Active packaging emerged in the last years in the food packaging area to prevent food spoilage and extend its shelf-life. Active packaging is produced with an active agent that interacts with the food. It is intended to prolong food shelf-life while preserving its organoleptic properties (appearance, aroma, consistency, texture, and flavor), i.e., to maintain food product quality and integrity, ensuring its safety. Active packaging systems can be divided into two groups: (1) active scavenging systems (or absorber systems) and (2) active releasing systems (or emitter systems). In the active scavenging systems, the active agent removes undesired substances from headspace, such as oxygen, moisture, carbon dioxide, ethylene, and odor, without going out of the packaging material. In the second group, the active agent is slowly released into the headspace to react with undesired substances from food products, such as reactive oxidizing species, or the active agent diffuses or migrates directly into the food.

Active packaging has been developed wherein the active agent can be added to the packaging as an independent device (e.g., pad or sachet), incorporated into the polymeric matrix (e.g., by extrusion), or applied on the film or packaging surface (e.g., by coating). Moreover, the active agent can also be firmly fixed or immobilized on the film surface using, e.g., super critical carbon dioxide technology. Since this review is focused on technologies that allow producing flexible active films with mono- or multilayer structure, the systems where the active agent is enclosed in an independent device are omitted. Figure 1 shows a general scheme of the structures used in the production of active packaging. In Figure 1A,B the active agent is incorporated in the polymeric matrix, and then a monolayer or multilayer film is produced, respectively. The addition of layers allows decreasing the diffusion of the active agent through the film and its subsequent evaporation during storage. Sometimes, the active agent is added in the adhesive layer, used between two incompatible polymer layers, rather than in polymeric layers. In Figure 1C, the active agent is immobilized on the mono- or multilayer film surface. Then, depending on how the active agents work, they will act as a scavenger (Figure 1D) (it does not migrate) or as an emitter, migrating to the food surface or headspace (Figure 1E). The scavenging systems are used mainly to control oxygen, moisture, and ethylene inside the packaging. The releasing systems are used to confer the antioxidant and antimicrobial capacity to the active packaging. Table 1 shows the most used active agents for food packaging, their mechanisms of action and potential benefits in food applications.

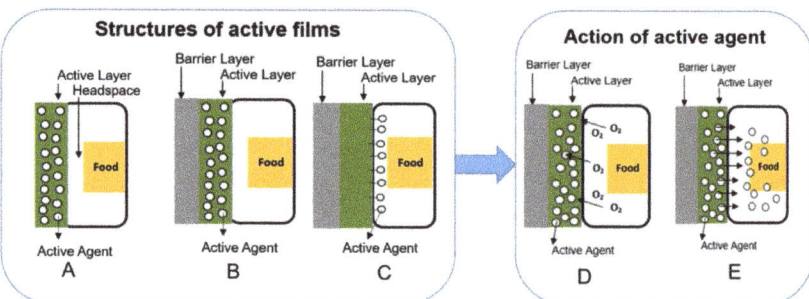

Figure 1. Structure of active film for active food packaging with headspace: (**A**) monolayer film with an active agent, (**B**) two-layer film with an active agent in the inner layer and (**C**) two-layer film with active substance immobilized or fixed on the surface of the film. (**D,E**) Schematic representation of active scavenging and releasing systems, respectively.

Table 1. Overview of active packaging technologies, active agents used, their mechanisms and potential benefits in food applications.

Active Scavenger System (or Absorber)	Classification	Materials	Mechanism	Potential Benefits	References
Oxygen scavenger	Metallic and metallic oxides	Iron, ferrous oxide, cobalt, zinc, copper, magnesium, aluminum, titanium	Oxidation of metals with the supply of moisture and action of an optional catalyst.	Prevention of discoloration; prevention of mold growth; retention of vitamin C content, prevention of browning; prevention of rancidity.	[22–26]
	Inorganic	Sulfite, thiosulfate, dithionite, hydrogen sulfite, titanium dioxide	Oxidation of inorganic substrate by UV light.		
	Organic	Ascorbic acid, tocopherol, gallic acid, hydroquinone, catechol, rongalit, sorbose, lignin, pyrogallol, glucose oxidase, laccase	Oxidation of organic substrate with metallic catalyst or alkaline substance.		
	Polymer-based	Polymer metallic complex	Oxidation of polymer components with metallic catalyst (mostly cobalt).		
Moisture absorber	Inorganic	Silica gel (SiO_2), potassium chloride (KCl), calcium chloride ($CaCl_2$), sodium chloride (NaCl), calcium sulfate ($CaSO_4$)	The common process is adsorption and absorption.	To control the moisture content in headspace of packaging and absorber of liquids.	[27–31]
	Organic	Sorbitol, fructose, cellulose and their derivatives (e.g., carboxymethylcellulose (CMC))			
	Polymer-based	Polyvinyl alcohol (PVOH) and sodium polyacrylate			
	Other synthesized	Synthesized attapulgite with acrylamide			
Ethylene scavenger	Minerals	Clays modified (e.g., MMT, organoclays, halloysite nanotubes (HNTs)) and zeolites, titanium dioxide (TiO_2)	Adsorption process and cation exchange.	Reduction in ripening and senescence of fruits and vegetables.	[32–35]
	Metallic and metallic oxides	Silver (Ag) and zinc oxide (ZnO)	Activated by either UV light, visible light or both.		

Active Releaser system (or emitter)	Classification	Materials	Mechanism	Potential benefits	References
Antioxidants	Organic	Tocopherol, carvacrol, quercetin, catechin, thymol, gallic acid, ascorbic acid, rosemary, green tea, oregano, cinnamon, sage leaf and bay leaf extracts, eugenol, olive leaf, mango leaf	Free radicals and peroxides react to retard or block the actual oxidation reactions.	Prevention of fat oxidation and food deterioration maintenance of nutritional quality, texture and functionality.	[36–38]
	Metallic and metallic oxides, and inorganic	Silver (Ag), copper (Cu), titanium dioxide (TiO_2) and zinc oxide (ZnO)	Catalytic function that reduces the rate oxidation.		
Antimicrobials	Organic	Allyl-isothiocyanate, cinnamaldehyde, carvacrol, thymol, eugenol, oregano, basil leaf, extract of allium, lauric arginate ester, sodium lactate, sorbic acid, citric acid	Metabolic and reproductive processes of microorganisms are blocked or inhibited. Cell wall conformation modification.	Inhibition of spoilage and retardation of pathogenic microorganism's growth.	[8,39,40]
	Polymers	Chitosan and ε-Polylysine			
	Enzymes, bacteriocins and antibiotics	Lysozyme, lactoferrin, nisin, lactocins, pediocin, enterocins			
	Metallic and metallic oxides, and inorganic	Silver (Ag), copper (Cu), titanium dioxide (TiO_2) and zinc oxide (ZnO)			

2.1. Scavenging Systems

The oxygen scavengers (OS) are used to remove any residual oxygen present inside the food packaging or to improve its barrier properties by acting as an active barrier to this gas. They are enclosed in sachets, or incorporated into the packaging materials using, for example, the extrusion or coating processes. In the extrusion process, as presented in this review, different types of materials have been used as oxygen scavenger agents; each one has different mechanisms to react with oxygen. Many efforts have been presented in the literature, targeting to understand the oxygen scavenging processes in polymeric films and to predict their performance [22–26]. These processes are complex and heterogeneous, and normally involve both physical and chemical phenomena. The physical phenomenon is related to the physical dissolution and diffusion of the gas through the polymer; the chemical phenomenon is related with the reaction of the active phase with oxygen. Table 1 provides some examples of materials used and their mechanisms of action. Inorganic and metallic materials are the most used as oxygen scavengers. These compounds are stable in extreme conditions, such as high temperatures and pressures, and some are considered nontoxic. In the past decades, organic compounds extracted from natural resources, such as plants, fruits, and vegetables, have also been used as oxygen scavengers, but most of them are sensitive to extrusion temperatures. In the last years, some companies developed oxygen scavengers for incorporation in packaging materials, such as Avient (product name Amosorb™, Avon Lake, OH, USA), IPL plastics (product name $ZerO_2$, Edmundston, NB, Canada), Sealed Air (product name Cryovac, Charlotte, NC, USA), and Crowne (product name Oxbar, Yardley, PA, USA).

Another type of active agent is the moisture absorber. It is used to absorb the fluids that exudate from fresh products, or to control the relative humidity inside the food packaging. The materials used for this function are commonly placed into packages in the form of sachets or pads, but there are already solutions where these agents are incorporated into the polymeric matrix used to produce trays or films. Taikous (product name Pichitto/Pichit, Gardena, CA, USA), Kyoto Printing (product name MoistCatchTM, Kyoto, Japan), and Aptar CSP Technologies (product name Active FilmTM, Auburn, AL, USA) are some examples of companies with commercial absorbing films. The most used materials in these commercial products are inorganic materials and synthetic polymers, such as silica gel and polyacrylate sodium. However, the food industry requests the development and application of natural materials or biodegradable polymers [27,29,30]. Table 1 provides examples of the natural materials most used as moisture absorbers in polymeric food packaging films. The common process used for moisture absorption in food packages is physical adsorption, but the absorption process can also happen in some cases.

The ethylene scavengers are used to remove the ethylene released during fruit ripening, thereby enhancing the fruit quality and shelf-life. Once ripening is underway, it triggers the production of more ethylene to continue the process of ripening. So, ethylene scavengers slow down the ripening process and senescence. The most used material is potassium permanganate ($KMnO_4$). This is usually included in the food packaging as a sachet, but has also been incorporated into the polymeric matrix [41]. The zeolites are another good candidate to be used as ethylene absorbers, mainly due to their porous 3-dimensional structure with cation exchange, adsorption, and molecular sieving properties. These unique properties opened the possibility of using zeolites in industrial and agricultural applications, in particular as an ethylene-absorbing additive that can be used in packaging materials [33–35]. Other examples of ethylene scavenger materials are active clays, metals, and metallic oxides, such as titanium dioxide (TiO_2), silver (Ag), and zinc oxide (ZnO) [32]. In Table 1, other examples are provided. There are commercial ethylene scavenging packaging materials produced by companies such as Evert Fresh (product name green bags, Katy, TX, USA) and PEAKfresh (product name PEAKfresh bag, Lake Forest, CA USA).

2.2. Releasing Systems

Antioxidant agents are used in food packaging to inhibit or slow down the oxidation reactions that affect food quality. They react with reactive oxidizing species (e.g., peroxides, superoxide, and hydroxyl radical) retarding or blocking the oxidation reactions of food products. The active agents are released from packaging material to headspace, by vaporization, or diffuse or migrate into the food [42,43]. Antioxidant packaging materials can be produced in the form of sachets, pads, and labels, or added directly into the films [44]. It has been shown that synthetic and natural antioxidant agents can be added to packaging systems resulting in active systems; however, for selecting an antioxidant, food characteristics and regulatory and safety issues should be considered. Synthetic antioxidants, such as butylated hydroxytoluene (BHT), butylated hydroxyanisole (BHA) and tert-butylhydroquinone (TBHQ) are already widely used by the food industry, but they are also associated with adverse side effects to human health [45,46]. Currently, the focus is on the incorporation of natural antioxidants extracted from plants and fruits, and essential oils from herbs and spices. Some examples of natural antioxidants compounds, as well as the mechanisms and potential benefits in food applications, are presented in Table 1.

Antimicrobial agents are applied to inhibit the growth of microorganisms that can cause food spoilage. Microorganisms, some of them foodborne pathogens, are the main ones responsible for food spoilage, especially in the case of fresh products (e.g., meat, fruits, and vegetables). Therefore, antimicrobial agents help to extend the shelf-life of a wide range of food products. The mechanism of action depends on the antimicrobial compounds, which can either inhibit the metabolic and reproductive processes of microorganisms, or modify the conformation of the cell wall. They are put directly in the food product, or in the food packaging as sachets and absorbent pads, or in the polymer matrix. Nowadays, antimicrobial packaging is already being used in the market. For example, the Prexelent® (Rajamäki, Finland) company produces antimicrobial plastics. Recently, flexible film packaging with halloysite nanotubes (HNTs) with antimicrobial essential oils was developed in the NanoPack project, carried out in the European Union (EU). This film allows increasing the shelf-life of bread, yellow cheese, and cherries, and maintains food quality and safety standards. In the past, the most used antimicrobial agents for food packaging were synthetic materials, such as ethylene diamine tetra-acetic acid (EDTA), and metallic or metallic oxide, such as Ag, Cu, TiO_2, and ZnO [47–49]. However, current developments in the industrial and scientific areas indicate that natural materials, such as chitosan, lysozyme, and citric acid, are efficient and safe for food contact [39,50,51]. More examples of this type of materials can be found in Table 1.

3. Methodologies for the Production of Active Flexible Packaging

3.1. Direct Incorporation of Active Agents in the Polymer Film Matrix

Film extrusion is the most used process for the production of plastic packaging. It consists of an extrusion line that can include one extruder, or several (in the case of co-extrusion), a die and equipment to stretch/blow, cool down, cut or wind the extrudate. The process starts at the extruder throat where the material, in granular form, is fed to the screw. The rotating screw forces the polymer granules ahead in the extruder barrel, which has several controlled heating zones (Figure 2).

The material experiences progressive heating and pressure until melting. This allows the polymeric granules to slowly melt, reducing the risk of overheating, which can result in polymer degradation. After melting, the polymer passes through the die, which shapes the melt into an initial high thickness geometry (annulus, in the case of blown film). Afterwards, and in the blown film case, this thick annulus is stretched in the two main directions: in the machine or longitudinal direction, by the action of the pulling rolls; in the transversal or circumferential direction, by the effect of compressed air inflated through the die. The cooling of the resulting film bubble is carried out by forced air, blown through the air ring (Figure 3A). When films are directly produced in the flat shape, the extrusion die has a rectangular shape, and the film is only oriented in the machine direction; in this case, the cooling process is promoted by direct contact with the chill rolls, as presented in Figure 3B. The conventional extruders used in these extrusion lines are usually of the single screw type (Figure 2). Another type of extruders, having two screws, usually co-rotating and intermeshed (twin-screw extruders), are used for compounding purposes (e.g., to produce polymer blends, to incorporate additives/active agents, to prepare composites and masterbatches). This type of extruders are, therefore, used to prepare the compounds that are fed to single screw extruders, for the production of the final film.

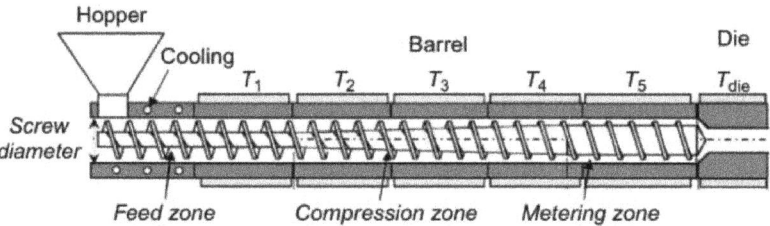

Figure 2. Schematic of a conventional extruder—single screw extruder. Reprinted from Covas, J., & Hilliou, L. (2018). Chapter 5—Production and Processing of Polymer-Based Nanocomposites. In M. Â. P. R. Cerqueira, J. M. Lagaron, L. M. P. Castro, & A. A. M. de O. S. Vicente (Eds.), Nanomaterials for Food Packaging (pp. 111–146) [10]. Copyright (2018), with permission from Elsevier.

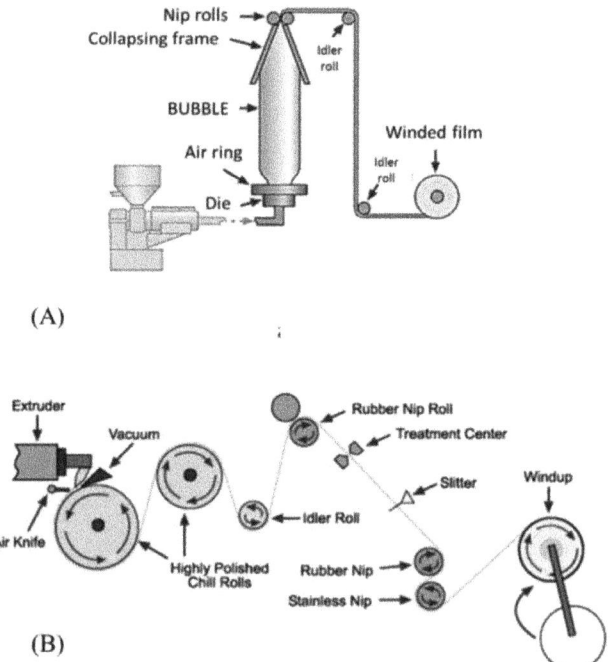

Figure 3. Schematic of the film extrusion/co-extrusion lines: (**A**) blown film extrusion line and (**B**) cast film extrusion line. Reprinted from Covas, J., & Hilliou, L. (2018). Chapter 5—Production and Processing of Polymer-Based Nanocomposites. In M. Â. P. R. Cerqueira, J. M. Lagaron, L. M. P. Castro, & A. A. M. de O. S. Vicente (Eds.), Nanomaterials for Food Packaging (pp. 111–146) [10]. Copyright (2018), with permission from Elsevier.

The extrusion process allows the production of blown or flat films with one or more layers (mono and multilayer films, respectively). In the multilayer case, the process used is called co-extrusion, and involves more than one extruder (at least one for each different polymer system to be included in the structure of the film). The various polymers are extruded through a single die, constituting a single multilayer structure at its exit. The remaining downstream equipment is similar to that of a conventional extrusion line.

Another process used to produce multilayer films for flexible packaging is the lamination process, illustrated in Figure 4. This process combines individual films, which can be polymeric or non-polymeric, into a multilayer structure. Polymeric adhesives (water or solvent-based) are used to bond the different layers. This process can be used in or outside (Figure 4A) an extrusion line. Another possibility is the lamination by extrusion, wherein the extruder provides the molten film that acts as adhesive (Figure 4B). In order to improve the adhesion of the substrates, a treatment on the film surface, such as corona, can be incorporated in the extrusion line.

Multilayer films have been used for increased barrier properties. Due to the multilayer structure, they can reduce the permeation of gases through the film, and thus avoid changing the headspace composition of the package over time. However, according to the type of food, this strategy is not always the best solution to increase food shelf-life (e.g., fresh and oxygen-sensitive foods). When this strategy is not enough, the incorporation of active agents in one of the polymeric layers can be a solution. This section reports works where the direct addition of active agents to the polymer matrix was employed as a solution. Table 2 summarizes the details of some studies, such as materials used, the function of the developed packaging, amount of active agent added, and their main effects. The amount of the active substances migrated are also mentioned. Below, these works will be reported, emphasizing the conditions used in the extrusion processes and the parameters that can influence the activity of the active agent, such as the thickness of the films, dispersion of the active agent, and processing temperatures.

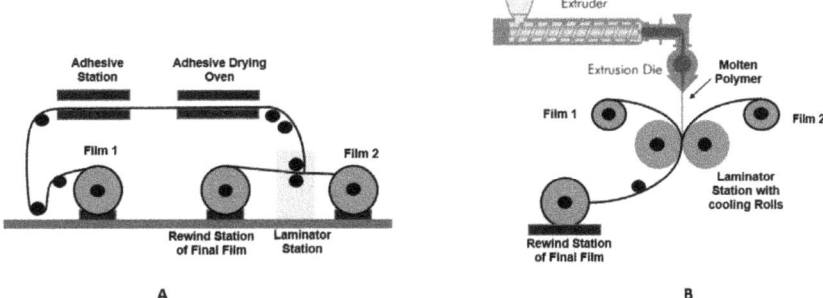

Figure 4. Schematic of the (**A**) lamination process outside the extrusion line and (**B**) lamination extrusion.

For example, Di Maio et al. [23] studied the effect of adding a polymeric oxygen scavenger (OS) (unsaturated hydrocarbon dienes—Amosorb DFC 4020) in a multilayer film using the same polymeric material, namely PET, to apply in fresh fruit. The films were produced by a co-extrusion process using a laboratory cast film extruder with a temperature profile of 285–280 °C. The OS was incorporated in the core layer and the pure PET was kept in the outer layers. Multilayer films with different thicknesses were produced. The authors reported that the active films developed showed a good oxygen scavenging capacity and a longer duration of activity time when the active internal layer presented a higher thickness. On the other hand, the oxygen scavenging rate was consistently lower when the external neat PET layers presented a higher thickness. The results are explained by the diffusion of oxygen through these layers, which needs more time for thicker samples before reacting with the active film. On the other hand, active monolayer films, also considered in this study, were saturated in a few days, which was explained by the fast reaction of the oxygen with the active compounds.

In 2013, Sängerlaub et al. [25] studied a multilayer film with different materials, namely PET, Al, and PE films, to apply in oxygen-sensitive food. The iron was used as an OS and was added to the PE inner layer. The PE active film was produced by extrusion (but the temperatures used were not specified). Afterwards, the lamination process was used to join the PET and aluminum (Al) films using an adhesive. The final structure obtained was PET/Adhesive/Al/Adhesive/PE-containing active agent without and with a sealing layer of PE. They also studied the effect of the OS layer and sealing layer thickness, showing that the thickness of these layers influenced the OS activity. In addition, they studied the effect of the addition of OS on sealing defects, such as small pinholes up to a diameter of 10 and 17 mm, and showed that the OS was able to compensate the sealing defects. Granda-Restrepo, Peralta, Troncoso-Rojas, & Soto-Valdez [52] studied the antioxidant properties of PE-HD/EVOH/PE-LD multilayer films produced by a blown film co-extrusion process (the temperatures used were not specified). Different active agents, such as BHA, BHT, and α-TOC, were added to the inner layer (PE-LD). TiO_2 was added to the outer layer (PE-HD) to prevent light transmission through the films, avoiding the use of an Al layer in the film structure. They packaged whole milk powder to perform migration tests and the PE-LD layer with antioxidant agent was put in direct contact with the product. They showed that the structure of these films avoided the loss of antioxidants to the environment and favored the migration to the product. However, they also showed that during the extrusion process the concentration of BHA, BHT, and α-TOC decreased approximately 17, 41, and 23%, respectively, which was explained by the processing temperatures.

Soysal et al. [8] studied the antimicrobial activity of PE-LD/PA/PE-LD multilayer packaging film using different antimicrobial agents, such as nisin, chitosan, potassium sorbate, or silver substituted zeolite (AgZeo). In addition, the authors selected different polymers to combine different barrier properties, namely a good barrier to water vapor that comes from PE-LD and a good barrier to gases that comes from PA. The multilayer film was produced by a blown film co-extrusion process (temperatures used were not specified). The drumsticks were the product selected for the study and singly vacuum-packaged in active films developed, it means that the active agents acted by direct contact with product. The results showed that the incorporation of the antimicrobial agent was an asset to avoid the microbial growth and increase the shelf life of product. They showed that the nisin and chitosan were among all those that reduced the levels of antimicrobial activity. However, the addition of these antimicrobials increased the cost of food packaging (not more than 2%), but they mentioned that this could be compensated by the benefits of increasing the food shelf-life.

Nowadays, the use of multilayer films is common among food packaging solutions, but there are few publications about active packaging based on this type of films. This is probably related to the lack of laboratory co-extrusion lines or to the difficulty in using the high throughput industrial co-extrusion lines. On the other side, there are a lot of studies about active packaging in monolayer films; some of them are presented below and in Table 2. There are also studies where the extrusion process was only used to produce compounds, being the films produced by compression molding in a hydraulic press. Since these films are not representative of the industrial ones, these studies were not considered in the present review. As with all multilayer active films, most of these works evaluated the development of antimicrobial and antioxidant activity of films.

Beigmohammadi et al. [47] studied the antimicrobial activity of PE-LD film loaded with Ag, Cu, and ZnO. The authors selected metallic nanoparticles to study their effect on microorganisms' growth in cheese, since these NPs have significantly reduced the microbial population in other products. PE-LD/NPs blends were produced using a twin-screw extruder, and different compounds were developed. Afterwards, the compounds were processed in a cast film extrusion line and active monolayer films were produced, using a temperature profile of 185–239 °C. Cheese samples were packaged with these films. However, the type of the packaging used (direct contact or headspace) was not specified. Of all active films developed, the one incorporating CuO was the one showing the lower coliform load of the cheese, and not showing any toxicity. Moreover using metallic nanoparticles, Li et al. [49] studied the antimicrobial capacity of PE-LD with Ag/TiO$_2$ nanopowder against *Aspergillus flavus* and the mildew. First, they prepared a masterbatch incorporating Ag/TiO$_2$ in a PE-LD matrix, using a twin-screw extruder. This masterbatch was later diluted in more PE-LD for the production of a flexible film, by the blown film extrusion process. The temperature profiles used in these extrusion processes were not provided. After, they studied the antimicrobial activity of the films and performed the migration test of Ag$^+$ ion using rice. The type of packaging used (direct contact or headspace) was not specified. The results showed that the small amount of silver migrated from the active films inhibited the *A. flavus* significantly and reduced the mildew of rice during storage. Emamifar et al. [53] also studied the effect loading different particles in a PE-LD film, such as P105 powder (with TiO$_2$ + Ag NPs) and ZnO NPs, on antimicrobial activity in packaging of fresh orange juice (packaging with direct contact with product). They produced the compounds in a twin-screw extruder, after they used a blown film extrusion line to produce a monolayer flexible film using a temperature profile of 60–175 °C. They reported that increasing the ZnO NPs concentration up to 1 wt.% caused the NPs agglomeration during the processing and this decreased the antimicrobial activity of the film. To reduce the tendency for NPs agglomeration, Emamifar & Mohammadizadeh [54] used a compatibilizer, namely polyethylene-grafted with maleic anhydride (PE-g-MA), in the preparation of the blends. They obtained a better dispersion of NPs even increasing their concentration for 3 and 5 wt.% This procedure resulted in a considerable increase in the antimicrobial activity of the film.

When the active agents are sensitive to the temperature and easily released, such as natural extracts or essential oils, some authors have tried some specific strategies. For example, Zhu, Lee, & Yam [55] incorporated α-TOC (3000 mg/kg) into the PE-LD/PP blends using a single-screw extruder, at 221 °C, and produced the PE-LD/PP blends monolayer films with antioxidant proprieties. They reported that 90% of α-TOC incorporated into the films was retained after the extrusion process. Concerning the release of α-TOC, the results showed that the higher the PP ratio in the blend the slower was the α-TOC release. The authors explained that this happened likely due to the more tightly packed structure and higher crystallinity of PP when compared to LD-PE. Graciano-Verdugo et al. [56] added 20 and 40 mg/g of α-TOC into pure PE-LD. First, they pre-mixed manually the component at room temperature, and then the blown film was produced at 165 °C, using a pilot size single-screw extruder. Even without using high temperature during mixing, it was not possible to avoid the losses of 5 and 25% α-TOC in films with 20 and 40 mg/g, respectively, after the extrusion process. However, and in both cases, the antioxidant capacity of the film was still observed in corn oil, where the active packaging acted by direct contact with the food product.

Biodegradable polymers have also been used to develop active film packaging, since many of them can already be used in conventional polymer processing technologies. For example, Llana-Ruiz-Cabello et al. [40] developed an active film with PLA containing Proallium as an active agent to produce films with antioxidant and antimicrobial properties. They made bags with the developed films and stored the iceberg salad within a modified atmosphere in some studies. Different concentrations of Proallium were incorporated into the PLA matrix and active films were obtained by extrusion using a twin-screw extruder at temperatures ranging between 200 and 205 °C. The Proallium was introduced into the extruder through a lateral barrel port where the polymer matrix was already molten to reduce its possible volatilization and degradation. They reported that Proallium alone lost around 80% of weight at temperatures up to 150 °C, but when Proallium was added to PLA, no films were formed. The results showed a great antimicrobial activity with the highest concentration of Proallium, such as 6.5 wt.%, and did not show antioxidant activity. Concerning the optical properties of the film produced, it was observed that the Proallium reduced its transparency, but no significant visual differences were observed. The authors did not mention if the films have the characteristic odor of Proallium. Manzanarez-López,

Soto-Valdez, Auras, & Peralta [57] used also PLA as polymeric matrix and added 3% w/w of α-TOC to produce a film with antioxidant properties. After the production of the compounding with a twin-screw extruder, the film was produced by blown extrusion process (pilot plant size extruder) using the same temperature profile of 165–170 °C. The concentration of α-TOC decreased to 2.58 wt.% after compounding, but after the film production the authors did not observe any loss. This happened because the film was immediately cooled after blowing, while the filament of the compounding was cooled at room temperature during 10–15 min. The PLA film produced with α-TOC showed a yellowish appearance. This difference was not perceptible to the naked eye in the single film, but perceptible in the film rolls. The authors did not explain the origin of the yellow color, but it probably originated from the high concentration of α-TOC used. Cestari et al. [58] developed an active biodegradable film with the addition of oregano essential oil (OEO) and potassium sorbate into TPS and PBAT (commercial name Ecoflex®). These mixtures were made using a twin-screw extruder with five heating zones (with a temperature profile of 90 and 120 °C). Films were produced by blown film extrusion using a temperature profile of 115–120 °C. Then the antimicrobial and antioxidant effects were studied in frozen chicken steaks stored with the film developed, but the authors did not specify the type of packaging (direct contact or headspace). They reported that the films reduced the risk of pathogen contamination, delayed the oxidation process of chicken meat, and extended its shelf-life.

Studies on ethylene scavenger and moisture absorber systems, with direct incorporation into polymer matrix and applied in monolayer packages, are scarce in the literature. For example, Tas et al. [32] studied the ethylene scavenging capacity of halloysite nanotubes HNTs-loaded PE-LD films. The incorporation of different concentrations of HNTs into PE-LD was performed using a twin-screw extruder. The film was produced in a blown film extrusion line with a temperature profile of 165–185 °C. To study the effect of these films, some products, such as bananas and tomatoes, were selected and tested, but the type of packaging (direct contact or headspace) was not specified. The authors observed that HNTs had an effect on the slowdown of the ripening process of bananas and on the retention of the firmness of tomatoes. Another example was presented by Sängerlaub et al. [59] that developed an active film with moisture absorber properties. They blended NaCl crystals with PP polymer using a twin-screw extruder with a temperature profile of 180–250 °C. The blend was used as masterbatch where the concentration of NaCl was 60% in weight. Afterwards, the masterbatch was blended (diluted) with neat PP and monolayer films were produced using a single screw extruder with a temperature profile of 180–230 °C. The results showed that the NaCl crystals incorporated in the film were able to avoid water vapor condensation in areas of reduced temperature. Moreover, the films showed an absorption capacity of water vapor around 80%.

The incorporation of active compounds through the extrusion processes can bring several advantages, namely in the production of films at the industrial typical high extrusion rates. However, extrusion is not adequate for some active compounds, such as the ones based on natural compounds, since it uses relatively high temperatures during the process. These high temperatures may lead to the degradation of the active agents, resulting in a loss of activity and change of color. Therefore, different alternative strategies have been explored to incorporate the active compounds. These will be presented in the following sections.

Table 2. Active agent incorporated directly in the polymer matrix.

Active Agent (AA)	Material/Matrix	Packaging Function	Processes Used	Food Product Tested/Packaging Type	Active Agent Amount	Main Effects Compared Control Film	Amount of AA Migrated *	References
Amosorb DFC 4020	PET/PET—containing AA/PET	Oxygen scavenger	Cast film co-extrusion (Temperature profile: 285–280 °C)	Fresh apple slices	10 g/100 g polymer	The multilayer films with higher thickness in internal active layer reduced the browning of fresh apple slices packaged after 15 days storage at 8 °C. This packaging also allowed preserving the initial values of the acidity and sugar content of apples.	nd	[23]
Iron	PET/Adhesive/Al/Adhesive/PE—containing AA/PE	Oxygen scavenger	Film extrusion and lamination (temperatures not specified)	Salami in a baked bread roll	-	The food samples stored 30 days at 23 °C with active film and with sealing defects of 10 mm, showed that the presence of OS was advantageous in the permanence of color of product, when compared to the packaging without OS.	nd	[25]
α-TOC and synthetic materials (BHA and BHT)	PE-HD—containing TiO$_2$/EVOH/PE-LD—containing antioxidant	Antioxidant activity	Blown film co-extrusion (temperatures not specified)	Whole milk powder/direct contact	4 g of α-TOC, 4 g of α-TOC mix with 1.5 g of BHA, 1.5 g of BHT and 1.5 g of BHA (all by 100 g polymer)	The multilayer film with α-TOC in contact with whole the milk powder showed a more gradual release of α-TOC during the 30 days storage (26.8% at 30 days). In addition, this film contributed to protect vitamin A degradation presents in whole milk powder.	α-TOC–63 ± 2 µg/g α-TOC mix with BHA–64 ± 0.6 µg/g (Product stored during 30 days at 30 °C) Regulation (EU) allows a maximum of 60 mg/kg of α-TOC	[52]
Nis, Chit, PSorbate or AgZeo	PE-LD/PA/PE-LD -containing AA	Antimicrobial activity	Blown film extrusion (temperatures not specified)	Chicken drumsticks/direct contact	2 g/100 g polymer	The results indicated that the use of active bags with nisin and chitosan reduced the levels of total aerobic mesophilic bacteria (APC) and total coliform in chicken drumsticks storage during 6 days at 5 °C.	nd	[8]
NPs Ag, CuO and ZnO	PE-LD film	Antimicrobial activity	Film extrusion (Temperature profile: 180–239 °C)	Cheese/ns	1 g metal nanoparticles/100 g polymer	All active films with metal NPs showed a decline of the number of coliform bacteria of 4.21 log cfu/g after 4 weeks of storage at 4 ± 0.5 °C. The effect of each individual NPs on decreasing coliform load had the following order: CuO > ZnO > Ag.	CuO–0.23 ± 0.005 mg/kg (it was used the simulant B at 40 °C for 10 days) EFSA1 legislation allows a maximum of 10 mg of Cu/kg of food	[47]
Ag/TiO$_2$ NPs	PE-LD film	Antimicrobial activity	Blown film extrusion (temperatures not specified)	Rice/ns	9 g/100 g polymer	Reduction from 7.15 to 5.48 log CFU/g in rice stored with active packaging after one month.	Ag$^+$–0.0035 mg/kg (product stored 35 days at 37 °C and relative humidity of 70%) EFSA1 legislation allows a maximum of silver migration of 0.05 mg of Ag$^+$/kg of food.	[49]
P105 powder (TiO$_2$ + Ag NPs) and ZnO NPs	PE-LD film	Antimicrobial activity	Film extrusion (Temperature profile: 60–160 °C)	Fresh orange juice/direct contact	1.5 and 5 g of P105 powder (TiO$_2$ + Ag NPs) and 0.25 and 1 g of ZnO NPs (all by 100 g polymer)	Nanocomposite film containing nano-Ag showed higher antimicrobial activity than films with nano-ZnO when they are used to pack orange juice.	5 g of P105 (Ag)–0.15 ± 0.002 µg/L 0.25 g ZnO–0.68 ± 0.002 µg/L 1 g ZnO–0.54 ± 0.005 µg/L (product stored at 40 °C for 112 days) EFSA1 legislation allows a maximum of 10 ppm of Ag Regulation (EU) allows a maximum of 25 mg of Zn/kg of food	[53]

Table 2. Cont.

Active Agent (AA)	Material/Matrix	Packaging Function	Processes Used	Food Product Tested/Packaging Type	Active Agent Amount	Main Effects Compared Control Film	Amount of AA Migrated *	References
α-TOC	PE-LD/PP blend film	Antioxidant activity	Film extrusion (Temperature profile: 221 °C)	-	3000 mg/kg	The PE-LD/PP blend films with higher PP ratio showed a longer induction period of oxidation against linoleic acid oxidation (6 days) due to the low releasing of TOC in LDE/PP blend films, allowing an antioxidant effect for more time.	nd	[55]
α-TOC	PE-LD film	Antioxidant activity	Film extrusion (Temperature profile: 165 °C)	Corn oil/direct contact	20 and 40 mg/g	Increase of shelf life of corn oil from 12 to 16 weeks stored at 30 °C.	nd	[56]
Proallium	PLA film	Antioxidant and antimicrobial activity	Film extrusion (Temperature profile: 200–205 °C)	Salad/ns	2, 5 and 6.5 g/100 g polymer	The films developed showed no significant antioxidant activity; however, they showed effectiveness during the storage time (7 days) against all microorganisms studied, except for aerobic bacteria.	nd	[40]
α-TOC	PLA film	Antioxidant activity	Blown film extrusion (Temperature profile: 165–170 °C)	-	3 g/100 g polymer	Diffusion of α-TOC to fractioned coconut oil was slower than to ethanol with 5.1–12.9% of release. Diffusion of α-TOC to soybean oil was able to decrease the induction of the oxidation at 20 and 30 °C, but not at 40 °C.	nd	[57]
PSorbate or/and OEO	TPS/PBAT-Ecoflex® blend film	Antioxidant and antibacterial activity	Blown film extrusion (Temperature profile: 90–120 °C)	Chicken steaks frozen/ns	0.5 and 1 g/100 g polymer	Active film showed a reduction of 50% in TBARS values and a delay in microbial development when using the film with OEO and PS.	nd	[58]
HNTs	PE-LD film	Ethylene scavenger	Blown film extrusion (Temperature profile: 165–185 °C)	Bananas and tomatoes/ns	1, 3 and 5 g/100 g polymer	The results showed that the presence of 5% w/w HNTs improved the ethylene adsorption capacity of PE films by 20%. Active films slowed down the ripening process of bananas during 8 days and tomatoes only decreased their firmness 16% after 10 days of storage.	nd	[32]
NaCl crystals	PP film	Moisture absorber	Cast film extrusion (Temperature profile: 180–250 °C)	-	0.03 g or 0.06 g per 1 g of film	The PP film developed with NaCl crystals showed an absorption capacity of water vapor around 0.8 g water/g film at 97% relative humidity.	nd	[31]

Legend: **nd**—not determined. **ns**—not specified if the active agent reacts by direct contact or headspace. * Amount of AA migrated from packaging to food product tested or food simulants.

3.2. Incorporation of the Active Agents by Coating

A coating is a thin layer formed from a single or multiple layers spread over the surface of a substrate that gives the substrate aesthetic and physical properties derived from the coating material. The most common types of coating used in the food packaging industry are varnishes applied on the outer surface of the food packaging, used to impart a clear and glossy surface. The most conventional techniques to apply coatings are spray, gap coating, slot die coating, roll coating, and gravure coating. Most of these techniques were used/are used as conventional printing techniques, but they can be adapted for other aims, such as applying active coatings on a pre-formed film.

The conventional production process of an active coating is based on the dissolution or dispersion of an active compound in a solvent or matrix that is then applied on the surface of a substrate, and dried by evaporation or crosslinking. The crosslinking can involve curing by oxidation, temperature treatment, and ultraviolet light. Before applying the coatings, it is usually needed to evaluate the application in terms of uniformity, stability, retention of the compound selected, and application cost. In addition, when the method of application is being selected, the size and shape of the substrate surface and the substrate nature must be taken into consideration. Sometimes a treatment stage is necessary (e.g., plasma, corona and ultraviolet (UV) treatments) in order to modify the substrate's surface in order to improve the adhesion of the coating to the surface. These techniques are commonly used in gravure printing or label printing, to increase the adhesion of inks, varnishes, and adhesives to plastic food packaging [60].

This review reports studies on active coatings applied on the surface of flexible films, and also applied between film layers together with an adhesive. Concerning the techniques used to apply the active coatings, the most used are plate coater (manual or automatic), spraying, or simply spreading the coating over the film's surface by brushing (Figure 5). These techniques are essentially laboratory, but they can be easily up-scaled for industrial applications [61,62]. Table 3 presents recent works focused on the use of coatings as a strategy to incorporate active agents and their most important findings. Below, these works will be reported with an emphasis on the conditions of coatings/adhesives preparation, application processes, and main results regarding the activity of active agents.

Bolumar, LaPeña, Skibsted, & Orlien [37] tested PE-LD films coated with rosemary extract. The coating was prepared using a solution of commercial rosemary extract containing 4.5% (w/v) of carnosic acid in ethanol. They applied the coating using a brush and reached a final concentration of rosemary extract of 0.45 mg/cm^2. The ability to counteract lipid oxidation was studied in pork patties, after 60 days of storage at 5 °C, packed in vacuum, packed (direct contact packaging) with the produced films. The rosemary extract-based active packaging showed effectiveness against the lipid oxidation when compared with the oxygen scavenging system.

Barbosa-Pereira et al. [63] studied the antioxidant activity of a coating produced with a natural extract obtained from a brewery residual waste. Different coating formulations were produced using different concentrations (3, 10, and 20% w/v) of natural extract added to a polyvinylic resin. Then the coating was applied on the PE-LD films' surface, by plate coater, using a rod of 40 μm. The final weight was 3.2 g/m^2. The authors studied the antioxidant activity of these coatings and compared them with a commercial rosemary extract and two synthetic antioxidants (the butylated hydroxytoluene (BHT) and propyl gallate). The effect of the natural extracts was evaluated using a headspace packaging enclosing a beef sample. The results showed that active films coated with natural extracts had an inhibitory effect on lipid oxidation and, therefore, it was concluded that they might be used to replace synthetic antioxidants.

Gaikwad, Singh, & Lee [64] studied the oxygen scavenging capacity of PE-LD films loaded with pyrogallol (PG) (a natural phenolic compound). The coating was prepared with different concentrations of PG (5, 10, and 20%, w/v) added to an ethyl acetate solution and different amounts of polyurethane. Then, the coating was applied to the PE-LD film using a plate stripe coater with a thickness of about 60—62 μm. The oxidative stability of soybean oil packed with the new films, stored during 30 days at 5, 23, and 60 °C and 95 ± 2% RH, was studied. The soybean oil showed better stabilization when packaged with PE-LD coated with 10 and 20% of PG and stored at 23 and 60 °C.

Guo, Jin, & Yang [65] developed several formulations of chitosan-based coatings (2 and 5% w/v) and tested the effect of different organic acids (acetic acid, citric acid, lactic acid, and levulinic acid, or their mixtures) in combination with antimicrobial agents, such as lauric arginate ester (LAE), sodium lactate (NaL), and sorbic acid (SA). Then, the coatings were applied on a PLA film using a brush or spray, obtaining different weights, such as 0.39 and 1.94 mg/cm^2 of chitosan, 1.94 and 3.89 μg/cm^2 of LAE, 0.78, 1.56, 3.8, and 7.78 mg/cm^2 of NaL, and 0.12 and 0.23 mg/cm^2 of SA. The antimicrobial efficacy in a microbial culture and ready-to-eat meat vacuum-packaged (direct contact packaging) was studied. In general, the results showed that the PLA films coated with chitosan containing multiple organic acids and other antimicrobials had an antimicrobial effect against *Listeria innocua*, *L. monocytogenes*, and *Salmonella Typhimurium*, showing a significant inhibition of microbial growth during 48 h at 22 °C. However, the active films developed showed to be more effective against the microorganisms in microbial culture than in RTE meat.

As mentioned above, the surface treatments of films can be used to improve the bonding of the coating to the substrate surface. For example, Al-Naamani, Dutta, & Dobretsov [66] developed a PE-LD film coated with 2% (w/v) of chitosan solution and 0.1 and 2% (w/v) of ZnO/chitosan solution. Before coating, they used plasma treatment to provide a hydrophilic PE-LD film surface. Then, they applied the chitosan solution and the chitosan/ZnO nanocomposite solution, by spray, on the PE-LD surface, and dried it at room temperature. They evaluated the shelf-life of packed okra samples during 12 days at room temperature (25 °C), but did not specify the type of packaging (contact direct or headspace). They observed that the coating developed with chitosan/ZnO reduced the fungal and bacterial growth more than the coating prepared only with chitosan. Moreover, a significant reduction in bacterial growth was observed in the samples stored in treated PE-LD film compared to the control one. Joerger, Sabesan, Visioli, Urian, & Joerger [67] applied a 2% (w/v) chitosan coating, and 5 and 4% (w/v) Ag/chitosan coating. The film, based on ethylene copolymer (EVA), was corona treated to create a reactive surface. Then the coating was spread onto the film surface using a plate coater with 28 wire wound rod. The antimicrobial activity, without and with the incorporation of Ag particles, using beef and chicken meat exudates (type of packaging not specified) was analyzed. They reported that the activity of the chitosan coating was more effective when silver was incorporated. However, the authors did not mention the effect of corona treatment.

Artibal company (Sabiñánigo, Spain) developed an active coating, described in the European Patent EP 1 477 519 A1 [68], that consists of adding the active agent to a varnish that is then applied to the films by rollers, tampography, serigraphy, or spraying systems. Using this technology, Lorenzo, Batlle, & Gómez [61] compared the effect of two natural antioxidants, such as OEO and green tea extract. These were applied at 1.5 and 2.0 g/m^2 to a PET/PE/EVOH/PE multilayer film used to package foal steaks. Camo, Lorés, Djenane, Beltrán, & Roncalés [62] also used an oregano coating, produced by Artibal company, but applied on a PP film. The aim was to evaluate the effect of the oregano extract concentrations (0.5, 1, 2, and 4%, w/v) on the quality characteristics of stored fresh beef. In both studies, the type of packaging used was not specified. In the first study they reported that the OEO was more effective to increase the shelf-life of fresh products than the green tea extract. In the second case, they demonstrated that it is possible to increase the shelf-life of fresh products from 14 to 23 days using 2% w/v concentration of oregano extract.

Lee, Park, Yoon, Na, & Han [69] developed a multilayer active film, with a five layers structure, containing PP, active coating, PET, active coating, and PE-LD. The star anise essential oil (SAEO) and thymol oil (TH) were used as an insect repellent and antimicrobial agent, respectively. Different coating formulations with 25% (v/v) of SAEO and TH were prepared using ethyl alcohol, polyurethane dispersed in distilled water, and a silicone surfactant. Those active coatings were applied on different sides of the PET film using an automatic control coater, then PP and PE-LD were laminated on the SAEO and TH coating layers, respectively. The insect repellent and antimicrobial activities were evaluated using slices of bread as a food model, but the authors did not specify the type of packaging (direct contact or headspace). The film developed demonstrated efficiency against the insects and impeded the growth of microorganisms in packed sliced bread.

In 2016, Carrizo, Taborda, Nerín, & Bosetti [42] studied the performance of the flexible films (commercial bags) produced with two layers of oriented polypropylene (OPP). A special adhesive incorporating green tea extract (GTE) (system under patent protection) was used to build the multilayer. This antioxidant multilayer packaging was produced at the industrial level using a lamination process. Then, the antioxidant activity of the films and their capacity for extending the shelf-life was tested for 16 months with real food, with headspace packaging, under atmospheric pressure, in the presence of oxygen. It was demonstrated that the packaging developed protected the food against the oxidation process, significantly reducing the rancidity and, in this way, extending the shelf-life of packaged food. The migration results showed that any of the compounds of GTE, such as catechins, diffused through the polymer. This means that the free radicals released in the oxidation process diffused through the polymeric layer and arrived at the catechins location, where they were trapped and consumed. The sensorial analysis demonstrated that the packaged food was not affected by GTE. Oudjedi, Manso, Nerin, Hassissen, & Zaidi [43] also studied the antioxidant activity of two active agents (from extracts of bay and sage leaves) added to an adhesive applied between the PE-LD/PET multilayer films. First, the extracts were dissolved into isopropanol (concentration not specified) and then incorporated at 10% (w/w) into the adhesive. After, the adhesive was spread on the PET sheet using a coating machine (KK coater, RK print), resulting in a final concentration of the extracts ranging between 0.025 and 0.03 g/m^2. Then the PE-LD film was laminated to the PET coated film using a lamination process. The authors evaluated the capacity of the active packaging with headspace to prevent lipid oxidation of fried potatoes, incubated at 40 °C for 20 days. They demonstrated that active components of extracts of bay and sage leaves also acted without migration from the material packaging to headspace, improving the shelf-life of fried potatoes. They also observed that the color of the packaging was not affected by the incorporation of the extracts.

Azlin-Hasim, Cruz-Romero, Cummins, Kerry, & Morris [70] used an innovative process to apply the active coatings. They treated PE-LD films with UV/ozone process and coated them using a layer-by-layer (LbL) technique. During the process, poly(ethyleneimine) (PEI) and poly(acrylic acid) (PAA) solutions loaded with Ag NPs were alternately deposited, producing an antimicrobial film. Another innovative solution already studied is atomic layer deposition (ALD). For example, Vähä-Nissi et al. [71] developed an active film with antimicrobial properties using ZnO and aluminum oxide (Al_2O_3) that were deposited at low temperature with ALD on biaxially oriented polymer films, namely BO-PLA and BO-PP.

Table 3. Active agents added directly to coatings.

Active Agent	Packaging Material	Packaging Function	Processes Used	Food Product Tested/Packaging Type	Active Agent Amount	Main Effects Compared Control Film	References
Rosemary extract	PE-LD films	Antioxidant activity	Brushing	Pork patties/direct contact	0.45 mg/cm²	The results demonstrated that PE-LD film coated with rosemary was the most effective active packaging to protect pork patties storage during 60 days at 5 °C.	[37]
Natural extract obtained from a brewery residual	PE-LD films	Antioxidant activity	Plate coater	Beef/headspace	3.2 g/m²	The results showed that active antioxidant films coated with natural extracts decreased lipid oxidation by up to 90% during 17 days stored at 4 °C.	[63]
Pyrogallol (PG) (a natural phenolic compound)	PE-LD films modified with sodium carbonate	Oxygen scavengers	Plate stripe coater	Soybean oil/headspace	Thickness 60–62 µm	The soybean oil samples packed with PE-LD/PG films coated with 10 and 20% PG and storage at 23 °C and 60 °C showed a better stabilizing effect during 30 days than oil packaged with pure PE-LD.	[64]
Chitosan, lauric arginate ester (LAE), sodium lactate (NaL), and sorbic acid (SA)	PLA films	Antimicrobial activity	Brushing or spraying	Ready-to-eat meat (RTE)/direct contact	0.39 and 1.92 mg/cm² of Chitosan; 1.94 and 3.89 µg/cm² of LAE; 0.78, 1.56, 3.89 and 7.78 mg/cm² of NaL and 0.12 and 0.23 mg/cm² of SA	The results showed that PLA films containing LAE were those that most significantly inhibited the growth of the tested microorganisms. The PLA films coated with NaL and SA showed to reduce significantly the growth of L. innocua but were less effective against Salmonella.	[65]
Chitosan and ZnO/chitosan	PE-LD films	Antimicrobial activity	Spraying (and plasma treatment)	Okra/ns	Not specified	The results showed that total bacterial concentrations in films coated with chitosan/ZnO coatings were reduced by 63%.	[72]
Chitosan and Ag/chitosan	Ethylene copolymer (EVA) film	Antimicrobial activity	Plate coater (and corona treatment)	Beef and chicken meat exudates/ns	Not specified	The film coated with chitosan reduced colony counts of E. coli 25922 and of L. monocytogenes Scott A by 5 and 2–3 log10, respectively, after 24 h exposure. However, this activity was increased when silver ions were incorporated into the films that, for example, originated the complete killing of E. coli O157:H7 DD3795.	[67]
Oregano essential oil and green tea extract	Multilayer film: PET/PE/EVOH/PE	Antioxidant activity	Rollers, tampograph, serigraphy or spraying systems	Foal steaks/ns	1.5 and 2.0 g/m²	The active films with essential oregano oil were significantly more efficient than those with green tea extract in case of extended fresh odor and color from 7 to 14 days, compared to the control.	[10]
Oregano extract	PP film	Antioxidant activity	Rollers, tampograph, serigraphy or spraying systems	Fresh beef//ns	Not specified	The results showed to be efficient in extending the fresh odor and color from 14 to 23 days. However, the addition of oregano should be around 1% due to the unacceptable oregano odor when the concentration is higher.	[62]
Star anise essential oil (SAEO) and thymol (TH)	Multilayer film PP/SAEO/PET/TH/PE-LD	Insect repellent and antimicrobial activity	Automatic control coater	Slices bread/ns	Thickness of active coating was 13.20 ± 1.72 µm	The developed film showed a strong and sustained insect repellent activity, lower microbial counts and better visual appearance of bread after 14 days of storage.	[69]
Green tea extract	Multilayer film: OPP/OPP	Antioxidant activity	Lamination	Dark chocolate peanuts and milk chocolatecereals/headspace	Not specified	The results demonstrated that it is possible to increase the shelf life of these products from 9 to 18 months without active agent migration from packaging.	[42]
Sage leaf (SL) and Bay leaf (BL) extracts	Multilayer film: PET/PE-LD	Antioxidant activity	Coating machine (KK coater, RK print)	Fried potatoes/headspace	0.025 and 0.03 g/m²	The results showed a strong antioxidant activity of SL and BL, either evaluated alone or as food packaging for fried potatoes. For example, in case of the malondialdehyde (MDA) the SL extract was more efficient, showing a reduction of 40% of MDA compared to the control, while BL showed a reduction of 31%.	[43]

Legend: nd—not determined. ns—not specified if the active agent reacts by direct contact or headspace.

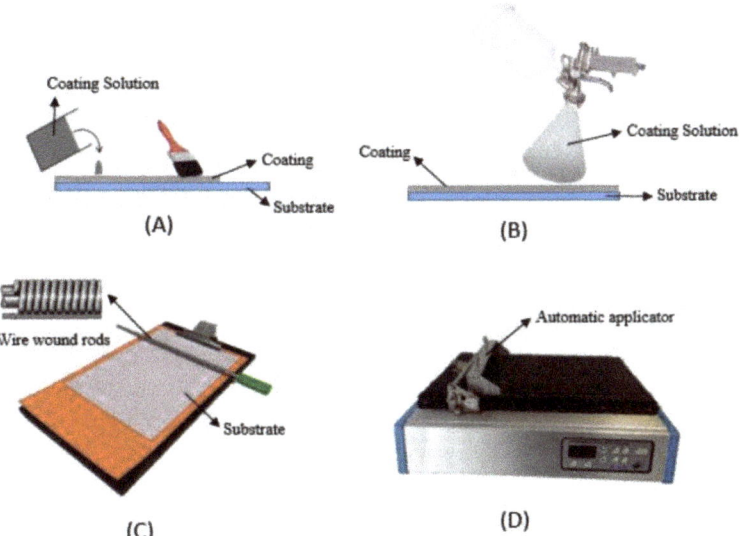

Figure 5. Techniques used to apply the actives coatings: (**A**) brushing, (**B**) spraying, (**C**) manual coater, and (**D**) automatic coater (or plate stripe coater).

3.3. Incorporation of Active Agents by Supercritical Impregnation

The supercritical impregnation process is a recent technique used to impregnate active agents in packaging materials. The supercritical fluids (SCFs) have been used as solvent, anti-solvent or plasticizer, in different applications, such as extraction, dyeing, cleaning, fractionation, polymerization, polymer processing, encapsulation, among others [73,74]. Supercritical solvent impregnation has been recently used to develop active packaging, as an alternative to conventional techniques. This process allows applying active agents at near ambient temperatures avoiding the degradation of the substances that are easily destroyed or deactivated by heat. Moreover, the supercritical fluid can work as a carrier of the active agent and change the matrix properties (e.g., swell), thus ensuring the impregnation success. Carbon dioxide (CO_2) is one of the most used SCFs, since it offers several advantages, such as low toxicity, low cost, non-flammability, environmental sustainability, and it is chemically inert under many conditions [74,75].

The impregnation process using CO_2 fluid involves three main steps (Figure 6). Briefly, first the active agent is dissolved in supercritical fluid CO_2 (dissolution step). Then, the dissolved active agent is transported to the surface of the material and subsequently penetrates and diffuses into the swollen polymer matrix (impregnation step). Finally, the CO_2 molecules are removed by the shrinking polymer (depressurization step), and the impregnated molecules are trapped inside the polymer matrix. The impregnation process depends on the pressure, temperature, depressurization rate and time, variables that should be optimized [74,75].

The impregnation process can be carried out in the static, dynamic, and semi-dynamic modes. In the static mode, the active agent and film are placed in a reactor and are physically separated (e.g., metal or paper filter) avoiding the direct contact between them. Then, the dissolution of the active agent and the sorption of the mixture (active agent and SC-CO_2) are carried out simultaneously. For the dynamic process, the dissolution of the active agent is done in the same impregnation reactor or in a previous dissolution reactor. In this case the sorption of the active agent in the film starts when the saturated mixture enters the impregnation cell. In the semi-dynamic mode, the static and dynamic modes work by time intervals. The active agent is dissolved in the CO_2-phase in the static period, and in the following dynamic mode period the dissolved agent diffuses into the polymer matrix [75].

In the past few years, some works have shown the effectiveness of the supercritical impregnation of natural active substances, such as extracts and essential oils, in food packaging. In most of the cases, the authors studied the effect of different operation conditions, namely pressure, contact time, temperature and depressurization rate, on impregnation yield and distribution of the active substance in flexible films. In addition, they studied possible changes in the polymer thermal and mechanical properties that could result from the process, such as the high pressure and the presence of active agents. In this review, we will report the effects of the operation conditions on antimicrobial and antioxidant agents. Table 4 summarizes some of the works performed using this methodology, the materials used, and the main effects on the activity of the films.

Below, these works will be reported, mainly emphasizing the production conditions. For example, Goñi, Gañán, Strumia, & Martini [76] produced an antioxidant active film, impregnating eugenol into a PE-LLD film, and studied the effects of different operation conditions, namely pressure and depressurization rate, on impregnation yield; also the antioxidant activity of eugenol was evaluated. They reported that the best penetration of eugenol in the film was observed when high pressure and slow depressurization (pressure released) were used. However, the results of antioxidant activity of impregnated films showed a strong activity, regardless of the eugenol loading. In the case of Cejudo Bastante, Casas Cardoso, Fernández-Ponce, Mantell Serrano, & Martínez de la Ossa [77] a multilayer structure with PET/PP with antioxidant and antimicrobial agent, namely olive leaf extract (OLE) was developed. Several parameters (pressure, temperature, depressurization rate, time and presence of a modifier (ethanol) to increase the solubility of the compound in CO_2) were studied. Then, the optimized parameters were used to impregnate the active agent OLE. Then, they applied these active films to cherry tomatoes (not specified the type of packaging applied) and evaluated them for 50 days of storage. They reported that the shelf-life of cherry tomatoes increased in the case of the impregnated films. The antioxidant and antimicrobial properties of the film were slightly higher when the OLE/polymer mass ratio of 1 and percentage of solvent of 7% ethanol was used.

Belizón, Fernández-Ponce, Casas, Mantell, & Martínez De La Ossa-Fernández [78] studied the effect of incorporation of polyphenols extracted from mango leaf in a similar multilayer film of PET/PP, and then studied the effect of their antioxidant activity in perishable food. They used the methyl gallate to determine the best conditions, namely the pressure, temperature, impregnation time, and stirring mode of the impregnation process; afterwards they used the optimized conditions for applying the mango leaf extract. Then, lettuce and tangerine were stored with the films produced (not specified the type of packaging applied) and their shelf-life was evaluated. The results showed that the films developed increased the shelf-life of the tested foods, showing that prevention of microbial contamination and organoleptic deterioration is possible.

Franco et al. [20] studied the effect of impregnation of α-TOC in single layer PP and PET, and in PP/PET multilayer films. To optimize the impregnation of α-TOC, a corona discharge treatment on PET film surface was performed. Results showed that PP films were the best option to produce a controlled-release packaging, with high values of loaded α-TOC, when compared with PET surface with corona discharge treatment. They reported that this happened due to the low penetration of TOC into the treated PET surface due to a low affinity with the additional polar groups formed during the treatment.

There are some authors that studied the effect of nanocomposite films on the impregnation process and release of an active substance. For example, Rojas et al. [79] studied PE-LD nanocomposites films with 2.5 and 5% (w/w) concentrations of an organo-modified montmorillonite (OM-MMT C20A) impregnated with thymol, using supercritical impregnation process. The aim was to study the effect of OM-MMT C20A content on the impregnation and the release of thymol, targeting its application in antimicrobial packaging. This work showed that the presence of OM-MMT C20A improved thymol incorporation in the PE-LD film, from 0.45 to 1.19% (w/w), depending on the impregnation conditions, and decreased its diffusion coefficient during the release tests, allowing its sustained release over time.

Some authors also studied this impregnation process using biodegradable polymers. For example, Villegas et al. [80] impregnated cinnamaldehyde (Ci) into PLA biodegradable films. They reported that Ci was successfully incorporated into PLA using SC-CO_2, at higher pressure and slower depressurization rate. Results showed that PLA films impregnated with Ci had better thermal, structural, and mechanical properties than neat PLA films. In addition, the films showed strong antibacterial activity against the tested microorganisms.

From the existing examples, it can be said that supercritical impregnation is promising from an industrial point of view, and a good strategy to produce active film for food applications.

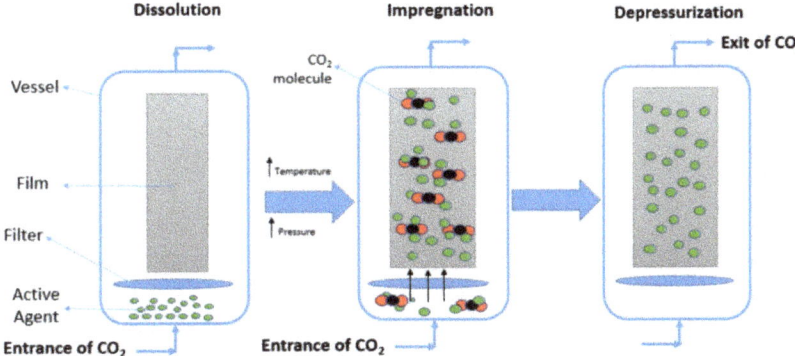

Figure 6. Schematic of the use of supercritical fluid carbon dioxide (CO_2) in the impregnation process during active film production.

Table 4. Active agent sorption on the surface of the flexible films using the SC-CO_2 impregnation process.

Active Agent	Material/Matrix	Packaging Function	Tested Product/Packaging Type	Active Agent Amount	Main effects Compared Control Film	References
Eugenol	PE-LLD film	Antioxidant activity	-	0.5 and 6 wt.%	The results of antioxidant activity showed approximately 80% inhibition after 96 h, regardless the eugenol loading.	[76]
Olive leaf extract (OLE)	PET/PP multilayer film	Antioxidant and antimicrobial activity	Cherry tomatoes/ns	2–5.5 mg/g film	The results showed that the tomatoes packed with the impregnated film did not show any physical change for the first 30 days and their appearance remained the same as at the initial moment of the experiment.	[77]
Polyphenols extracted of mango leaf extract	PET/PP multilayer film	Antioxidant activity	Lettuce and tangerine/ns	36–40 mg of total polyphenols (TP)/100 g film	The results showed that the films increase the shelf-life of lettuces for 14 days and tangerines until 39 days by preventing microbial infections and organoleptic deterioration.	[78]
α-TOC	PET film, PP film and PET/PP multilayer film	Antioxidant activity	-	2.66–3.20 mg/cm^2 film	The results showed that it is possible to produce a multilayer film with controlled releasing of α-TOC until 14 h.	[20]
Thymol	PE-LD film/Cloisite 20A nanocomposite film	Antimicrobial activity	-	0.3–1 wt.%	They reported that the presence of nanoclays makes the release of thymol from PE-LD film difficult allowing a sustained release over time of the active compound.	[81]
Cinnamaldehyde essential oil (Ci)	PLA film	Antibacterial activity	-	72–162 mg/g film	The films impregnated with 13% of Ci showed strong antibacterial activity against E. coli and S. aureus, where no microorganism was detected.	[80]

Legend: ns—not specified if the active agent reacts by direct contact or headspace.

4. Processes to Incorporate the Active Agents into Carriers

In general, carriers of active compounds are materials used to transport and protect the active compounds. In the past few years, carriers loaded with active compounds have been considered an attractive method in different areas such as the pharmaceutic, cosmetic, and food industries. However, despite the wide applicability and advantages, the production processes have found little space in the food industry, mainly in food packaging area, due to its high production cost. While the pharmaceutical and cosmetic sectors often support the use of high-cost methodologies, the food industry works with lower profit margins, and therefore, production costs should be maintained low [74]. As a consequence, the use of carriers loaded with active compounds and their incorporation in food packaging has not been widely studied. Table 5 summarizes some of the works found in the literature, namely the main characteristics of the systems used, materials, amount of active agent added, and main effects on the packaging developed. Below, these works will be reported, emphasizing the production process of carriers loaded with active agents and the main outcomes.

Regarding the processes to incorporate the active agents into carriers, most of the work has been focused on the absorption in porous media materials and integration through ultrathin fibers. More rarely, the encapsulation of active compounds in hollow bodies of polymeric materials has also been studied.

In the case of the absorption in porous media materials, essential minerals, such as aluminosilicates and mesoporous silica, have been used. For example, Wrona et al. [82] used crystalline microporous aluminosilicates and incorporated green tea extract using an absorption, adsorption, or ion exchange process, as mentioned in the EP1564242B1 patent. The microporous existing on the shell facilitate the incorporation and protection of the active agents. The active materials were then incorporated into polyethylene (PE) films (not specified the type of PE), and the antioxidant activity in minced pork meat vacuum packaged (direct contact packaging) was studied. The active PE film was developed by NUREL company (Zaragoza, Spain) using an extrusion process. The results demonstrated that inorganic carriers protected the green tea extract from the adverse effects of the high temperature used in the extrusion process. Moreover, it was reported that the referred active packaging extended the shelf-life of meat and preserved its red color. The results showed that the evaluated shelf-life parameters of packed foods were better in samples packaged with active films than the ones packed with the control film. In the case of Gargiulo et al. [83] the impregnation method to load mesoporous silica with α-TOC was used. Briefly, they dissolved α-TOC in ethanol and added mesoporous silica to the solution. Then they dried at 35 °C for 16 h under reduced pressure to obtain a solid powder. Then, mesoporous silica loaded with α-TOC was added into PE-LD matrix by using an internal mixer. The pellets obtained were then fed into a co-rotating laboratory twin-screw extruder equipped with a sheet die, to produce the active antioxidant films. The results showed a slower antioxidant release of the tocopherol loaded into a silica substrate when compared to the samples containing free tocopherol. However, it also was demonstrated that the release of α-TOC depends on the size of the silica mesoporous.

Sun, Lu, Qiu, & Tang [84] reported a similar study using PE-LD films with α-TOC impregnated into silica mesoporous. They reported similar effects of α-TOC impregnated onto silica mesoporous on the release behavior of the antioxidant. Melendez-Rodriguez et al. [85] developed mesoporous silica nanoparticles loaded with eugenol by vapor absorption process. This process consisted in mixing 100 mg of eugenol with 100 mg of the mesoporous silica in a tightly closed vial. The mixture was incubated in an oven at 40 °C for 24 h while being continuously shaken. Thereafter, the loaded nanoparticles were incorporated into poly(3-hydroxybutyrate-co-3-hydroxyvalerate) (PHBV) by electrospinning, and electrospun composite fibers were produced. These fibers were subjected to annealing in a 4122-model press at 155 °C for 5 s, without pressure, and PHBV active films were produced. The antimicrobial performance against foodborne bacteria was evaluated, whereas it showed to inhibit bacterial growth successfully against *S. aureus* and *E. coli*.

Alkan, Sehit, Tas, Unal, & Cebeci [86] report the study of halloysite nanotubes (HNTs) loaded with carvacrol producing an aqueous solution and applying ultrasonication process. Then, carvacrol loaded halloysite was mixed with chitosan solution (positively charged) and poly (sodium-4-styrene sulfonate) (PSS) solution (negatively charged), and applied on PE film (not specified the type of PE) surface treated with plasma, using the spray LbL deposition technique. Then, antimicrobial activity in chicken meat samples packaged with the coated PE films was evaluated (type of packaging not specified). The results demonstrated that the microbiological quality of the samples packaged with PE films coated with HNTs loaded with carvacrol were significantly improved, when compared to samples packaged with neat PE films. This improvement in the microbiological quality means an increase in shelf-life for the food product.

Another option that has been used to incorporate active compounds are cyclodextrins (CDs). They are cyclic oligosaccharides capable of forming inclusion complexes with many organic compounds, including essential oils and volatiles, or their components. They can lead to changes in the solubility and reactivity of the added molecules. For example, Oliveira et al. [87] produced β-CS loaded with sorbic acid (SA) by molecular inclusion, which were added in the formulation of starch and poly (butylene adipate co-terephthalate) (PBAT). Then, the active films were produced using an extrusion process. They reported that CDs protected SA during film preparation, reducing its sublimation at high temperatures, and provided a controlled release of SA. However, the color of the film presented changes, and the concentration of SA was not efficient in controlling microorganisms.

Another strategy used is the incorporation, in packaging materials, of active compounds into fibers that are electrospun. The fibers are essentially produced by electrospinning with diameters that can range from nanometers to microns. To produce these fibers, a wide range of synthetic polymers, biopolymers, and biodegradable polymers are used. For example, Quiles-Carrillo et al., 2019, [38] studied PLA films coated with PLA electrospun structures loaded with gallic acid, used as an antioxidant agent. The PLA films were produced by cast film extrusion and the fibers loaded with gallic acid were deposited on surface of PLA film by the electrospinning process. They produced a PLA film coated with fibers (referred to as bilayer film) and also another film with fibers. Then they sandwiched both films with another PLA film on the other side (called multilayer). The antioxidant activity of gallic acid from PLA films was evaluated by the 2,2-diphenyl-1-picrylhydrazyl radical (DPPH) methodology. The results showed that during the first days, the bilayer PLA films showed more antioxidant activity than the multilayer films. However, the PLA films containing the electrospun multilayer achieved a stronger DPPH inhibition in a longer time.

Another example was presented by Figueroa-Lopez et al. [88] that produced an electrospun antimicrobial layer made of PHBV that was deposited on a previously produced PHA-based film. They combined OEO and zinc oxide nanoparticles (ZnO NPs) as antimicrobial agents in concentration of 2.5 and 2.25 wt.%, respectively, and produced the PHBV nanofibers. Then, they produced different active multilayer films, and the antimicrobial and antioxidant activity were evaluated. The multilayer films showed a significant inhibition of the antimicrobial activity (R) (R $\geq$ 1 and <3) against *S. aureus* and *E. coli* after 15 days. Concerning the antioxidant activity after 15 days, the multilayer films showed a reduction of DPPH inhibition due to the reduced release of the active compounds.

Chiralt, Tampau, & Gonz [89] developed electrospun fiber mats of PCL loaded with carvacrol. Afterwards, these fibers were deposited in between starch films in order to obtain multilayer films with a starch film/PCL loaded/starch film structure. The starch films were produced by melt blending of starch-glycerol-water at 160 °C and 8 rpm, for 30 min., in a two-roll mill. The multilayer structure was obtained using a thermocompression process. The barrier properties and antimicrobial activity were analyzed. The results showed that the film has an antimicrobial effect against *Escherichia coli* and a better barrier to water vapor.

In 2020, Figueroa-Lopez et al. [90] loaded OEO into α- and γ-CDs and incorporated them during electrospinning into poly(3-hydroxybutyrate-co-3-hydroxyvalerate) (PHBV) fibers. Afterwards, the loaded PHBV electrospun mats were subjected to annealing, and transparent films were produced. The antioxidant and antimicrobial activity were evaluated. The authors reported that antioxidant and antimicrobial activity for γ-CD inclusion complex was higher than for α-CD, because the OEO has higher solubility in the γ-CD and has bigger pore size.

The production of small capsules or nanoparticles for the incorporation of active agents is also possible by using different methodologies and materials. For example, Glaser et al. [91] used chitosan to produce nanocapsules loaded with resveratrol, by the ionic gelation technique. This technique allows the production of nano- and microparticles by ionic interactions between the two materials of different charges. Then, the solution with nanoparticles loaded with the active agent was applied to PE (not specified the type of PE) and PP surface (untreated and oxygen plasma treated), in two layers, by the LbL deposition technique. The effect of plasma treatment on films' properties, and on antimicrobial and antioxidant activity, was evaluated. They observed that the O_2 plasma treatment on the surface of PP and PE films before the coating improved the interaction between the two films and the coating with nanoparticles of chitosan. The results of active films functionalized with plasma treatment showed better properties and superior antimicrobial and anti-oxidative activity.

In 2011, Guarda, Rubilar, Miltz, & Jose [92] studied the antimicrobial activity of a BO-PP film coated with microcapsules produced by an emulsion of oil in water (O/W) containing thymol and carvacrol. For the production of microcapsules an aqueous solution of Arabic gum and a mixture of four different concentrations of the active agents with soybean oil (used as oil phase) was used. Then, the emulsion was homogenized at ambient temperature with high shear equipment. Before applying the coating on the BO-PP, the film surface was treated by corona, but the authors do not refer to the coating process used. The results demonstrated that antimicrobial activity was not altered by the microencapsulation, but the release rate of the active agents was lower and controlled when compared to the case where the active agents were incorporated directly into the matrix.

Table 5. Applications of carriers of the active agents added into flexible films.

Active Agent	Materials of Carrier	Material/Matrix	Packaging Function	Tested Product/Packaging Type	Loading of AA	Active Agent Amount	Main Effects Compared Control Film	References
Green tea	Crystalline microporous aluminosilicates	PE film	Antioxidant activity	Fresh minced meat/direct contact	6.4–12.8 mg/g of carrier material	20 and 40 wt.%	The results showed that active packaging developed extended the shelf life of fresh minced meat for 3 days when compared to a control sample.	[82]
α-TOC	Mesoporous silica	PE-LD film	Oxygen scavengers	-	α-TOC/silica weight ratio was 0.42–0.73	3 wt.%	The results demonstrated a slower release of α-TOC into a silica substrate (decrease about 60%) when compared to films samples with free tocopherol.	[83]
α-TOC	Mesoporous silica	PE-LD film	Antioxidant activity	-	Not specified	1 wt.%	The results exhibited radical scavenging activity of the active film, which increased from 28.45% to 46.50% during 24 h of DPPH test.	[84]
Eugenol	Mesoporous silica (MCM-41)	PHBV films	Antimicrobial activity	-	500 mg/g of MCM-41	2.5, 5, 7.5 and 10 wt.%	The electrospun PHBV films incorporated mesoporous silica nanoparticles with eugenol showed antimicrobial activity after 15 days.	[85]
Carvacrol	Halloysite nanotubes	PE film	Antimicrobial activity	Chicken meat/ns	-	15 wt.%	The results showed that the samples packaged with films developed with HNTs loaded with carvacrol decreased 85% in the viability of cells, demonstrating a strong bactericidal effect against *A. hydrophila*.	[86]
Sorbic acid (SA)	beta-cyclodextrin (β-CD)	PBAT film	Antimicrobial activity	-	100 mg SA/1 g β-CD	1 wt.%	The results showed that active films developed were not efficient in the control microorganisms, due to the low concentration (1% w/w) of active agent used in the film formulation.	[87]
Gallic acid	PLA fibers	PLA film	Antioxidant activity	-	40% based on the PLA weight	Not specified	The results showed that the PLA films containing the electrospun GA-loaded interlayer have a sustained release of the active agent for 10 weeks.	[38]
Oregano essential oil (OEO) and ZnO NPs	PHBV fibers	PHA film	Antimicrobial activity	-	Not specified	2.5 wt.% of OEO and 2.25 wt.% of ZnO NPs	The multilayer films developed showed a high antimicrobial and antioxidant activities in both open and closed systems for up to 15 days.	[88]
Carvacrol	PCL fibers	Starch film	Antimicrobial activity	-	12 g/100 g fibers	15 wt.%	The active film developed showed the antimicrobial effect against *E. coli*, but was not effective at controlling the growth of *Listeria innocua*.	[89]
Oregano essential oil (OEO)	alfa- and gamma-cyclodextrin (α- and γ-CD)	PHBV	Antioxidant and antimicrobial activity	-	Weight ratios of α-CD:OEO and γ-CD:OEO were 80:20 wt/wt and 85:15 wt/wt, respectively	10, 15, 20, 25, and 30 wt.%	The activity of films was evaluated during storage and it was observed that they are stable up to 15 days, which was explained by the protection offered by the developed system.	[90]
Resveratrol	Chitosan	PE and PP film	Antimicrobial and antioxidant activity	-	Not specified	2 wt.%	The active films showed over 90% reduction of *S. aureus* and over 77% reduction of *E. coli* as compared to untreated samples and increase antioxidant activity for over a factor of 10.	[91]
Carvacrol and thymol	Oil-in-water emulsion	BO-PP film	Antimicrobial activity	-	Not specified	1, 2, 5 and 10 wt.%	The results demonstrated that thymol and carvacrol microencapsulated and added on surface film were able to act for fresh food preservation against microorganisms, such as *E. coli* O157:H7, *S. aureus*, *L. innocua*, *Saccharomyces cerevisiae*, and *Aspergillus niger*.	[92]

Legend: **ns**—not specified if the active agent reacts by direct contact or headspace.

5. Conclusions and Future Trends

Active food packaging has been seen as a way to extend shelf-life and to maintain the sensory properties, quality, and safety of packaged food. In some cases, this type of packaging can also help to decrease the addition of preservative agents into foods. Among the active agents used to add this capability to the packaging, metallic particles already showed their capacity, while natural compounds still face several technological changes (e.g., sensitivity to temperature). Among the industrial production processes, incorporating active agents by extrusion is a challenge since the homogenous distribution of the active components in the matrix and the high temperatures used can limit the active agents' performance. To overcome some of these issues, the scientific community and the industry have developed other strategies such as incorporating the active agents by coating technologies, which in most cases are commonly used by the industry. Nevertheless, some new methodologies, such as LbL technique and ultrasonic atomization, are expected to rise in the coming years as a solution for the development of active packaging. There are also more innovative strategies such as the supercritical fluid carbon dioxide or the use of carrier systems. For the last case, it is expected that micro and nanoencapsulation could bring some advantages. In fact, loading of active compounds in some of those systems can help to stabilize the compounds during the extrusion process, or the controlled release of the compounds when they are in contact with food.

It is also important to check the current legislation and the cost of all these new and innovative materials and technologies to assess if they can be easily implemented. For example, some of the solutions presented in this review are still at the research and development level, and their commercialization requires, in some cases, regulatory approval.

Author Contributions: Conceptualization, A.G.A. and M.A.C.; writing—original draft preparation, A.G.A.; writing—review and editing, A.G.A., O.S.C. and M.A.C., writing—review, A.G.A., C.B., S.M., A.V.M., O.C., B.S., M.S., A.S.S., L.P., O.S.C. and M.A.C.; funding acquisition, A.V.M., B.S., A.S.S., L.P., O.S.C. and M.A.C. All authors have read and agreed to the published version of the manuscript.

Funding: Conducted under the project "MobFood-Mobilizing scientific and technological knowledge in response to the challenges of the agri-food market" (POCI-01-0247-FEDER-024524), by "MobFood" Consortium, and financed by European Regional Development Fund (ERDF), through the Incentive System to Research and Technological development, within the Portugal2020 Competitiveness and Internationalization Operational Program. IPC researchers acknowledge also funding by National Funds through FCT-Portuguese Foundation for Science and Technology, References UIDB/05256/2020 and UIDP/05256/2020.

Institutional Review Board Statement: Not applicable.

Informed Consent Statement: Not applicable.

Data Availability Statement: Not applicable.

Conflicts of Interest: The authors declare no conflict of interest.

Abbreviations

ALD	Atomic layer deposition
Al_2O_3	Aluminum oxide
Ag	Silver
Ag NPs	Silver nanoparticles
AgZeo	Silver substituted zeolite
Al	Aluminum
BHT	Butylated hydroxytoluene
BHA	Butylated hydroxyanisole
BioPE	Biobased linear low-density polyethylene
BO-PLA	Biaxially oriented polylactic acid

BO-PP	Biaxially oriented polypropylene
CA	Citric acid
CaO	Calcium oxide
Chit	Chitosan
CuO	Copper oxide
DPPH	2,2-diphenyl-1-picrylhydrazyl radical
EC	European Commission
EU	European Union
EVA	Ethylene vinyl acetate copolymer
EVOH	Ethylene vinyl alcohol
PE-HD	High-density polyethylene
HNTs	Halloysite nanotubes
$KMnO_4$	Potassium permanganate
LA	Lactic acid
LbL	Layer-by-layer
PE-LD	Low-density polyethylene
LevA	Levulinic acid
PE-LLD	Linear low-density polyethylene
MAP	Modified atmosphere packaging
NPs	Nanoparticles
Nis	Nisin
OEO	Oregano essential oil
OPP	Bioriented polypropylene
OS	Oxygen scavenger
PA	Polyamide
PAA	Poly(acrylic acid)
PBAT	Poly (butylene adipate-coterephthalate)
PCL	Poly-(ε-caprolactone)
PE	Polyethylene
PEI	Polyethyleneimine
PEO	Poly(ethylene oxide)
PE-g-MA	Polyethylene-grafted maleic anhydride
PP	Polypropylene
PET	Polyethylene terephthalate
PHAs	Polyhydroxyalkanoates
PHBV	Poly(3-hydroxybutyrate-co-3-hydroxyvalerate)
PLA	Poly(lactic acid)
PS	Potassium sorbate
PSS	Poly (sodium-4-styrene sulfonate)
PVA	Poly(vinyl alcohol)
PVdC	Poly(vinylidene chloride)
R	Value of the antimicrobial activity
RH	Relative humidity
RTE	Ready-to-eat meat
$SC-CO_2$	Supercritical impregnation process by carbon dioxide
SFCs	Supercritical fluids
TBARS	Thiobarbituric acid reactive substance
TBHQ	Tert-butylhydroquinone
TPP	Sodium tripolyphosphate
TPS	Thermoplastic starch
TiO_2	Titanium dioxide
UV	Ultraviolet
ZnO	Zinc oxide
α-TOC	Alfa-tocopherol

References

1. Han, J.W.; Ruiz-Garcia, L.; Qian, J.P.; Yang, X.T. Food Packaging: A Comprehensive Review and Future Trends. *Compr. Rev. Food Sci. Food Saf.* **2018**, *17*, 860–877. [CrossRef] [PubMed]
2. Prasad, P.; Kochhar, A. Active Packaging in Food Industry: A Review. *IOSR J. Environ. Sci. Toxicol. Food Technol.* **2014**, *8*, 1–7. [CrossRef]
3. Yildirim, S.; Röcker, B.; Pettersen, M.K.; Nilsen-Nygaard, J.; Ayhan, Z.; Rutkaite, R.; Radusin, T.; Suminska, P.; Marcos, B.; Coma, V. Active Packaging Applications for Food. *Compr. Rev. Food Sci. Food Saf.* **2018**, *17*, 165–199. [CrossRef] [PubMed]
4. Ataei, S.; Azari, P.; Hassan, A.; Pingguan-Murphy, B.; Yahya, R.; Muhamad, F. Essential Oils-Loaded Electrospun Biopolymers: A Future Perspective for Active Food Packaging. *Adv. Polym. Technol.* **2020**, *2020*, 9040535. [CrossRef]
5. Barbosa-Pereira, L.; Angulo, I.; Lagarón, J.M.; Paseiro-Losada, P.; Cruz, J.M. Development of new active packaging films containing bioactive nanocomposites. *Innov. Food Sci. Emerg. Technol.* **2014**, *26*, 310–318. [CrossRef]
6. Pereira De Abreu, D.A.; Paseiro Losada, P.; Maroto, J.; Cruz, J.M. Natural antioxidant active packaging film and its effect on lipid damage in frozen blue shark (*Prionace glauca*). *Innov. Food Sci. Emerg. Technol.* **2011**, *12*, 50–55. [CrossRef]
7. Singh, P.; Wani, A.A.; Saengerlaub, S. Active packaging of food products: Recent trends. *Nutr. Food Sci.* **2011**, *41*, 249–260. [CrossRef]
8. Soysal, Ç.; Bozkurt, H.; Dirican, E.; Güçlü, M.; Bozhüyük, E.D.; Uslu, A.E.; Kaya, S. Effect of antimicrobial packaging on physicochemical and microbial quality of chicken drumsticks. *Food Control* **2015**, *54*, 294–299. [CrossRef]
9. Huang, T.; Qian, Y.; Wei, J.; Zhou, C. Polymeric Antimicrobial food packaging and its applications. *Polymers* **2019**, *11*, 560. [CrossRef]
10. Covas, J.; Hilliou, L. Chapter 5—Production and Processing of Polymer-Based Nanocomposites. In *Nanomaterials for Food Packaging*; Cerqueira, M.Â.P.R., Lagaron, J.M., Castro, L.M.P., Vicente, A.A.M.O.S., Eds.; Elsevier: Amsterdam, The Netherlands, 2018; pp. 111–146. ISBN 9780323512718.
11. Souza, V.G.L.; Fernando, A.L. Nanoparticles in food packaging: Biodegradability and potential migration to food—A review. *Food Packag. Shelf Life* **2016**, *8*, 63–70. [CrossRef]
12. Youssef, A.M.; El-sayed, S.M. Bionanocomposites materials for food packaging applications: Concepts and future outlook. *Carbohydr. Polym.* **2018**, *193*, 19–27. [CrossRef]
13. Wang, Y.; Padua, G.W. Water Sorption Properties of Extruded Zein Films. *J. Agric. Food Chem.* **2004**, *52*, 3100–3105. [CrossRef]
14. Chen, H.; Wang, J.; Cheng, Y.; Wang, C.; Liu, H.; Bian, H.; Pan, Y.; Sun, J.; Han, W. Application of protein-based films and coatings for food packaging: A review. *Polymers* **2019**, *11*, 2039. [CrossRef]
15. Li, H.; He, Y.; Yang, J.; Wang, X.; Lan, T.; Peng, L. Fabrication of food-safe superhydrophobic cellulose paper with improved moisture and air barrier properties. *Carbohydr. Polym.* **2019**, *211*, 22–30. [CrossRef]
16. Deshwal, G.K.; Panjagari, N.R.; Alam, T. An overview of paper and paper based food packaging materials: Health safety and environmental concerns. *J. Food Sci. Technol.* **2019**, *56*, 4391–4403. [CrossRef]
17. Cazón, P.; Vázquez, M. Bacterial cellulose as a biodegradable food packaging material: A review. *Food Hydrocoll.* **2021**, *113*, 106530. [CrossRef]
18. Saedi, S.; Garcia, C.V.; Kim, J.T.; Shin, G.H. Physical and chemical modifications of cellulose fibers for food packaging applications. *Cellulose* **2021**, *28*, 8877–8897. [CrossRef]
19. Jiang, Z.; Ngai, T. Recent Advances in Chemically Modified Cellulose and Its Derivatives for Food Packaging Applications: A Review. *Polymers* **2022**, *14*, 1533. [CrossRef]
20. Franco, P.; Incarnato, L.; De Marco, I. Supercritical CO_2 impregnation of α-tocopherol into PET/PP films for active packaging applications. *J. CO_2 Util.* **2019**, *34*, 266–273. [CrossRef]
21. Vilela, C.; Kurek, M.; Hayouka, Z.; Röcker, B.; Yildirim, S.; Antunes, M.D.C.; Nilsen-Nygaard, J.; Pettersen, M.K.; Freire, C.S.R. A concise guide to active agents for active food packaging. *Trends Food Sci. Technol.* **2018**, *80*, 212–222. [CrossRef]
22. Ahn, B.J.; Gaikwad, K.K.; Lee, Y.S. Characterization and properties of LDPE film with gallic-acid-based oxygen scavenging system useful as a functional packaging material. *J. Appl. Polym. Sci.* **2016**, *133*, 44138. [CrossRef]
23. Di Maio, L.; Scarfato, P.; Galdi, M.R.; Incarnato, L. Development and oxygen scavenging performance of three-layer active PET films for food packaging. *J. Appl. Polym. Sci.* **2015**, *132*, 41465. [CrossRef]
24. Matche, R.S.; Sreekumar, R.K.; Raj, B. Modification of linear low-density polyethylene film using oxygen scavengers for its application in storage of bun and bread. *J. Appl. Polym. Sci.* **2011**, *122*, 55–63. [CrossRef]
25. Sängerlaub, S.; Gibis, D.; Kirchhoff, E.; Tittjung, M.; Schmid, M.; Müller, K. Compensation of Pinhole Defects in Food Packages by Application of Iron-based Oxygen Scavenging Multilayer Films. *Packag. Technol. Sci.* **2013**, *26*, 17–30. [CrossRef]
26. Shin, Y.; Shin, J.; Lee, Y.S. Preparation and characterization of multilayer film incorporating oxygen scavenger. *Macromol. Res.* **2011**, *19*, 869–875. [CrossRef]
27. Chen, C.W.; Xie, J.; Yang, F.X.; Zhang, H.L.; Xu, Z.W.; Liu, J.L.; Chen, Y.J. Development of moisture-absorbing and antioxidant active packaging film based on poly(vinyl alcohol) incorporated with green tea extract and its effect on the quality of dried eel. *J. Food Process. Preserv.* **2018**, *42*, e13374. [CrossRef]
28. Lee, J.H.; Lee, S.G. Preparation and swelling behavior of moisture-absorbing polyurethane films impregnated with superabsorbent sodium polyacrylate particles. *J. Appl. Polym. Sci.* **2016**, *133*, 43973. [CrossRef]

29. Chand, K.; Kumar, S. Effect of Active Packaging and Coating Materials on Quality Parameters of Jaggery Cubes. *Int. J. Eng. Res.* **2018**, *7*, 4–9. [CrossRef]
30. Choi, H.Y.; Lee, Y.S. Characteristics of moisture-absorbing film impregnated with synthesized attapulgite with acrylamide and its effect on the quality of seasoned laver during storage. *J. Food Eng.* **2013**, *116*, 829–839. [CrossRef]
31. Sängerlaub, S.; Seibel, K.; Miesbauer, O.; Pant, A.; Kiese, S.; Rodler, N.; Schmid, M.; Müller, K. Functional properties of foamed and/or stretched polypropylene-films containing sodium chloride particles for humidity regulation. *Polym. Test.* **2018**, *65*, 339–351. [CrossRef]
32. Tas, C.E.; Hendessi, S.; Baysal, M.; Unal, S.; Cebeci, F.C.; Menceloglu, Y.Z.; Unal, H. Halloysite Nanotubes/Polyethylene Nanocomposites for Active Food Packaging Materials with Ethylene Scavenging and Gas Barrier Properties. *Food Bioprocess Technol.* **2017**, *10*, 789–798. [CrossRef]
33. Srithammaraj, K.; Magaraphan, R.; Manuspiya, H. Modified Porous Clay Heterostructures by Organic–Inorganic Hybrids for Nanocomposite Ethylene Scavenging/Sensor Packaging Film. *Packag. Technol. Sci.* **2012**, *25*, 63–72. [CrossRef]
34. Boonruang, K.; Chonhenchob, V.; Singh, S.P.; Chinsirikul, W.; Fuongfuchat, A. Comparison of Various Packaging Films for Mango Export. *Packag. Technol. Sci.* **2012**, *25*, 107–118. [CrossRef]
35. Esturk, O.; Ayhan, Z.; Gokkurt, T. Production and Application of Active Packaging Film with Ethylene Adsorber to Increase the Shelf Life of Broccoli (*Brassica oleracea* L. var. *Italica*). *Packag. Technol. Sci.* **2014**, *27*, 179–191. [CrossRef]
36. Busolo, M.A.; Lagaron, J.M. Antioxidant Polyethylene Films Based On A Resveratrol Containing Clay Of Interest In Food Packaging Applications | Elsevier Enhanced Reader. Available online: https://reader.elsevier.com/reader/sd/pii/S2214289415300119?token=90922F3D2440CAF6A65DC36ED590788F26E271C821005623F69B89F780919755B7EBF0E8195E246085D9FA6EEB1C8DA6&originRegion=eu-west-1&originCreation=20210729160631 (accessed on 29 July 2021).
37. Bolumar, T.; LaPeña, D.; Skibsted, L.H.; Orlien, V. Rosemary and oxygen scavenger in active packaging for prevention of high-pressure induced lipid oxidation in pork patties. *Food Packag. Shelf Life* **2016**, *7*, 26–33. [CrossRef]
38. Quiles-Carrillo, L.; Montanes, N.; Lagaron, J.M.; Balart, R.; Torres-Giner, S. Bioactive Multilayer Polylactide Films with Controlled Release Capacity of Gallic Acid Accomplished by Incorporating Electrospun Nanostructured Coatings and Interlayers. *Appl. Sci.* **2019**, *9*, 30533. [CrossRef]
39. Ahmed, M.Y.; EL-Sayed, S.M.; EL-Sayed, H.S.; Salama, H.H.; Dufresne, A. Enhancement of Egyptian Soft White Cheese Shelf Life Using A Novel Chitosan/Carboxymethyl Cellulose/Zinc Oxide Bionanocomposite Film | Elsevier Enhanced Reader. Available online: https://reader.elsevier.com/reader/sd/pii/S0144861716305367?token=548191AFA89F297C72DE875E19C4823F9D1F0D6F6193FC787E6A2A30480AC67AC044A4FE7407DE22019A5BECCFB81005&originRegion=eu-west-1&originCreation=20210729135351 (accessed on 29 July 2021).
40. Llana-Ruiz-Cabello, M.; Pichardo, S.; Bãnos, A.; Núñez, C.; Bermúdez, J.M.; Guillamón, E.; Aucejo, S.; Cameán, A.M. Characterisation and evaluation of PLA films containing an extract of Allium spp. to be used in the packaging of ready-to-eat salads under controlled atmospheres. *LWT Food Sci. Technol.* **2015**, *64*, 1354–1361. [CrossRef]
41. Syamsu, K.; Warsiki, E.; Yuliani, S.; Widayanti, S.M. Nano Zeolite-kmno4 as Ethylene Adsorber in Active Packaging of Horticulture Products (*Musa Paradisiaca*). *Int. J. Sci. Basic Appl. Res.* **2016**, *30*, 93–103.
42. Carrizo, D.; Taborda, G.; Nerín, C.; Bosetti, O. Extension of shelf life of two fatty foods using a new antioxidant multilayer packaging containing green tea extract. *Innov. Food Sci. Emerg. Technol.* **2016**, *33*, 534–541. [CrossRef]
43. Oudjedi, K.; Manso, S.; Nerin, C.; Hassissen, N.; Zaidi, F. New active antioxidant multilayer food packaging films containing Algerian Sage and Bay leaves extracts and their application for oxidative stability of fried potatoes. *Food Control* **2019**, *98*, 216–226. [CrossRef]
44. Barbosa-Pereira, L.; Cruz, J.M.; Sendón, R.; Rodríguez Bernaldo de Quirós, A.; Ares, A.; Castro-López, M.; Abad, M.J.; Maroto, J.; Paseiro-Losada, P. Development of antioxidant active films containing tocopherols to extend the shelf life of fish. *Food Control* **2013**, *31*, 236–243. [CrossRef]
45. Wang, W.; Kannan, K. Quantitative identification of and exposure to synthetic phenolic antioxidants, including butylated hydroxytoluene, in urine. *Environ. Int.* **2019**, *128*, 24–29. [CrossRef]
46. Ousji, O.; Sleno, L. Identification of in vitro metabolites of synthetic phenolic antioxidants BHT, BHA, and TBHQ by LC-HRMS/MS. *Int. J. Mol. Sci.* **2020**, *21*, 49525. [CrossRef]
47. Beigmohammadi, F.; Peighambardoust, S.H.; Hesari, J.; Azadmard-Damirchi, S.; Peighambardoust, S.J.; Khosrowshahi, N.K. Antibacterial properties of LDPE nanocomposite films in packaging of UF cheese. *LWT Food Sci. Technol.* **2016**, *65*, 106–111. [CrossRef]
48. Cutter, C.N.; Willett, J.L.; Siragusa, G.R. Improved antimicrobial activity of nisin-incorporated polymer films by formulation change and addition of food grade chelator. *Lett. Appl. Microbiol.* **2001**, *33*, 325–328. [CrossRef]
49. Li, L.; Zhao, C.; Zhang, Y.; Yao, J.; Yang, W.; Hu, Q.; Wang, C.; Cao, C. Effect of stable antimicrobial nano-silver packaging on inhibiting mildew and in storage of rice. *Food Chem.* **2017**, *215*, 477–482. [CrossRef]
50. Rollini, M.; Nielsen, T.; Musatti, A.; Limbo, S.; Piergiovanni, L.; Munoz, P.H.; Gavara, R.; Barringer, S. Antimicrobial Performance of Two Different Packaging Materials on the Microbiological Quality of Fresh Salmon. *Coatings* **2016**, *6*, 6. [CrossRef]
51. Pang, Y.-H.; Sheen, S.; Zhou, S.; Liu, L.; Yam, K.L. Antimicrobial Effects of Allyl Isothiocyanate and Modified Atmosphere on Pseduomonas Aeruginosa in Fresh Catfish Fillet under Abuse Temperatures. *J. Food Sci.* **2013**, *78*, M555–M559. [CrossRef]

52. Granda-Restrepo, D.; Peralta, E.; Troncoso-Rojas, R.; Soto-Valdez, H. Release of antioxidants from co-extruded active packaging developed for whole milk powder. *Int. Dairy J.* **2009**, *19*, 481–488. [CrossRef]
53. Emamifar, A.; Kadivar, M.; Shahedi, M.; Soleimanian-Zad, S. Effect of nanocomposite packaging containing Ag and ZnO on inactivation of Lactobacillus plantarum in orange juice. *Food Control* **2011**, *22*, 408–413. [CrossRef]
54. Emamifar, A.; Mohammadizadeh, M. Preparation and Application of LDPE/ZnO Nanocomposites for Extending Shelf Life of Fresh Strawberries. *Food Technol. Biotechnol.* **2015**, *53*, 488–495. [CrossRef] [PubMed]
55. Zhu, X.; Lee, D.S.; Yam, K.L. Release property and antioxidant effectiveness of tocopherol-incorporated LDPE/PP blend films. *Food Addit. Contam. Part A* **2012**, *29*, 461–468. [CrossRef] [PubMed]
56. Graciano-Verdugo, A.Z.; Soto-Valdez, H.; Peralta, E.; Cruz-Zárate, P.; Islas-Rubio, A.R.; Sánchez-Valdes, S.; Sánchez-Escalante, A.; González-Méndez, N.; González-Ríos, H. Migration of α-tocopherol from LDPE films to corn oil and its effect on the oxidative stability of soybean oil. *Food Res. Int.* **2010**, *43*, 1073–1078. [CrossRef]
57. Manzanarez-López, F.; Soto-Valdez, H.; Auras, R.; Peralta, E. Release of α-Tocopherol from Poly(lactic acid) films, and its effect on the oxidative stability of soybean oil. *J. Food Eng.* **2011**, *104*, 508–517. [CrossRef]
58. Cestari, L.A.; Gaiotto, R.C.; Antigo, J.L.; Scapim, M.R.S.; Madrona, G.S.; Yamashita, F.; Pozza, M.S.S.; Prado, I.N. Effect of active packaging on low-sodium restructured chicken steaks. *J. Food Sci. Technol.* **2015**, *52*, 3376–3382. [CrossRef]
59. Pant, A.F.; Sangerlaub, S.; Muller, K. Gallic acid as an oxygen scavenger in bio-based multilayer packaging films. *Materials* **2017**, *10*, 489. [CrossRef]
60. Tyuftin, A.A.; Kerry, J.P. Review of surface treatment methods for polyamide films for potential application as smart packaging materials: Surface structure, antimicrobial and spectral properties. *Food Packag. Shelf Life* **2020**, *24*, 100475. [CrossRef]
61. Lorenzo, J.M.; Batlle, R.; Gómez, M. Extension of the shelf-life of foal meat with two antioxidant active packaging systems. *LWT Food Sci. Technol.* **2014**, *59*, 181–188. [CrossRef]
62. Camo, J.; Lorés, A.; Djenane, D.; Beltrán, J.A.; Roncalés, P. Display life of beef packaged with an antioxidant active film as a function of the concentration of oregano extract. *Meat Sci.* **2011**, *88*, 174–178. [CrossRef]
63. Barbosa-Pereira, L.; Aurrekoetxea, G.P.; Angulo, I.; Paseiro-Losada, P.; Cruz, J.M. Development of new active packaging films coated with natural phenolic compounds to improve the oxidative stability of beef. *Meat Sci.* **2014**, *97*, 249–254. [CrossRef]
64. Gaikwad, K.K.; Singh, S.; Lee, Y.S. A new pyrogallol coated oxygen scavenging film and their effect on oxidative stability of soybean oil under different storage conditions. *Food Sci. Biotechnol.* **2017**, *26*, 1535–1543. [CrossRef]
65. Guo, M.; Jin, T.Z.; Yang, R. Antimicrobial Polylactic Acid Packaging Films against Listeria and Salmonella in Culture Medium and on Ready-to-Eat Meat. *Food Bioprocess Technol.* **2014**, *7*, 3293–3307. [CrossRef]
66. Al-Naamani, L.; Dobretsov, S.; Dutta, J. Chitosan-zinc oxide nanoparticle composite coating for active food packaging applications. *Innov. Food Sci. Emerg. Technol.* **2016**, *38*, 231–237. [CrossRef]
67. Joerger, R.D.; Sabesan, S.; Visioli, D.; Urian, D.; Joerger, M.C. Antimicrobial activity of chitosan attached to ethylene copolymer films. *Packag. Technol. Sci.* **2009**, *22*, 125–138. [CrossRef]
68. Lardiés, O.G.; Nerin de la Puerta, C.; Garcia, J.A.B.; Rabinal, P.R. Antioxidant Active Varnish. European Patent EP1477519A1, 17 November 2004. pp. 1–8. Available online: https://data.epo.org/gpi/EP1477519A1-Antioxidant-active-varnish (accessed on 29 July 2021).
69. Lee, J.S.; Park, M.A.; Yoon, C.S.; Na, J.H.; Han, J. Characterization and Preservation Performance of Multilayer Film with Insect Repellent and Antimicrobial Activities for Sliced Wheat Bread Packaging. *J. Food Sci.* **2019**, *84*, 3194–3203. [CrossRef]
70. Azlin-Hasim, S.; Cruz-Romero, M.C.; Cummins, E.; Kerry, J.P.; Morris, M.A. The potential use of a layer-by-layer strategy to develop LDPE antimicrobial films coated with silver nanoparticles for packaging applications. *J. Colloid Interface Sci.* **2016**, *461*, 239–248. [CrossRef]
71. Vähä-Nissi, M.; Pitkänen, M.; Salo, E.; Kenttä, E.; Tanskanen, A.; Sajavaara, T.; Putkonen, M.; Sievänen, J.; Sneck, A.; Rättö, M.; et al. Antibacterial and barrier properties of oriented polymer films with ZnO thin films applied with atomic layer deposition at low temperatures. *Thin Solid Film.* **2014**, *562*, 331–337. [CrossRef]
72. Al-Naamani, L.; Dutta, J.; Dobretsov, S. Nanocomposite zinc oxide-chitosan coatings on polyethylene films for extending storage life of okra (*Abelmoschus esculentus*). *Nanomaterials* **2018**, *8*, 479. [CrossRef]
73. Nalawade, S.P.; Picchioni, F.; Janssen, L.P.B.M. Supercritical carbon dioxide as a green solvent for processing polymer melts: Processing aspects and applications. *Prog. Polym. Sci.* **2006**, *31*, 19–43. [CrossRef]
74. Tadesse Abate, M.; Ferri, A.; Guan, J.; Chen, G.; Nierstrasz, V. Impregnation of materials in supercritical CO_2 to impart various functionalities. In *Advanced Supercritical Fluids Technologies*; IntechOpen: London, UK, 2020; pp. 1–14. [CrossRef]
75. Rojas, A.; Torres, A.; José Galotto, M.; Guarda, A.; Julio, R. Supercritical impregnation for food applications: A review of the effect of the operational variables on the active compound loading. *Crit. Rev. Food Sci. Nutr.* **2020**, *60*, 1290–1301. [CrossRef]
76. Goñi, M.L.; Gañán, N.A.; Strumia, M.C.; Martini, R.E. Eugenol-loaded LLDPE films with antioxidant activity by supercritical carbon dioxide impregnation. *J. Supercrit. Fluids* **2016**, *111*, 28–35. [CrossRef]
77. Cejudo Bastante, C.; Casas Cardoso, L.; Fernández-Ponce, M.T.; Mantell Serrano, C.; Martínez de la Ossa, E.J. Supercritical impregnation of olive leaf extract to obtain bioactive films effective in cherry tomato preservation. *Food Packag. Shelf Life* **2019**, *21*, 100338. [CrossRef]
78. Belizón, M.; Fernández-Ponce, M.T.; Casas, L.; Mantell, C.; Martínez De La Ossa-Fernández, E.J. Supercritical impregnation of antioxidant mango polyphenols into a multilayer PET/PP food-grade film. *J. CO_2 Util.* **2018**, *25*, 56–67. [CrossRef]

79. Rojas, A.; Torres, A.; Añazco, A.; Villegas, C.; Galotto, M.J.; Guarda, A.; Romero, J. Effect of pressure and time on scCO$_2$-assisted incorporation of thymol into LDPE-based nanocomposites for active food packaging. *J. CO$_2$ Util.* **2018**, *26*, 434–444. [CrossRef]
80. Villegas, C.; Torres, A.; Rios, M.; Rojas, A.; Romero, J.; de Dicastillo, C.L.; Valenzuela, X.; Galotto, M.J.; Guarda, A. Supercritical impregnation of cinnamaldehyde into polylactic acid as a route to develop antibacterial food packaging materials. *Food Res. Int.* **2017**, *99*, 650–659. [CrossRef] [PubMed]
81. Rojas, A.; Torres, A.; Martínez, F.; Salazar, L.; Villegas, C.; Galotto, M.J.; Guarda, A.; Romero, J. Assessment of kinetic release of thymol from LDPE nanocomposites obtained by supercritical impregnation: Effect of depressurization rate and nanoclay content. *Eur. Polym. J.* **2017**, *93*, 294–306. [CrossRef]
82. Wrona, M.; Nerín, C.; Alfonso, M.J.; Caballero, M.Á. Antioxidant packaging with encapsulated green tea for fresh minced meat. *Innov. Food Sci. Emerg. Technol.* **2017**, *41*, 307–313. [CrossRef]
83. Gargiulo, N.; Attianese, I.; Giuliana, G.; Caputo, D.; Lavorgna, M.; Mensitieri, G.; Lavorgna, M. Microporous and Mesoporous Materials a -Tocopherol release from active polymer films loaded with functionalized SBA-15 mesoporous silica. *Microporous Mesoporous Mater.* **2013**, *167*, 10–15. [CrossRef]
84. Sun, L.; Lu, L.; Qiu, X.; Tang, Y. Development of low-density polyethylene antioxidant active films containing a -tocopherol loaded with MCM-41 (Mobil Composition of Matter No. 41). *Food Control* **2017**, *71*, 193–199. [CrossRef]
85. Melendez-Rodriguez, B.; Figueroa-Lopez, K.J.; Bernardos, A.; Cabedo, L.; Torres-Giner, S. Electrospun Antimicrobial Films of Poly(3-hydroxybutyrate-co-3-hydroxyvalerate) Containing Eugenol Essential Oil Encapsulated in Mesoporous Silica. *Nanoparticles* **2019**, *9*, 227. [CrossRef]
86. Alkan, B.; Sehit, E.; Tas, C.E.; Unal, S.; Cebeci, F.C. Carvacrol loaded halloysite coatings for antimicrobial food packaging applications. *Food Packag. Shelf Life* **2019**, *20*, 100300. [CrossRef]
87. De Oliveira, C.M.; de Gomes, B.O.; Batista, A.F.P.; Mikcha, J.M.G.; Yamashita, F.; Scapim, M.R.S.; de Bergamasco, R.C. Development of sorbic acid microcapsules and application in starch-poly (butylene adipate co-terephthalate) films. *J. Food Process. Preserv.* **2021**, *45*, e15459. [CrossRef]
88. Figueroa-Lopez, K.J.; Torres-Giner, S.; Angulo, I.; Pardo-Figuerez, M.; Escuin, J.M.; Bourbon, A.I.; Cabedo, L.; Nevo, Y.; Cerqueira, M.A.; Lagaron, J.M. Development of active barrier multilayer films based on electrospun antimicrobial hot-tack food waste derived poly(3-hydroxybutyrate-co-3-hydroxyvalerate) and cellulose nanocrystal interlayers. *Nanomaterials* **2020**, *10*, 2356. [CrossRef]
89. Chiralt, A.; Tampau, A.; Gonz, C. Food Hydrocolloids Release kinetics and antimicrobial properties of carvacrol encapsulated in electrospun poly-(ε-caprolactone)nano fibres. Application in starch multilayer films. *Food Hydrocoll.* **2018**, *79*, 158–169. [CrossRef]
90. Figueroa-Lopez, K.J.; Enescu, D.; Torres-Giner, S.; Cabedo, L.; Cerqueira, M.A.; Pastrana, L.; Fuciños, P.; Lagaron, J.M. Development of electrospun active films of poly(3-hydroxybutyrate-co-3-hydroxyvalerate) by the incorporation of cyclodextrin inclusion complexes containing oregano essential oil. *Food Hydrocoll.* **2020**, *108*, 106013. [CrossRef]
91. Glaser, T.K.; Plohl, O.; Vesel, A.; Ajdnik, U.; Ulrih, N.P.; Hrnčič, M.K.; Bren, U.; Zemljič, L.F. Functionalization of polyethylene (PE) and polypropylene (PP) material using chitosan nanoparticles with incorporated resveratrol as potential active packaging. *Materials* **2019**, *12*, 2118. [CrossRef]
92. Guarda, A.; Rubilar, J.F.; Miltz, J.; Jose, M. The antimicrobial activity of microencapsulated thymol and carvacrol. *Int. J. Food Microbiol.* **2011**, *146*, 144–150. [CrossRef]

Article

Lignin Nanoparticles for Enhancing Physicochemical and Antimicrobial Properties of Polybutylene Succinate/Thymol Composite Film for Active Packaging

Angel Jr Basbasan [1], Bongkot Hararak [2], Charinee Winotapun [2], Wanwitoo Wanmolee [3], Wannee Chinsirikul [3], Pattarin Leelaphiwat [1,4], Vanee Chonhenchob [1,4] and Kanchana Boonruang [4,5,*]

1. Department of Packaging and Materials Technology, Faculty of Agro-Industry, Kasetsart University, Bangkok 10900, Thailand
2. National Metal and Materials Technology Center, National Science and Technology Development Agency, Pathum Thani 12120, Thailand
3. National Nanotechnology Center, National Science and Technology Development Agency, Pathum Thani 12120, Thailand
4. Center for Advanced Studies for Agriculture and Food, Kasetsart University, Bangkok 10900, Thailand
5. Department of Horticulture, Faculty of Agriculture, Kasetsart University, Bangkok 10900, Thailand
* Correspondence: kanchana.boon@ku.th

Abstract: The natural abundance, polymer stability, biodegradability, and natural antimicrobial properties of lignin open a wide range of potential applications aiming for sustainability. In this work, the effects of 1% (w/w) softwood kraft lignin nanoparticles (SLNPs) on the physicochemical properties of polybutylene succinate (PBS) composite films were investigated. Incorporation of SLNPs into neat PBS enhanced T_d from 354.1 °C to 364.7 °C, determined through TGA, whereas T_g increased from −39.1 °C to −35.7 °C while no significant change was observed in T_m and crystallinity, analyzed through DSC. The tensile strength of neat PBS increased, to 35.6 MPa, when SLNPs were added to it. Oxygen and water vapor permeabilities of PBS with SLNPs decreased equating to enhanced barrier properties. The good interactions among SLNPs, thymol, and PBS matrix, and the high homogeneity of the resultant PBS composite films, were determined through FTIR and FE-SEM analyses. This work revealed that, among the PBS composite films tested, PBS + 1% SLNPs + 10% thymol showed the strongest microbial growth inhibition against *Colletotrichum gloeosporioides* and *Lasiodiplodia theobromae*, both in vitro, through a diffusion method assay, and in actual testing on active packaging of mango fruit (cultivar "Nam Dok Mai Si Thong"). SLNPs could be an attractive replacement for synthetic substances for enhancing polymer properties without compromising the biodegradability of the resultant material, and for providing antimicrobial functions for active packaging applications.

Keywords: lignin nanoparticles; thymol; polybutylene succinate; antimicrobial packaging; biodegradable materials

Citation: Basbasan, A.J.; Hararak, B.; Winotapun, C.; Wanmolee, W.; Chinsirikul, W.; Leelaphiwat, P.; Chonhenchob, V.; Boonruang, K. Lignin Nanoparticles for Enhancing Physicochemical and Antimicrobial Properties of Polybutylene Succinate/Thymol Composite Film for Active Packaging. *Polymers* 2023, 15, 989. https://doi.org/10.3390/polym15040989

Academic Editors: Victor G. L. Souza, Lorenzo M. Pastrana and Ana Luisa Fernando

Received: 19 January 2023
Revised: 10 February 2023
Accepted: 13 February 2023
Published: 16 February 2023

Copyright: © 2023 by the authors. Licensee MDPI, Basel, Switzerland. This article is an open access article distributed under the terms and conditions of the Creative Commons Attribution (CC BY) license (https:// creativecommons.org/licenses/by/ 4.0/).

1. Introduction

The research of natural and biodegradable materials is of great interest, especially now that many industries, such as the packaging industry, are aiming for sustainability. Currently, biopolymers are the most attractive alternative for nonbiodegradable packaging materials, such as conventional plastics. However, previous studies have shown that biopolymers are still not equal to conventional plastics in terms of packaging performance.

Compared with conventional plastics, biopolymers have disadvantages, such as sensitivity to heat, humidity, and shear stress, that could lead to early partial thermal and mechanical degradation during the processing stage, and possess inferior mechanical properties—e.g., brittleness, rigidity, and low tensile strength—and inappropriate barrier properties—e.g., high water and oxygen permeabilities, resulting in limited application possibilities [1,2].

Innovations have been produced and a great amount research is ongoing to improve the functional performance and widen the applications of biopolymers. The most common and technically feasible method is adding another compound into the biopolymeric matrix to enhance material properties, such as mechanical, barrier, and thermal stability, and to provide active functions, including antimicrobial and antioxidant.

Recently, lignin has drawn distinct attention as a natural additive to polymeric matrices because of its capability to modify material properties. Lignin is a natural aromatic polymer exhibiting a complex, high molecular weight and a highly random structure. It is found in the cell walls of plants and constitutes the most stable component of biomass [3,4]. Lignin is an abundant agro-industrial waste, mainly from the papermaking and biorefinery industries. Its natural abundance and low cost open diverse potential applications in various industries. Lignin can be used in the agriculture industry as an agrochemical, i.e., as fertilizer, pesticide, and plant growth regulator [5]. In the medical industry, promising findings were reported, which may improve human health, involving the potential applications of lignin [6].

Aside from being naturally biodegradable, lignin generally shows high strength and rigidity and satisfactory thermal resistance. As a result, thermoplastic polymers compounded with lignin, such as polyethylene (PE), polypropylene (PP), and polyvinyl chloride (PVC), have shown improved flowability and processing performance of the material, and have exhibited significant enhancement in material properties [7]. In this present work, the effects of incorporation of softwood kraft lignin nanoparticles (SLNPs) on the properties of the resultant material were investigated. To be consistent with the aimed biodegradability and for the biopolymer composite not to be compromised, polybutylene succinate (PBS) was used as the polymer matrix. PBS is among the most important emerging biodegradable polymers, with a wide range of potential applications due to its excellent processability and chemical resistance [8]. It is synthesized by polycondensation between succinic acid and butanediol in a two-step reaction process: esterification between the diacid and the diol, and then polycondensation under high temperature to form PBS with high molecular weight [9]. Interestingly, PBS can be produced both from monomers derived from fossil-based sources and by the fermentation route of bacteria [10]. Moreover, it was reported that full biomass-based PBS was chemically synthesized from furfural that was produced from inedible agricultural cellulosic waste [11]. Even though PBS has excellent processability, some of its important properties, such as brittleness and low thermal stability, restrict its wide commercial implementation and applications. Hence, various techniques, such as copolymerization using the extrusion method coupled with a line for film casting [12], blending with another biopolymer through extrusion compounding followed by injection molding [13], and incorporation of natural additives [14], such as bamboo fiber, through extrusion followed by injection molding [15], have been performed to achieve improved properties. The thermal stability of the PBS polymer composite was enhanced with the addition of low sulfonate kraft lignin, but utilizing chemically modified low sulfonate lignin further improved the thermal stability of the resultant PBS composite [16]. It was reported that PBS and softwood kraft lignin have shown noncovalent interactions [17] which could possibly help in the enhancement of the thermal stability of the PBS composite. Contrasting results were presented when lignosulphonate (a classification of lignin) was blended in 70/30 PBS/polybutylene adipate-co-terephthalate (PBAT) with zinc nanoparticles, wherein the thermal stability and tensile strength of the hybrid composite decreased with the incorporation of lignosulphonate [18]. However, it is interesting to note that the lignosulphonate used in their study was not modified. In this present work, softwood kraft lignin was processed into nanoparticles to achieve the advantages of nano-size particles in terms of reactivity. Moreover, different classifications of lignin will produce different results when blended with a polymer, because the reaction is affected by the complexity of the lignin structure, which is dependent on the source and isolation process [19,20].

Aside from the effects that modify the properties of materials, lignin is a natural antimicrobial agent because it contains numerous functional groups [21] that are responsible for antimicrobial activities [22], such as phenolic and aliphatic hydroxyls, methyl, carboxylic, and carbonyl groups. However, despite its great potential to replace synthetic chemicals, only 2–5% of lignin in its macromolecular form is commercially utilized [23]. Considering that the worldwide paper and pulp industry discards lignin from its production, generating waste amounting to 50–70 million tons annually, the commercial utilization of lignin is considered very low.

Recent studies have shown the viability of lignin as an antimicrobial agent. For example, fractionated kraft lignin from bamboo obtained by organosolv fractionation can destroy, in vitro, the cell wall of *Escherichia coli*, *Salmonella*, *Streptococcus*, and *Staphylococcus aureus*, thus inhibiting the growth of these bacteria [24]. Similarly, it was reported that lignin derived from beechwood flour is comparable with existing antibacterial agents, such as chlorhexidine, in inhibiting the growth of *E. coli* and 189 *S. aureus* [25]. Moreover, lignin residue of corn stover from ethanol production exhibited antimicrobial activities against *Listeria monocytogenes*, *S. aureus*, and *Candida lipolytica* [26]. The authors [27] demonstrated that lignin from different sources, such as eucalyptus, acacia, sugarcane bagasse, corn, and cotton stalks, conferred antimicrobial activity against microorganisms mentioned above and several other microorganisms, including *Proteus vulgaris*, *Pseudomonas aeruginosa*, *Aspergillus niger*, *Bacillus subtilis*, and *Klebsiella pneumoniae*, among others.

In this present work, an innovative and sustainable manner of utilizing abundant agro-industrial waste lignin was presented by converting it into a high-value product that can be beneficial to the industry. Raw softwood kraft lignin was processed into lignin nanoparticles and used as a natural additive to enhance the material properties of PBS composite films. Additionally, while most previous published works dealt with bacteria, a potential novel work was demonstrated in this study by using the SLNPs as a natural antimicrobial agent against fungal species. *Colletotrichum gloeosporioides* and *Lasiodiplodia theobromae* are two major fungal species that cause significant postharvest decay—anthracnose and stem-end rot, respectively—to many economically important fruit crops such as mango, papaya, and citrus. The use of a natural additive, such as SLNPs, as an alternative to synthetic fungicides for postharvest decay control would not only be economically beneficial but also safer to humans and the environment.

The incorporation of essential oils into polymers as antimicrobial agents has widely been studied as a competitive alternative to chemical preservatives. Aside from being generally recognized as safe (GRAS), essential oils, which are hydrophobic liquids that are naturally present in various plant parts, demonstrate a wide range of antimicrobial activities, such as antibacterial and antifungal activities, and function as natural preservatives [28]. Their volatility enables them to diffuse out from the polymer matrix and perform antimicrobial functions, whereas the antimicrobial activities of SLNPs were restricted to the polymer matrices. The addition of SLNPs into polymer could enhance the antimicrobial efficacy of polymeric film. As thymol was found to be the most effective phenol against tested fungal species in our previous studies [29], SLNPs were added to thymol in the PBS composite. In our previous study, a high level of thymol, i.e., 20% (w/w), was used, whereas this present study aimed to utilize a lower amount of thymol with the addition of SLNPs.

Due to the reactivity of lignin as a function of the numerous functional groups found in it, lignin can interact with many polymers and change their properties [30]. According to this premise, SLNPs would produce interactions with PBS and thymol that could potentially enhance both the physicochemical and antimicrobial properties of the resultant PBS composite film. Therefore, this present work aimed to investigate the effects of the incorporation of SLNPs into PBS film containing thymol through characterizing the physicochemical properties of the resultant composite films in terms of morphology, Fourier-transform infrared spectroscopy (FTIR) analysis, and thermal, mechanical, and barrier properties. Moreover, the antimicrobial activities of SLNPs in PBS composite film containing thymol were investigated, both in vitro and in vivo.

2. Materials and Methods

2.1. Materials

Raw softwood kraft lignin powder (BioPiva™100, denoted as SR-lignin) was purchased from UPM Biochemicals, Helsinki, Finland. PBS pellets were obtained from PPT Global Chemical Public Company Limited, Bangkok, Thailand. Pure culture of *C. gloeosporioides* was obtained from the Plant Protection Research and Development Office of the Department of Agriculture, Bangkok, Thailand, while pure culture of *L. theobromae* was isolated from "Nam Dok Mai Si Thong" mango fruit.

2.2. Preparation of SLNPs

SR-lignin was processed into lignin nanoparticles (LNPs) by antisolvent precipitation assisted by ultrasonication, following the method of Hararak et al. [31]. It was purified first through fractionation method using acetone. In detail, 10 g of SR-lignin was pre-dried in a vacuum oven at 80 °C for 6 h and then placed in a 3-neck flask containing 100 mL acetone. A water condenser was attached to the top neck of the flask while a thermocouple and nitrogen bubbling at a flow rate of 10 mL min^{-1} were attached to the other two necks. The flask containing the acetone and dried raw lignin was heated up, maintained at 56 °C, and then stirred for 6 h using a heating mantle with a built-in stirrer. The acetone-soluble fraction was filtrated through a 1 μm glass filter and the obtained acetone-soluble lignin (denoted as S-lignin) was dried at 80 °C in a vacuum oven for 30 min and then ground into powder.

The obtained S-lignin was processed into LNPs (denoted as SLNPs) by ultrasonication. Briefly, 1 g of S-lignin was fully dissolved in 20 mL acetone at room temperature. Afterward, the 20 mL acetone containing dissolved S-lignin was poured into 200 mL deionized water at room temperature under ultrasonication (VCX 500 sonicator, 500 watts, 20 kHz, and 13 mm probe) for 5 min. The mixture was then centrifuged at a speed of 11,180 g-force for 20 min and the obtained SLNPs were dried at 80 °C in a vacuum oven for 6 h and then kept in a hermetically sealed container until use.

2.3. Preparation of PBS Composite Films

In total, 2 masterbatches of polymer blends, namely, PBS + 2% SLNPs and PBS + 20% thymol, were initially prepared. The first masterbatch (MB) was prepared by mixing 2 g SLNP powder and 98 g PBS pellets and then gradually loaded the mixture into a co-rotating twin screw extruder (LABTECH Twin-screw extruder; 20 mm screw diameter; length to diameter ratio of 32). The barrel temperature profile from feeding to the die zone was set at 110, 130, 140, 150, 154, 155, 155, and 160 °C with a screw speed of 0.112 g-force. The produced extrudate passed through a cooling water, was air-dried, and then palletized to around 2.5 mm long. The second MB was prepared using an internal mixer (HAAKE™, Rheomex OS, Bremen, Germany; 310 cm^2 capacity; 0.8 fill factor; roller rotor) to minimize the vaporization of the volatile thymol (vapor pressure of 53.33 Pa at 25 °C) and to produce good dispersion and homogeneity of the thymol in the PBS matrix [32,33]. The rotor speed was set at 0.038 g-force and the temperature of the mixing chamber was set at 120 °C. An amount of 250 g PBS pellets were gradually fed into the internal mixer until melt phase then a 62 g thymol powder was introduced into the mixer and allowed to mix for 12 min. The produced bulk polymer mixture was crushed into pellets. Proportions from the two masterbatches were pre-mixed to produce the final formulations, as reported in Table 1.

Table 1. PBS composite films' formulations.

Formulations	PBS (w/w)	SLNPs (w/w)	Thymol (w/w)
PBS	100	-	-
PBS + 10T	90	-	10
PBS + 1SLNPs	99	1	-
PBS + 1SLNPs + 1T	98	1	1
PBS + 1SLNPs + 5T	94	1	5
PBS + 1SLNPs + 7.5T	91.5	1	7.5
PBS + 1SLNPs + 10T	89	1	10

PBS: polybutylene succinate; SLNPs: softwood kraft lignin nanoparticles; T: thymol.

The mixtures were formed into composite films by using a blown film, single-screw extruder (Thermo Scientific HAAKE™ Rheomex OS, Bremen, Germany; 19 mm screw diameter; length to diameter ratio of 25) with a screw speed set at 0.038 g-force and temperature profile of 150, 155, 160, and 160 °C for the feeding, compression, metering, and die zone, respectively. Neat PBS film which served as the control for analyses was also formed using the same blown film, single-screw extruder with the same setting.

2.4. Physicochemical Characterization of PBS Composite Films

2.4.1. Morphology

The morphology image of the cryofractured neat PBS and (selected) PBS composite films was carried out through a field-emission scanning electron microscope (FE-SEM, model SU5000, Hitachi, Tokyo, Japan). All samples were coated with platinum by Quorum-Q150RS (UK) before imaging.

2.4.2. FTIR Analysis

The PBS composite films were determined for infrared absorption spectra using Fourier-transform infrared spectroscopy (FTIR, Tensor 27, Bruker Corporation, Bremen, Germany) with an attenuated total reflection (ATR) mode in the range of 500–4000 cm^{-1} with 64 scans.

2.4.3. Thermal Properties

A differential scanning calorimetry (DSC, Mettler Toledo, Greifensee, Switzerland) was used to determine the glass transition temperature (T_g) and the melting point temperatures (T_m) of neat PBS and PBS composite films. The first heating scan was performed from −80 to 180 °C; the cooling scan was performed from 180 down to −80 °C; and the second heating scan was performed from −80 to 180 °C. The heating and cooling rates were set at 10 °C min^{-1}, under a nitrogen atmosphere, with a flow rate of 50 mL min^{-1}. From the data gathered in the DSC scan, the crystallinity of neat PBS and PBS composite films was calculated using Equation (1) [34]:

$$\chi = \frac{\Delta H}{\Delta H_o(1 - mf)} \times 100 \qquad (1)$$

where ΔH is the heat of fusion of the sample obtained from the second heating scan, ΔH_o is the heat of fusion of 100% crystalline PBS theoretically equal to 200 J/g [35], and $(1 - mf)$ is the weight fraction of PBS in the composite films.

A thermogravimetric analysis (TGA, Mettler Toledo, TGA/SDTA 851e, Columbus, OH, USA) was used to determine the PBS composite films' thermal decomposition (T_d). Samples were heated from 30–600 °C at a heating rate of 10 °C min^{-1}, under a nitrogen atmosphere, with a flow rate of 50 mL min^{-1}. The remaining content of thymol in the composite films was also determined through the gathered data in TGA. It was calculated based on weight loss as a function of temperature.

2.4.4. Mechanical Properties

The film samples were pre-conditioned inside a chamber set at 25 ± 2 °C and 50 ± 5% RH for 48 h. Young's modulus (YM), tensile strength (TS), and elongation at break (EB) in machine direction were determined according to ASTM D882-09 using Instron universal testing machine (Model 5965, Instron, Norwood, MA, USA) equipped with 5 KN load cell and a cross-head speed of 10 mm min^{-1}. At least 5 film samples (2.5 cm × 10 cm) were tested for each treatment.

2.4.5. Barrier Properties

The oxygen transmission rate (OTR) was determined using an Oxygen Permeation Analyzer 8501 (Illinois Instruments, Inc., Johnsburg, IL, USA) in accordance with ASTM D3985-17. The oxygen permeability (OP), expressed in cm^3 m h^{-1} m^{-2} atm^{-1}, was calculated using the obtained OTR values (cm^3 m^{-2} day^{-1}), film thickness l (m), and difference in partial pressure between sides of the film ΔP (Pa), applied in Equation (2):

$$\text{OP} = \frac{OTR \times l}{\Delta P} \qquad (2)$$

The water vapor permeability (WVP) was determined using the desiccant method according to ASTM E96-95. Film samples were placed on metal cups filled with dehydrated silica gel. The water vapor transmission rate (WVTR, g m^{-2} day^{-1}) was determined from the slope of the plot of the weight change of cup versus time using linear regression, divided by the area of diffusion (m^2). The WVP (g m h^{-1} m^{-2} atm^{-1}) was calculated using Equation (3) [32], using the determined WVTR, thickness l (m), and water vapor pressure difference ΔP (Pa) between sides of the film samples:

$$\text{WVP} = \frac{WVTR \times l}{\Delta P} \qquad (3)$$

2.5. Investigation of Antimicrobial Activities of SLNPs

2.5.1. Isolation of Pure Culture of L. theobromae

The pure culture of L. theobromae was isolated from mango fruit cultivar "Nam Dok Mai Si Thong" following the method of Khan et al. [36]. Briefly, a small part of decaying mango fruit tissue near the stem end was cut using a sterile blade and placed on the potato dextrose agar (PDA) contained in Petri dishes. The Petri dishes were sealed with Parafilm and incubated in an ambient condition (30 ± 2 °C) until mycelial growth was observed. Subculturing was performed up to 3 times to isolate the pure culture of L. theobromae. Pure culture was verified using a light microscope and based on the colony and conidial characteristics.

2.5.2. Antimicrobial Activities of SR-Lignin and SLNPs In Vitro

The poisoned food method, employed according to Balouiri et al. [37], was used to determine the antimicrobial activities using PDA. A mycelial plug of 8 mm diameter extracted from the edge of an actively growing, pure culture of C. gloeosporioides or L. theobromae was placed at the center of the newly prepared PDA containing SR-lignin or SLNPs. The final concentration of SR-lignin in PDA was 0.2% (w/v), while that of SLNPs varied from 0.2, 0.5, 2.0, to 5.0% (w/v). A pure PDA growth medium was used as the control. All prepared samples were sealed immediately with Parafilm and incubated at 30 ± 2 °C. Mycelial growth was measured each sampling day using a digital Vernier caliper (Mitutoyo 0.01 mm resolution) by measuring the average of the diameters in a perpendicular direction. The mycelial growth data were transformed into percentage inhibition using Equation (4) [38]. In total, 3 experiments with 3 replications per treatment were conducted for the determination of the antimicrobial activities of lignin in PDA.

$$\text{Inhibition (\%)} = 100 - \frac{\varnothing \text{ with lignin}}{\varnothing \text{ control}} \times 100 \qquad (4)$$

2.5.3. Antimicrobial Activities of SLNPs Incorporated into PBS Composite Films In Vitro

The antimicrobial activities against *C. gloeosporioides* and *L. theobromae* of the SLNPs incorporated in PBS composite films containing thymol were determined using the vapor diffusion assay, according to Boonruang et al. [29], with slight modification. A film sample (7.5 cm diameter) was sterilized under UV light and then fixed inside the cover of a Petri dish containing solidified pure PDA inoculated with an 8 mm mycelial plug of pure culture of the tested fungal species. The antimicrobial activities were investigated at room temperature (30 ± 2 °C) and 12 ± 2 °C (optimum storage temperature for mango fruit). Neat PBS film served as the control. The incubation condition, measuring of mycelial growth, and calculation of percentage inhibition were the same as described in Section 2.5.2.

2.5.4. Antimicrobial Activities of PBS Composite Films In Vivo

The antimicrobial activities of the developed PBS composite films were tested on mango fruit (*Mangifera indica* cultivar "Nam Dok Mai Si Thong"). Export quality mangoes (85% maturity, 100 days from full blossom) were purchased from Chiang Mai, Thailand, and carefully delivered to the laboratory. The mango fruit were sorted further based on similarity in shape, size, color peel, and absence of visual defects. The selected mango fruit were washed with tap water, sanitized with 200 ppm NaOCl for 2 minutes, and then air-dried for 4 h at ambient room temperature before being packed in neat PBS, PBS + 10T, and PBS + 1SLNPs + 10T composite films (20 cm × 15 cm), with 3 samples each. Mango fruit without packaging served as the control. The prepared samples were stored in a cold chamber set at 12 ± 2 °C and 90 ± 5% RH. The total area of decay caused by both *C. gloeosporioides* and *L. theobromae* was quantified by measuring the average perpendicular diameters of the decay at the end of the storage time, using a digital Vernier caliper (Mitutoyo 0.01 mm resolution).

2.6. Statistical Analysis

SPSS Statistics was used to perform an analysis of variance (ANOVA) test on samples. Duncan's multiple range test (DMRT) was employed to determine differences among sample means at a significant level of $p \leq 0.05$.

3. Results

3.1. Physicochemical Characterization of PBS Composite Films

3.1.1. Morphology

Figure 1 shows the cryo-fractured cross-section image of neat PBS, PBS + 10T, PBS + 1SLNPs, and PBS + 1SLNPs + 10T films. The captured image of the neat PBS film showed a relatively smooth cross-section surface with fibrils forming web-like structures, indicating a semi-ductile characteristic of the PBS polymer [39]. The smoothness of the cross-section surface remains when 10% thymol was added into the PBS matrix. Furthermore, no visible phase separation was seen on the image with the presence of 10% thymol in the PBS + 10T composite film, an indication of the homogeneity of the PBS composite. The incorporated 1% SLNPs were very small relative to the PBS matrix; hence, only a single particle of the SLNPs was captured in the images of both PBS + 1SLNPs and PBS + 1 SLNPs + 10T. Moreover, no agglomeration of SLNPs was seen, an indication of good dispersion of the 1% SLNPs in the PBS matrix.

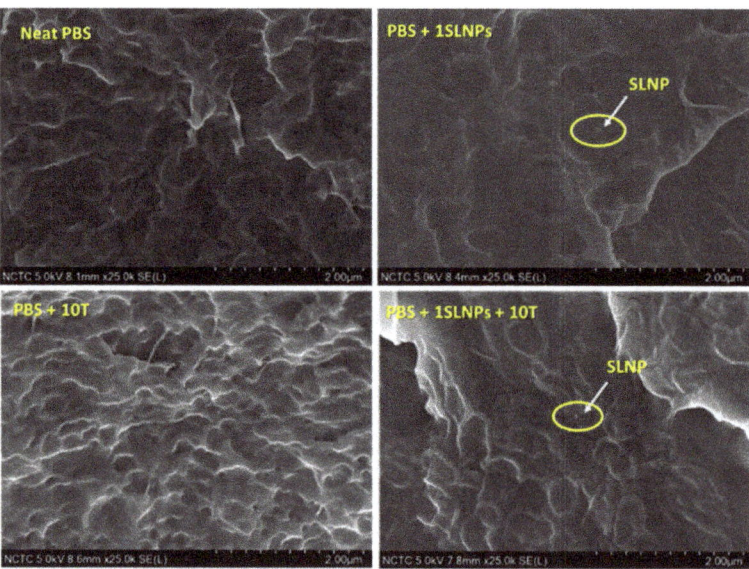

Figure 1. SEM images of cryo-fractured cross-section (side view) of neat polybutylene succinate (PBS) and PBS composite films with thymol (T) at a magnification of ×25 K. The size of softwood kraft lignin nanoparticles (SLNPs) ranged from 40–300 nm with an average particle size of 120 ± 18 nm and average polydispersity index (PDI) of 0.07 ± 0.01, determined by dynamic light scattering (DLS) technique at 25 °C (Malvern Zetasizer-4 Instrument, Malvern, UK).

3.1.2. FTIR Analysis

Interactions among 1% SLNPs, thymol, and PBS matrix were identified using ATR-FTIR analysis. Figure 2 shows the IR spectra of pure thymol, neat PBS film, and PBS composite films. The FTIR curves of the PBS composite films with 1% SLNPs + varied concentrations of thymol showed a similarity of patterns, indicating a high degree of homogeneity of the films. This could be attributed to the occurrence of intermolecular interactions, such as hydrogen bonding, between the phenolic hydroxyl groups in SLNPs and the terminal –OH or –COOH of PBS, π–π interactions between aromatic structures of lignin and thymol, and possible polar–polar interactions between the phenolic hydroxy group of SLNPs and carbonyl group of PBS [40] (Figure 3). The peaks at 2958–2868 cm^{-1} were ascribed to the symmetric and asymmetric stretching of C–H in thymol [41]. Meanwhile, the neat PBS film and all PBS composite films showed CH_2 vibration of the methylene group at 2947 cm^{-1}, which is typical for PBS [39]. The observed peak at 2349 cm^{-1} was assigned to asymmetric stretching of O=C=O, the peak at 1710 cm^{-1} was ascribed to C=O stretching of the carboxyl group, and the peak at 1335 cm^{-1} occurred due to C–O stretching of the ester group [42] in neat PBS. These peaks shifted slightly, to 2350–2352 cm^{-1}, 1711–1712 cm^{-1}, and 1332–1334 cm^{-1}, respectively, signaling interactions among 1% SLNPs, thymol, and PBS matrix. The broadening of the peak at 1152 cm^{-1} for PBS containing 1% SLNPs and thymol lower than 10% concentrations was due to the generation of intermolecular hydrogen bonding and Van der Waals force among PBS, SLNPs, and thymol [42]. Moreover, the observed 1155 cm^{-1} peak in neat PBS film, which was attributed to the ester C–O–C bond stretching [43], showed a slight change of wavenumbers, to 1154 and 1153 cm^{-1}, when 10% thymol (PBS + 10T) and 1% SLNPs (PBS + 1SLNPs + 10T) were added, respectively. The slight shift implied that there were some noncovalent interactions between PBS and thymol [17], and possibly among PBS, SLNPs, and thymol. The intense band spectra ring vibration of thymol seen at 804 cm^{-1} was attributed to out-of-plane C–H wagging vibrations, and the slight shifting of this band, to 806–807 cm^{-1}, indicated the presence of thymol in the PBS composites [41]. Meanwhile, the presence of intense bands of thymol at 588 cm^{-1} was ascribed to O–H out-of-plane deformation vibration of the content phenols [44].

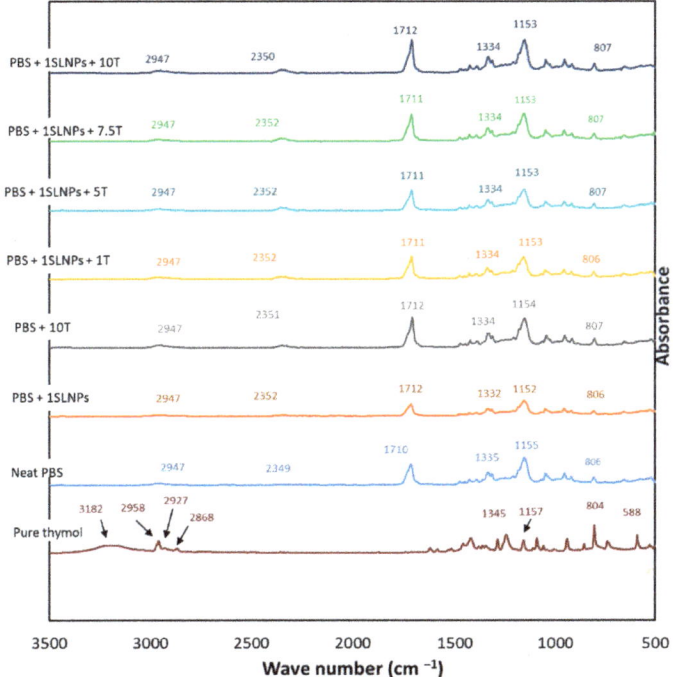

Figure 2. FTIR spectra of neat PBS, pure thymol, and PBS composites.

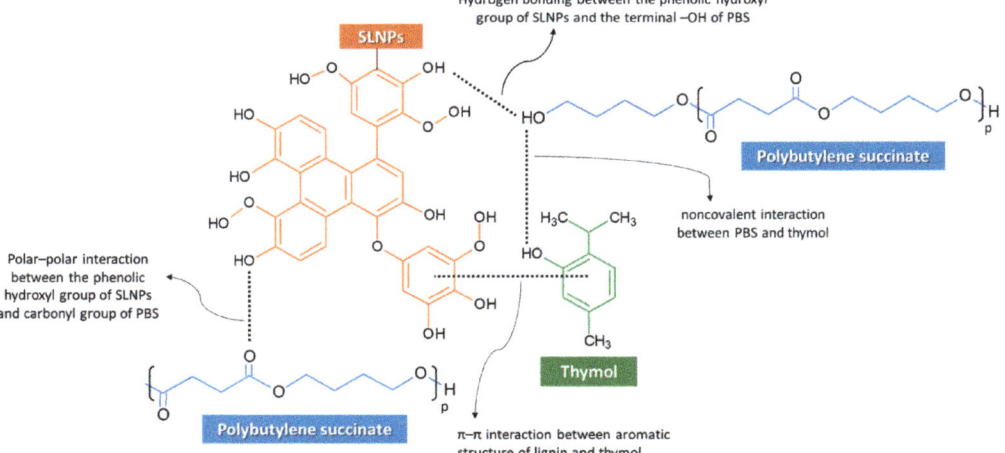

Figure 3. Representation of interactions among SLNPs, thymol, and PBS.

3.1.3. Thermal Properties

The thermal properties of neat PBS film and PBS composite films were reported in terms of glass transition temperature (T_g) and melting temperatures (T_{m1} and T_{m2}), determined through DSC analysis, and decomposition temperature (T_d), determined with the use of TGA. The results of the DSC analysis were obtained from the second heating scan of film samples, because this provided data on the ideal behavior of the material, since residual solvents and the thermal history of the material were removed in the first heating scan. Thermal properties and crystallinity of the neat PBS film and PBS composite films are summarized in Table 2. The T_g of the neat PBS film was determined to be −39.1 °C, and the addition of 1% SLNPs increased the T_g to −35.7 °C. The increase in T_g was an indicator of good interactions between 1% SLNPs and the PBS matrix as a result of strong intermolecular hydrogen bonds between polymers [45,46]. Meanwhile, the 10% (w/w) thymol functioned as a plasticizer to the PBS matrix, decreasing the T_g of the PBS + 10T film to −43.7 °C. The plasticizing effects of essential oils, such as thymol, can reduce the intermolecular interactions of the polymer chains, increasing mobility and leading to improved flexibility and ductility of the polymer [33,47,48]. The lower amount of thymol, i.e., 1 and 5% (w/w), had insignificant plasticization effects in the PBS + 1SLNPs + 1T and PBS + 1SLNPs + 5T films, since the T_g increased to −33.2 and −34.3 °C, respectively. However, it was observed that the T_g in PBS + 1SLNPs + 10T films decreased to −42.1 °C, owing to the plasticization effects of the higher amount of thymol in the PBS matrix. It was previously believed that lower thymol concentrations, i.e., of 1 and 5%, had no effect on the intermolecular bond between 1% SLNPs and PBS, yet it was observed that these concentrations produced a non-plasticizing effect in the PBS composites.

Table 2. Thermal properties of neat PBS and PBS composites.

Film Samples	T_g (°C)	T_{m1} (°C)	T_{m2} (°C)	T_c (°C)	T_d^o (°C)	T_d^p (°C)	X_c (%)
Neat PBS	−39.1	105.2	112.5	91.3	354.1	394.5	36.1
PBS + 1SLNPs	−35.7	105.2	112.7	90.8	364.7	405.0	38.5
PBS + 10T	−43.7	100.0	109.5	87.5	361.7	404.0	35.4
PBS + 1SLNPs + 1T	−33.2	104.8	112.8	90.3	364.6	405.5	34.1
PBS + 1SLNPs + 5T	−34.3	105.2	112.5	90.2	362.7	402.8	35.9
PBS + 1SLNPs + 7.5T	−40.9	104.5	112.5	90.0	362.9	400.2	37.0
PBS + 1SLNPs + 10T	−42.1	103.7	111.8	89.7	365.2	404.8	32.2

T_g: glass transition temperature; T_m: melting temperature; T_c: crystallization temperature; T_d^o: onset of thermal decomposition temperature; T_d^p: peak of thermal decomposition temperature; and X_c: crystallinity.

The melting peaks of neat PBS and PBS composite films are shown in Figure 4A. As illustrated, neat PBS and PBS composite films exhibited double melting peaks—T_{m1} and T_{m2}—due to the two different types of crystalline lamella present in PBS. The lower melting exotherm (T_{m1}) corresponded to the melting of the original crystallites, while the higher melting exotherm (T_{m2}) revealed the melting of the recrystallized crystals [39]. Generally, the T_m increases with the increase in crystallinity [49], due to the higher energy required for melting. The presence of 1% SLNPs in the PBS matrix had no significant effects both on the T_{m1} and T_{m2} of the PBS composites, the same as thymol <10%. This was reflected in the % crystallinity, where the calculated values showed no significant change. However, PBS composite films with 10% thymol, either with or without 1% SLNPs, showed noticeably reduced melting temperatures. The present work was in accordance with our previous study [33] which revealed that incorporation of 10 to 20% (w/w) thymol into antifungal PLA films led to the depletion of the T_m from 150 °C (neat PLA film) to 142 °C (PLA + 10% thymol film). Similar results were reported by Celebi and Gunes [50], where neat PLA and PLA containing 5% thymol had the same T_m at 154 °C, but the T_m decreased to 148 °C with 10% thymol in the polymer matrix. They explained that, due to the low molecular sizes of the plasticizers, thymol occupied the intermolecular spaces between the polymer chains, resulting in (1) reduced energy for molecular motion, (2) decreased formation of hydrogen bonding between polymer chains, and (3) increased free volume and molecular mobility. Nevertheless, comparing the 2 PBS composite films, PBS + 10T had lower T_{m1} and T_{m2} (100 and 109.5 °C, respectively) than PBS + 1 SLNPs + 10T (103.7 and 111.8 °C, respectively). It could be stated that the 1% SLNPs in the PBS composite helped to lessen the reduction of hydrogen bonding between PBS chains caused by the plasticization effects of thymol.

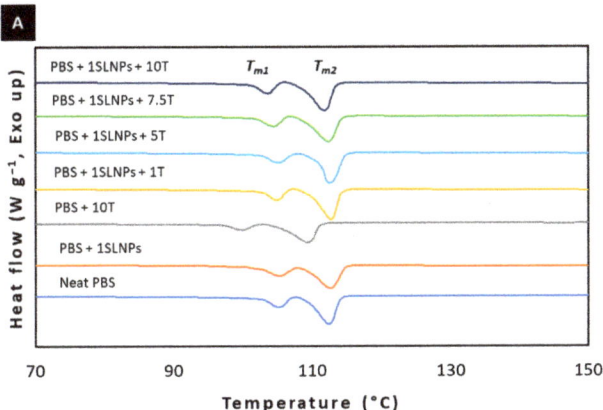
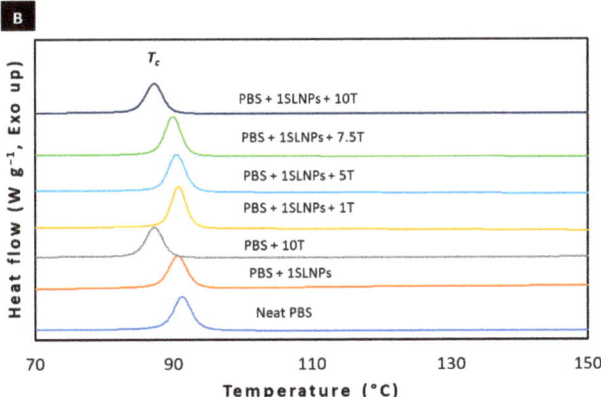
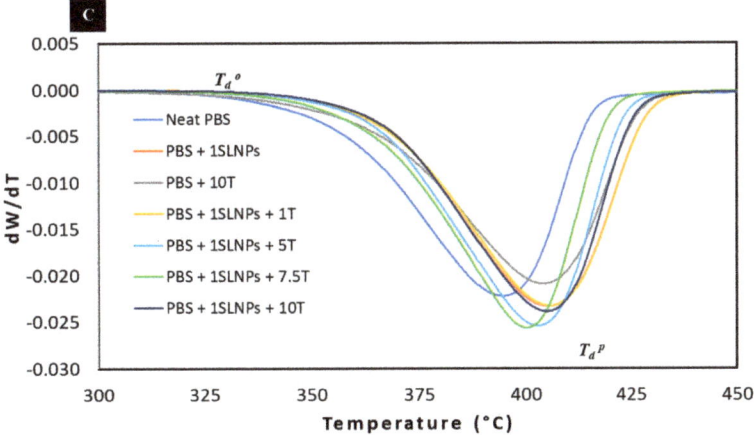

Figure 4. DSC thermograms of neat PBS and PBS composite films under nitrogen atmosphere—(**A**) second heating scan melting peaks and (**B**) crystallization peak; (**C**) derivative thermogravimetry of neat PBS and PBS composite films showing the onset and peak of thermal decomposition using TGA.

The T_c is the temperature where the molecules have enough energy to form into ordered arrangements [51]; therefore, certain molecules crystallize at this temperature [52]. Figure 4B shows the T_c observed in the cooling scan of the DSC thermogram, and the analyzed values are summarized in Table 2. Similar to the obtained results for the melting temperatures, the 1% SLNPs and thymol <10% had no significant effects on the T_c of the PBS composites. It was noticeable, however, that the lowest T_c was obtained in PBS + 10T at 87.5 °C; however, with the presence of 1% SLNPs, the obtained T_c was increased to 89.7 °C.

Thermal stability of neat PBS and PBS composites films was analyzed through TGA by determining the onset and peak of the T_d under a nitrogen atmosphere. The thermal stability of a polymer refers to the capability of the polymer to resist thermal action while maintaining its properties (strength, toughness, elasticity) at a given temperature [53]. Figure 4C shows the derivative thermogravimetry of neat PBS and PBS composite films; the onset and peak of T_d is shown. The onset of T_d was recorded at 354.1 °C for the neat PBS film. Incorporation of 1% SLNPs into PBS increased the onset T_d to 364.7 °C, and the incorporation of 10% thymol increased the T_d to 361.7 °C. The results showed that 1% SLNPs in the PBS enhanced the thermal stability of the PBS matrix more than the 10% thymol. Moreover, the combined effects of 1% SLNPs and 10% thymol in the PBS matrix enhanced further the thermal stability of the resultant PBS composite film, in which the recorded onset of T_d of PBS + 1SLNPs + 10T was 365.2 °C. Furthermore, as an effect of the presence of 1% SLNPs in the PBS composites, the recorded onset of T_d of PBS + 1SLNPs + T was 364.6, 362.7, and 362.9 °C for 1T, 5T, and 7.5T, respectively, which were generally higher than the onset of T_d of PBS + 10T (without 1% SLNPs). The peak of T_d revealed that 1% SLNPs could improve the T_d of PBS composite more than 10% thymol. Moreover, the peak of T_d revealed that thymol could help to improve the T_d of PBS composites, but the incorporation of 1% SLNPs into PBS + 10T could enhance the T_d to a greater degree. The obtained results showed that 1% SLNPs and thymol both enhanced the thermal stability of PBS, a finding that is in accordance with Mariën et al. [54] and Mousavioun et al. [55], who reported that lignin fragments strongly improved the thermal stability of modified silicone polymer and soda lignin improved the overall thermal stability of poly(hydroxybutyrate) or PHB, respectively. Meanwhile, Mohamad et al. [42] reported that the incorporation of thymol into PBS could help enhance the thermal stability of neat PBS.

Based on the TGA results, the remaining content of thymol in PBS + 1SLNPs + 1T, PBS + 1SLNPs + 5T, and PBS + 1SLNPs + 7.5T composite films was 0.99, 4.96, and 7.45%, respectively. Meanwhile, the remaining content of thymol in PBS + 10T was 9.58%; this was slightly increased to 9.72% in PBS + 1SLNPs + 10T, which could be an effect of the π–π interactions between aromatic structures of SLNPs and thymol in the PBS matrix (Figure 3).

The degree to which the polymer chains are aligned with one another is referred to as the crystallinity of a polymer [56]. The calculated % crystallinity of the neat PBS and PBS composites as affected by 1% SLNPs and thymol conformed to the obtained T_g, T_m, and T_d of the composite materials. As shown in Table 2, the % crystallinity of neat PBS was 36.1%, and was slightly increased to 38.5% when 1% SLNP was incorporated into the PBS matrix, due to the nucleation effects of lignin. Lignin can act as a nucleating agent [57,58]; hence, it can increase the crystallinity of a polymer. A similar result was reported by Sahoo et al. [45], who concluded that the % crystallinity of PBS increased with the addition of lignin up to 65% (w/w)—a much higher amount of lignin compared with the present study. Moreover, 10% thymol caused a decrease in the % crystallinity of PBS + 1SLNPs + 10T film, which confirmed the plasticizing effects affecting T_g and T_m of the PBS composite. This is in accordance with our previous study which showed the reduction of the degree of crystallinity of PLA after the addition of 10–20% (w/w) thymol [33]. A similar result, that PBS blended with 2–10% (w/w) thymol exhibited a lower degree of crystallinity compared with neat PBS, was reported by [59].

3.1.4. Mechanical Properties

The mechanical properties of a material describe its behavior upon the application of external force(s). Table 3 shows the effects of 1% SLNPs, thymol, and combined 1% SLNPs and thymol on the Young's modulus (YM), tensile strength (TS), and elongation at break (EB) of the PBS composite films.

Table 3. Young's modulus (YM), tensile strength (TS), and elongation at break (EB) in the machine direction of neat PBS and PBS composite films.

Film Samples	YM (MPa)	TS (MPa)	EB (%)
Neat PBS	593.4 ± 84.2 [bc]	34.3 ± 1.4 [bc]	11.6 ± 0.8 [a]
PBS + 1SLNPs	601.9 ± 57.9 [c]	35.6 ± 1.8 [c]	10.3 ± 0.8 [a]
PBS + 10T	491.9 ± 62.8 [ab]	28.3 ± 2.5 [a]	11.9 ± 0.4 [a]
PBS + 1SLNP + 1T	589.1 ± 63.9 [bc]	34.9 ± 1.2 [bc]	11.6 ± 0.9 [a]
PBS + 1SLNP + 5T	445.3 ± 97.3 [a]	34.0 ± 7.1 [bc]	11.6 ± 2.1 [a]
PBS + 1SLNP + 7.5T	466.5 ± 62.2 [a]	34.9 ± 3.2 [bc]	11.2 ± 1.7 [a]
PBS + 1SLNP + 10T	486.6 ± 104.4 [ab]	30.0 ± 4.6 [ab]	10.6 ± 0.6 [a]

Values are presented as mean ± standard deviation ($n = 5$). Means followed by the same superscript letters in columns are not significantly different at $p \leq 0.05$ using DMRT.

The YM conformed to the T_m of the film samples, in general. PBS composites with higher T_m showed higher YM. This relationship is associated with molecular mobility as a result of the strength of the atomic bonding of polymers in the composite [60]. Lignin is highly interactive, which can enhance the films' mechanical properties, due to the strong hydrogen bonds between its functional group and the polymer matrix [61]. This could be the reason for the increase in YM from 593.4 to 601.9 MPa, and in TS from 34.3 to 35.6 MPa, when 1% SLNP was incorporated into the PBS matrix. Similarly, the YM and TS of wheat gluten bioplastic increased with the presence of kraft lignin in the polymer, as reported in [62]. Sakunkittiyut et al. [63] showed that up to 30% (w/w) of kraft lignin significantly increased the YM and TS of a fish protein-based polymer by 50 and 300%, respectively. Moreover, the YM and TS of starch/lignin bio-composites increased with the addition of a concentration of lignin up to 3% (w/w) [64]. Similarly, the YM and TS of agar/lignin bio-composites increased [65], and the YM and TS of whey protein isolate/lignin bio-composites increased with the addition of a 0.5% (w/w) concentration of lignin [66]. Meanwhile, it was expected that the YM and TS of the PBS composites containing thymol would generally decrease, as thymol lessens the strong intermolecular bonding of PBS chains, resulting in higher mobility of the chain [59,67]. These studies dealt with a low level of lignin, from 0.5 to 3.0%, to attain an improved YM and TS. Other previous studies revealed that a higher amount of lignin in the polymer matrix resulted in decreased YM and TS. Aadil et al. [68] reported that high lignin concentrations, of 10 and 20%, decreased the TS of a lignin–alginate composite down to 0.5 and 0.1 MPa, respectively, compared to the TS of a neat alginate, which was 0.6 MPa. A similar result was shown by Izaguirre et al. [69]; lignin concentrations higher than 0.25% (w/w) did not increase further the YM and TS of chitosan films. The decrease in YM and TS of the polymer matrix incorporated with a higher level of lignin could be the result of poor interfacial interaction between the lignin and the polymer matrix, due to agglomeration of lignin particles in higher concentrations caused by strong hydrogen bonding between the functional groups present in lignin [70].

The use of 1% SLNPs in the present work seemed to be the appropriate level, as it enhanced the mechanical properties, in particular, the YM and TS, of the PBS composite films. For the EB, the results showed that the change was not significant ($p \leq 0.05$) as compared to the neat PBS.

3.1.5. Barrier Properties

Barrier properties of the neat PBS and PBS composite films are reported in terms of OP and WVP (Table 4). The calculated OP of neat PBS was 2.28×10^{-3} cm^3 m h^{-1} m^{-2} atm^{-1} and significantly reduced to 1.59×10^{-3} cm^3 m h^{-1} m^{-2} atm^{-1} with the incorporation of 1% SLNPs. This effect was due to the creation of a winding path by the lignin in the PBS matrix, resulting in difficulty for the movement of the diffusing O_2 molecules. The O_2 molecules navigated around the lignin dispersed in the PBS, resulting in a longer time of diffusion [71]. The reduced OP of the PBS + 1SLNPs was related to the increased crystallinity of the PBS composite films. Since lignin is a nucleating agent [57], the presence of the SLNPs in the PBS matrix could result in the development of crystal phases that are assumed to be impermeable [72]; this could hinder the movement of the diffusing O_2 molecules. Moreover, the reduced OP of PBS with the presence of 1% SLNPS could be due to the reinforcing effects of lignin on the biobased polymer matrix. The results were in accordance with Kovalcik et al. [73], who reported that the OP of poly(3-hydroxybutyrate-co-3-hydroxyvalerate) (PHBHV) bio-polyester decreased with the addition of 1% (w/w) kraft lignin, from 6.97×10^{-4} cm^3 m h^{-1} m^{-2} atm^{-1} down to 1.69×10^{-4} cm^3 m h^{-1} m^{-2} atm^{-1}. The effectiveness of the 1% SLNPs in reducing the OP of the PBS composite films can also be associated with the rigid nature of lignin particles in conjunction with the good dispersion of the SLNPs in the PBS matrix.

Table 4. Thickness, OP, and WVP of neat PBS and PBS composites films.

Film Samples	Thickness (μm)	Oxygen Permeability (cm^3 m h^{-1} m^{-2} atm^{-1})	Water Vapor Permeability (g m h^{-1} m^{-2} atm^{-1})
Neat PBS	29.8 ± 0.8 [a]	2.28×10^{-3} ± 0.0002 [b]	1.54×10^{-3} ± 0.0001 [d]
PBS + 1SLNPs	30.0 ± 1.6 [a]	1.59×10^{-3} ± 0.0002 [a]	1.41×10^{-3} ± 0.0001 [cd]
PBS + 10T	31.8 ± 4.9 [a]	2.37×10^{-3} ± 0.0002 [bc]	1.14×10^{-3} ± 0.0001 [ab]
PBS + 1SLNP + 1T	30.2 ± 2.7 [a]	2.40×10^{-3} ± 0.0004 [bc]	1.28×10^{-3} ± 0.0001 [bc]
PBS + 1SLNP + 5T	30.4 ± 2.7 [a]	2.08×10^{-3} ± 0.0001 [b]	1.15×10^{-3} ± 0.0002 [ab]
PBS + 1SLNP + 7.5T	33.4 ± 1.5 [a]	2.56×10^{-3} ± 0.0005 [bc]	1.28×10^{-3} ± 0.0003 [ab]
PBS + 1SLNP + 10T	33.4 ± 2.4 [a]	2.84×10^{-3} ± 0.0004 [c]	1.09×10^{-3} ± 0.0000 [ab]

Values are presented as mean ± standard deviation (n = 4). Means followed by the same superscript letters in columns are not significantly different at $p \leq 0.05$ using DMRT.

Meanwhile, the general increase of the OP of the PBS matrix incorporated with thymol was ascribed to the destabilization of the PBS chain due to the plasticizing effects of thymol leading to a less dense and less cohesive PBS polymer network (in the case of PBS + 10T) and reduced intermolecular interaction between SLNPs and PBS chain (in case of PBS + 1SLNPs + varied concentrations of thymol). The increased OP values in PBS composites with thymol were probably due to the changes in the PBS matrix structure caused by thymol, as proven by the FTIR analysis. These changes increased the diffusion of O_2 molecules through the PBS composite films with thymol. This was consistent with Othman et al. [74], who reported that the OP values of corn starch films increased as an effect of thymol present in the polymer matrix.

The WVP of the neat PBS reduced from 1.54×10^{-3} g m h^{-1} m^{-2} atm^{-1} to 1.41×10^{-3} g m h^{-1} m^{-2} atm^{-1} with the addition of 1% SLNPs because of the difficult path for the water vapor in the PBS matrix created by the lignin. This reduction could also be due to the good compatibility between 1% SLNPs and the PBS matrix leading to strong molecular interactions between polymers, as shown in Figures 2 and 3. Previous studies have shown that lignin enhanced the water vapor barrier of thermoplastic starch [75], sago starch-based food packaging film [64], agar composite films [65], and soy protein isolate film [30]. Meanwhile, the addition of 10% thymol in the PBS caused a significant decrease in the WVP of the PBS composites. This reduction in WVP was due to the hydrophobic property of thymol resulting in decreased affinity of the PBS composite with water. Moreover, the hydrogen and covalent interactions between the PBS network and the phenolic compound of thymol may reduce the availability of hydrogen groups to produce hydrophilic bonds with water [74].

3.2. Investigation of Antimicrobial Activities of SLNPs

3.2.1. Antimicrobial Activities of SR-Lignin and SLNPs In Vitro

The poisoned food method is a common technique used to evaluate the antimicrobial activities against fungi. In this method, the antimicrobial agent is mixed well into the growth medium at the desired final concentration [37]. The present study investigated the antimicrobial activities of SR-lignin and SLNPs against *C. gloeosporioides* and *L. theobromae* through the poisoned food method, at 30 ± 2 °C, using PDA as the growth medium. As shown in Figure 5A, the SR-lignin at 0.2% (w/v) concentration showed low growth inhibition against *C. gloeosporioides*, with 2.03, 4.03, 1.26, and 0.70% on days 3, 6, 9, and 12, respectively. There was no inhibition on day 15 because the mycelia on both SR-lignin samples and control reached the edge of the Petri dish. It was found that purifying the SR-lignin and processing it into LNPs significantly increased the antimicrobial activities, to 646.79%, at the same concentration of 0.2% (w/v). Moreover, the growth inhibition was significantly increased further when the concentration of SLNPs was increased from 0.2 to 0.5% (w/v). The growth inhibition of SLNPs at 0.5% (w/v) concentration was 352.06% higher than SLNPs at 0.2% (w/v) concentration. Therefore, it was expected that a concentration of SLNPs higher than 0.5% (w/v) would result in a stronger growth inhibition. However, the growth inhibition against *C. gloeosporioides* in 0.5, 2.0, and 5.0% (w/v) concentrations were not significantly ($p \leq 0.05$) different throughout the days of observation. The results were comparable with what was reported by Dominguez-Robles et al. [17], regarding the degree of adherence of *S. aureus* on the PBS composites containing different concentrations of lignin. PBS composites containing 2.5, 5, 10, and 15% (w/w) lignin showed a similar degree of bacterial adherence at $p \leq 0.05$, which suggests that the resistance to bacterial adherence was not directly dependent on lignin concentration.

SR-lignin at 0.2% (w/v) concentration showed a stronger growth inhibition against *L. theobromae*, as shown in Figure 5B. The growth inhibitions were 44.61 and 16.07% on days 3 and 6, respectively. With the same concentration of 0.2% (w/v), SLNPs showed 32.06% significantly stronger antimicrobial effects compared with SR-lignin. With an increase in the SLNPs' concentrations from 0.2 to 0.5% (w/v), the growth inhibition significantly increased, by 18.96%, and further increasing the concentrations, to 2.0 and 5.0% (w/v), produced no significant ($p \leq 0.05$) increase in the antimicrobial effects. The same level of growth inhibition at 0.5, 2.0, and 5.0% (w/v) concentrations could be due to the agglomeration of SLNPs which occurred, visible on the reverse side of the Petri dishes containing the growth medium (Figure 6). The agglomeration of SLNPs at higher concentration can hide a portion of the surface area of the SLNPs, affecting the full interactions against the tested fungal species. The agglomeration could be the result of high surface energy, a large number of hydrogen bonds, and Van der Waals forces between SLNP particles [76]. In comparison, the growth inhibition of SR-lignin and SLNPs, both at 0.2% (w/v) concentration, was higher than that of *C. gloeosporioides* against *L. theobromae*. However, it could be observed that the antimicrobial effects against *L. theobromae* remained up to 6 days, compared with the effects against *C. gloeosporioides*, which persisted for 15 days. The difference in the antimicrobial activities was the outcome of the faster growth of *L. theobromae* at 30 °C [77].

The presence of SLNPs in the culturing media could inhibit growth and cause changes in the morphology of the microorganism. In particular, the presence of many functional groups, especially the phenolic hydroxyl group, is responsible for the antimicrobial effects of lignin. This functional group can modify the physiological process of the microorganism, leading to damaging the cell membrane and eventually causing the loss of functionalities of the microorganism [78,79]. It was suggested that the phenolics in lignin can disrupt the cell wall, resulting in altered physiological processes and, eventually, dysfunction of parts of the microorganism and growth inhibition [20].

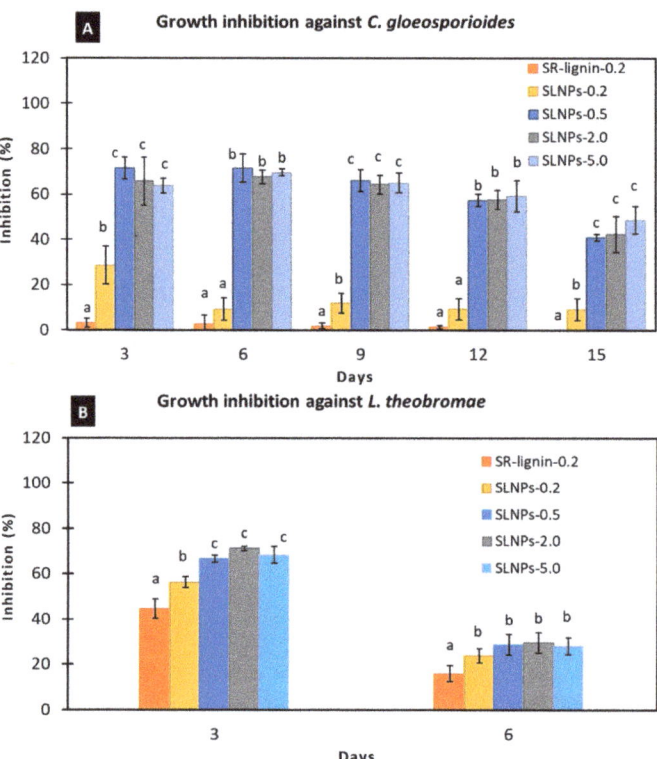

Figure 5. Antimicrobial effects of raw softwood kraft lignin (SR–lignin) and SLNPs in PDA against (**A**) *Colletotrichum gloeosporioides* and (**B**) *Lasiodiplodia theobromae*. Values are presented as mean ($n = 3$) with the standard deviation represented by vertical bars. Means followed by the same letter between treatments per day are not significantly different at $p \leq 0.05$ using DMRT. SR–lignin: raw softwood kraft lignin.

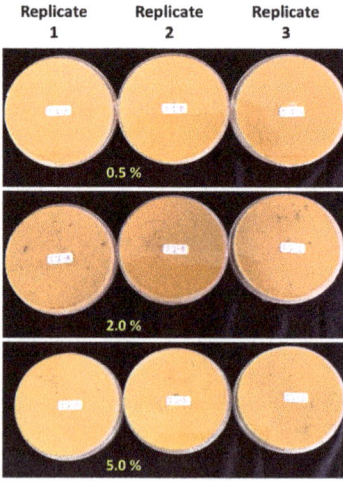

Figure 6. Reverse side of the Petri dishes containing PDA growth medium. The black spots visible at 2.0 and 5.0% were agglomerated SLNPs.

3.2.2. Antimicrobial Activities of SLNPs Incorporated into PBS Composite Films In Vitro

Figure 7 shows the antimicrobial activities against *C. gloeosporioides* and *L. theobromae* of 1% SLNPs incorporated into the PBS composite films containing thymol, determined using the vapor diffusion method. Thymol can inhibit the mycelial growth of the tested fungi either at 30 or 12 °C, and the inhibition was significantly reduced as the concentration of thymol decreased. At 30 °C, 10% (w/w) thymol in PBS was sufficient to fully inhibit the growth of *C. gloeosporioides* for at least 8 days; the % inhibition decreased to 95.1 and 88.94% on days 10 and 15, respectively. However, with the presence of 1% SLNPs in the PBS + 10T, mycelial growth was not observed until the last day of observation, which equates to 100% growth inhibition. Meanwhile, both PBS + 10T and PBS + 1SLNPs + 10T fully inhibited the mycelial growth of *L. theobromae* until day 2. The % inhibition decreased over time with 95.44, 87.40, 69.91, 40.88, and 17.53% on days 4, 6, 8, 10, and 15, respectively, for PBS + 10T, and 96.21, 90.68, 78.77, 47.03, and 25.42%, respectively, for PBS + 1SLNPs + 10T. Although the calculated % inhibition in PBS + 1SLNPs + 10T was generally higher than in PBS + 10T, the significant ($p \leq 0.05$) difference was determined starting on day 10 for *C. gloeosporioides* and on day 6 for *L. theobromae*. It is interesting to note that with the presence of 1% SLNPs, the antimicrobial effects against the studied fungal species increased, as well as the duration of effectiveness.

The observation for the growth inhibition against the tested fungi at 12 °C was started on day 8 because no significant ($p > 0.05$) mycelial growth was observed before that day. Like the observations at 30 °C, the inhibition against *C. gloeosporioides* and *L. theobromae* reduced as the concentrations of thymol decreased, but both PBS + 10T and PBS + 1SLNPs + 10T produced 100% inhibition against the tested fungi for 32 days. Since temperature directly affects fungal growth [80] and the fungal metabolic activities important for growth are slowed down at low temperatures, making the fungi dormant until optimum temperature [81], samples were moved out to ambient temperature (30 ± 2 °C) for further observation. After 10 days in ambient temperature, there was an average of 3.67 mm mycelial growth of *C. gloeosporioides* in PBS + 10T, and the calculated % inhibition was 95.24%; however, mycelial growth was still not observed in PBS + 1SLNPs + 10T. Thus, the inhibition remained at 100%. Meanwhile, for *L. theobromae*, the mycelia grew up to the edge of the Petri dishes after ten days in PBS + 10T, resulting in zero inhibition. However, an average of 39.42 mm mycelial growth was recorded in PBS + 1SLNPs + 10T, resulting in 48.81% growth inhibition. The results signify that the presence of 1% SLNPs in the PBS composite films containing thymol could significantly enhance the antimicrobial effects, particularly in the final days of observation. This could be due to the molecular interaction of SLNPs and thymol, particularly the substantial π-stacking between aromatic compounds in thymol and aromatic units of lignin [82], which could retain thymol in the PBS matrix for a longer period, leading to longer antimicrobial effectiveness.

3.2.3. Antimicrobial Activities of PBS Composite Films In Vivo

The antimicrobial activities on "Nam Dok Mai Si Thong" mango fruit in PBS composite films containing 1% SLNPs and 10% thymol were quantified by determining the total area of decay on the last day of observation, day 33. Mango samples were unpacked on this day due to the unmarketable appearance of control mangoes and samples packed in neat PBS. There were black-brown sunken circular spots observed on these mango samples, a manifestation of the mango disease anthracnose, caused by *C. gloeosporioides* [83]. Additionally, these fruit samples showed a dark-brownish area on the surface around the base of the fruit's stem end, an indicator of stem-end rot disease caused by *L. theobromae* [84]. Conversely, no visible fruit diseases were observed in the early stage of storage, e.g., on days 3 and 6, as shown in Figure 8.

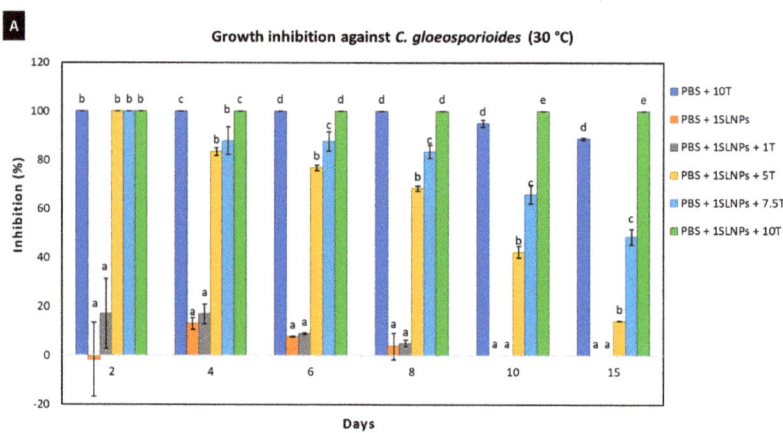

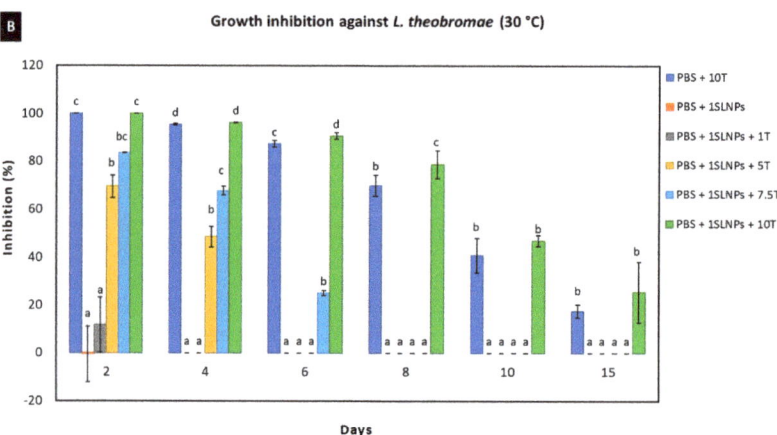

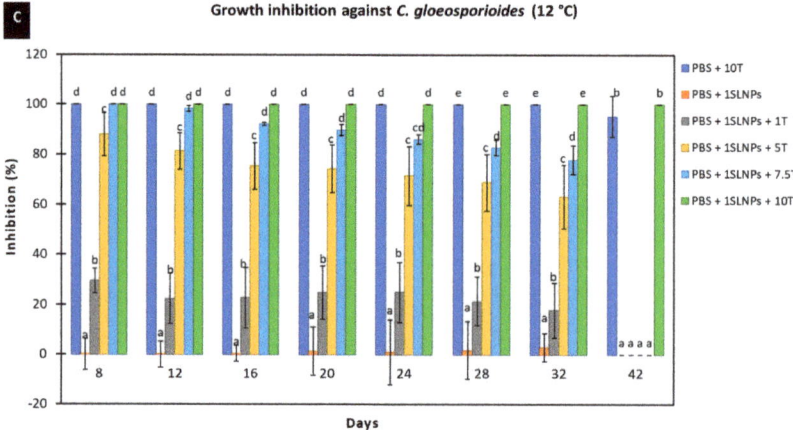

Figure 7. *Cont.*

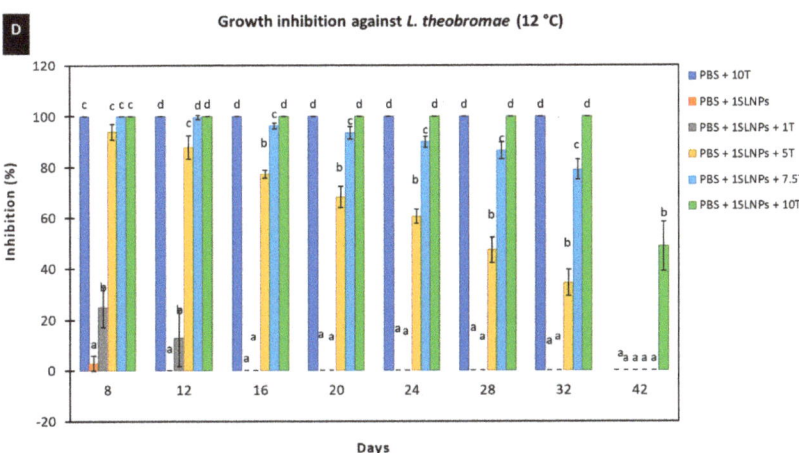

Figure 7. Growth inhibition of PBS composite films containing 1% SLNPs at 30 °C against (**A**) *C. gloeosporioides* and (**B**) *L. theobromae*; at 12 °C against (**C**) *C. gloeosporioides* and (**D**) *L. theobromae*. Values are presented as mean (*n* = 3) with the standard deviation represented by vertical bars. Means followed by the same letter between treatments per day are not significantly different at $p \leq 0.05$ using DMRT.

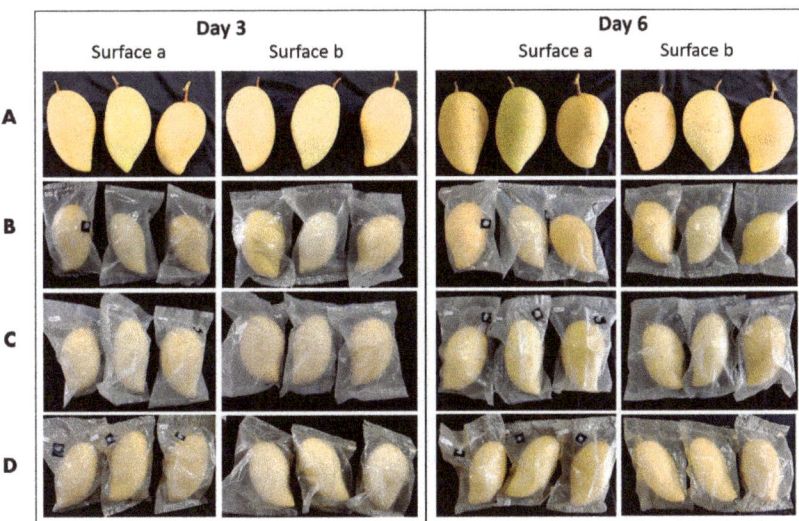

Figure 8. Mango fruit samples on days 3 and 6, stored at 12 °C and 90 ± 5% RH. (**A**) Control, (**B**) neat PBS, (**C**) PBS + 10T, and (**D**) PBS + 1SLNPs + 10T.

Table 5 is the summary of the calculated total decay of the mango samples. Mango fruit packed in neat PBS exhibited the largest total area of decay, with 17.2 cm^2, followed by control samples, with 16.0 cm^2. This could be due to the saturated RH inside the package (100% RH) compared to the % RH of the cold storage chamber (90 ± 5%). As shown in the work of Dannemiller et al. [85], the fungal growth rate at 100% equilibrium RH was higher than at 85% equilibrium RH. Meanwhile, 10% thymol in the PBS matrix significantly inhibited the growth of the decay-causing fungal species, as shown by the low total area of decay of 5.2 cm^2. However, with the presence of 1% SLNPs in the PBS + 10T composite, the total area of decay was significantly ($p \leq 0.05$) reduced further to 1.1 cm^2. The results were associated with what was obtained in the in vitro study.

Table 5. Total area of decay on day 33.

Treatment	Decay Area (cm^2)
CONTROL	16.0 ± 2.1 [c]
Neat PBS	17.2 ± 0.3 [c]
PBS + 10T	5.2 ± 0.7 [b]
PBS + 1SLNPs + 10T	1.1 ± 0.1 [a]

Means followed by the same superscript letter between treatments are not significantly different at $p \leq 0.05$ using DMRT.

4. Conclusions

The incorporation of SLNPs into the PBS film containing thymol was shown to increase the tensile strength, barrier against oxygen, thermal decomposition temperature, and antifungal activities. FTIR results confirmed the good interactions among SLNPs, thymol, and PBS matrix, and SEM corroborated the homogeneity of the PBS composite film. The synergistic effects of SLNPs and thymol in the PBS matrix showed the strongest microbial growth inhibition against *C. gloeosporioides* and *L. theobromae*, two major fungal species that cause anthracnose and stem-end rot diseases in mango fruit, respectively. This work showed that SLNPs could be an attractive natural alternative to synthetic substances for enhancing polymer properties without compromising the biodegradability of the resultant material. Furthermore, the results gathered are important and recommended for extending the shelf life of many economically important fruit crops that are susceptible to anthracnose and stem-end rot diseases. In addition, this natural alternative could potentially be applied as an antimicrobial packaging for other food products.

Author Contributions: Conceptualization, A.J.B., B.H., C.W., W.W., W.C., P.L., V.C. and K.B.; methodology, A.J.B., B.H., C.W., W.W., P.L., V.C. and K.B.; formal analysis, A.J.B., B.H., C.W. and K.B.; investigation, A.J.B., B.H. and C.W.; resources, V.C., B.H., C.W., W.W., P.L. and K.B.; data curation, B.H., C.W., P.L. and K.B.; writing—original draft preparation, A.J.B.; writing—review and editing, B.H., C.W., W.W., P.L., V.C. and K.B.; visualization, A.J.B.; supervision, K.B., V.C. and W.C.; funding acquisition, K.B., V.C. and W.C. All authors have read and agreed to the published version of the manuscript.

Funding: Kasetsart University Research and Development Institute, KURDI (Project No. FF (KU) 22.66).

Institutional Review Board Statement: Not applicable.

Informed Consent Statement: Not applicable.

Data Availability Statement: All data are contained in the article.

Acknowledgments: This research was supported by the Kasetsart University Research and Development Institute, KURDI (Project No. FF (KU) 22.66). The authors wish to acknowledge the Department of Packaging and Materials Technology, Faculty of Agro-Industry, Kasetsart University, Bangkok 10900, Thailand, for the scholarship grant for A.J.B.

Conflicts of Interest: The authors declare no conflict of interest.

References

1. Sikora, J.W.; Majewski, Ł.; Puszka, A. Modern biodegradable plastics—Processing and properties part II. *Materials* **2021**, *14*, 2523. [CrossRef] [PubMed]
2. Abe, M.M.; Martins, J.R.; Sanvezzo, P.B.; Macedo, J.V.; Branciforti, M.C.; Halley, P.; Botaro, V.R.; Brienzo, M. Advantages and disadvantages of bioplastics production from starch and lignocellulosic components. *Polymers* **2021**, *13*, 2484. [CrossRef] [PubMed]
3. Misra, M.; Vivekanandhan, S.; Mohanty, A.; Denault, J. Nanotechnologies for Agricultural Bioproducts. In *Comprehensive Biotechnology*, 3rd ed.; Elsevier: Amsterdam, The Netherlands, 2011; pp. 119–127. [CrossRef]

4. Brunner, G. Processing of Biomass with Hydrothermal and Supercritical Water. In *Supercritical Fluid Science and Technology*; Elsevier: Amsterdam, The Netherlands, 2014; pp. 395–509. [CrossRef]
5. Ahmad, U.M.; Ji, N.; Li, H.; Wu, Q.; Song, C.; Liu, Q.; Ma, D.; Lu, X. Can lignin be transformed into agrochemicals? Recent advances in the agricultural applications of lignin. *Ind. Crop. Prod.* 2021, *170*, 113646. [CrossRef]
6. Vinardell, M.P.; Mitjans, M. Lignins and Their Derivatives with Beneficial Effects on Human Health. *Int. J. Mol. Sci.* 2017, *18*, 1219. [CrossRef]
7. Huang, J.; Fu, S.; Gan, L. Lignin-modified thermoplastic materials. In *Lignin Chemistry and Applications*; Huang, J., Fu, S., Gan, L., Eds.; Elsevier: Amsterdam, The Netherlands, 2019; pp. 135–161.
8. Rudnik, E. Compostable polymer properties and packaging applications. In *Plastic films in food packaging*; Ebnesajjad, S., Ed.; William Andrew Publishing: Norwich, NY, USA, 2013; pp. 217–248.
9. Jiang, L.; Zhang, J. Biodegradable and biobased polymers. In *Applied plastics engineering handbook*, 2nd ed.; Kutz, M., Ed.; William Andrew Publishing: Norwich, NY, USA, 2017; pp. 127–143.
10. Aliotta, L.; Seggiani, M.; Lazzeri, A.; Giganate, V.; Cinelli, P. A brief review of Poly(Butylene Succinate) (PBS) and Its Main Copolymers: Synthesis, Blends, Composites, Biodegradability, and Applications. *Polymers* 2022, *14*, 844. [CrossRef]
11. Tachibana, Y.; Masuda, T.; Funabashi, M.; Kunioka, M. Chemical Synthesis of Fully Biomass-Based Poly(butylene succinate) from Inedible-Biomass-Based Furfural and Evaluation of Its Biomass Carbon Ratio. *Biomacromolecules* 2010, *11*, 2760–2765. [CrossRef]
12. Soccio, M.; Dominici, F.; Quattrosoldi, S.; Luzi, F.; Munari, A.; Torre, L.; Lotti, N.; Puglia, D. PBS-Based Green Copolymer as an Efficient Compatibilizer in Thermoplastic Inedible Wheat Flour/Poly(butylene succinate) Blends. *Biomacromolecules* 2020, *21*, 3254–3269. [CrossRef]
13. Jordá-Reolid, M.; Ibáñez-García, A.; Catani, L.; Martínez-García, A. Development of Blends to Improve Flexibility of Biodegradable Polymers. *Polymers* 2022, *14*, 5223. [CrossRef]
14. Dintcheva, N.T.; Infurna, G.; Baiamonte, M.; D'Anna, F. Natural Compounds as Sustainable Additives for Biopolymers. *Polymers* 2020, *12*, 732. [CrossRef] [PubMed]
15. Pivsa-Art, S.; Pivsa-Art, W. Eco-friendly bamboo fiber-reinforced poly(butylene succinate) biocomposites. *Polym. Compos.* 2021, *42*, 1752–1759. [CrossRef]
16. Saffian, H.A.; Hyun-Joong, K.; Tahir, P.; Ibrahim, N.A.; Lee, S.H.; Lee, C.H. Effect of Lignin Modification on Properties of Kenaf Core Fiber Reinforced Poly(Butylene Succinate) Biocomposites. *Materials* 2019, *12*, 4043. [CrossRef] [PubMed]
17. Domínguez-Robles, J.; Larrañeta, E.; Fong, M.L.; Martin, N.K.; Irwin, N.J.; Mutjé, P.; Tarrés, Q.; Delgado-Aguilar, M. Lignin/poly(butylene succinate) composites with antioxidant and antibacterial properties for potential biomedical applications. *Int. J. Biol. Macromol.* 2019, *145*, 92–99. [CrossRef] [PubMed]
18. Mtibe, A.; Hlekelele, L.; Kleyi, P.E.; Muniyasamy, S.; Nomadolo, N.E.; Ofosu, O.; Ojijo, V.; John, M.J. Fabrication of a Polybutylene Succinate (PBS)/Polybutylene Adipate-Co-Terephthalate (PBAT)-Based Hybrid System Reinforced with Lignin and Zinc Nanoparticles for Potential Biomedical Applications. *Polymers* 2022, *14*, 5065. [CrossRef]
19. Calvo-Flores, F.G. Lignin: A renewable raw material. In *Encyclopedia of Renewable and Sustainable Materials*; Hashmi, S., Choudhury, I.A., Eds.; Elsevier: Amsterdam, The Netherlands, 2020; Volume 5, pp. 102–118.
20. Basbasan, A.J.; Hararak, B.; Winotapun, C.; Wanmolee, W.; Leelaphiwat, P.; Boonruang, K.; Chinsirikul, W.; Chonhenchob, V. Emerging challenges on viability and commercialization of lignin in biobased polymers for food packaging: A review. *Food Packag. Shelf Life* 2022, *34*, 100969. [CrossRef]
21. Liu, R.; Smeds, A.; Tirri, T.; Zhang, H.; Willför, S.; Xu, C. Influence of Carbohydrates Covalently Bonded with Lignin on Solvent Fractionation, Thermal Properties, and Nanoparticle Formation of Lignin. *ACS Sustain. Chem. Eng.* 2022, *10*, 14588–14599. [CrossRef]
22. Alzagameem, A.; Klein, S.E.; Bergs, M.; Do, X.T.; Korte, I.; Dohlen, S.; Hüwe, C.; Kreyenschmidt, J.; Kamm, B.; Larkins, M.; et al. Antimicrobial Activity of Lignin and Lignin-Derived Cellulose and Chitosan Composites against Selected Pathogenic and Spoilage Microorganisms. *Polymers* 2019, *11*, 670. [CrossRef]
23. Morena, A.G.; Bassegoda, A.; Natan, M.; Jacobi, G.; Banin, E.; Tzanov, T. Antibacterial Properties and Mechanisms of Action of Sonoenzymatically Synthesized Lignin-Based Nanoparticles. *ACS Appl. Mater. Interfaces* 2022. [CrossRef]
24. Yun, J.; Wei, L.; Li, W.; Gong, D.; Qin, H.; Feng, X.; Li, G.; Ling, Z.; Wang, P.; Yin, B. Isolating High Antimicrobial Ability Lignin From Bamboo Kraft Lignin by Organosolv Fractionation. *Front. Bioeng. Biotechnol.* 2021, *9*. [CrossRef]
25. Gregorova, A.; Redik, S.; Sedlarik, V.; Stelzer, F. Lignin-containing polyethylene films with antibacterial activity. In Proceedings of the 3rd International Conference on Thomson Reuters of NANOCON, Brno, Czech Republic, 21–23 September 2011.
26. Dong, X.; Dong, M.; Lu, Y.; Turley, A.; Jin, T.; Wu, C. Antimicrobial and antioxidant activities of lignin from residue of corn stover to ethanol production. *Ind. Crop. Prod.* 2011, *34*, 1629–1634. [CrossRef]
27. Lobo, F.; Franco, A.; Fernandes, E.; Reis, R. An Overview of the Antimicrobial Properties of Lignocellulosic Materials. *Molecules* 2021, *26*, 1749. [CrossRef] [PubMed]
28. Pandey, A.K.; Kumar, P.; Singh, P.; Tripathi, N.N.; Bajpai, V.K. Essential Oils: Sources of Antimicrobials and Food Preservatives. *Front. Microbiol.* 2017, *7*, 2161. [CrossRef] [PubMed]
29. Boonruang, K.; Kerddonfag, N.; Chinsirikul, W.; Mitcham, E.J.; Chonhenchob, V. Antifungal effect of poly(lactic acid) films containing thymol and R-(-)-carvone against anthracnose pathogens isolated from avocado and citrus. *Food Control.* 2017, *78*, 85–93. [CrossRef]

30. Zadeh, E.M.; O'Keefe, S.F.; Kim, Y.-T. Utilization of Lignin in Biopolymeric Packaging Films. *ACS Omega* **2018**, *3*, 7388–7398. [CrossRef]
31. Hararak, B.; Wanmolee, W.; Wijaranakul, P.; Prakymoramas, N.; Winotapun, C.; Kraithong, W.; Nakason, K. Physicochemical properties of lignin nanoparticles from softwood and their potential application in sustainable pre-harvest bagging as transparent UV-shielding films. *Int. J. Biol. Macromol.* **2023**, *229*, 575–588. [CrossRef]
32. Suwanamornlert, P.; Kerddonfag, N.; Sane, A.; Chinsirikul, W.; Zhou, W.; Chonhenchob, V. Poly (lactic acid)/poly (butylene-succinate-co-adipate)(PLA/PBSA) blend films containing thymol as alternative to synthetic preservatives for active packaging of bread. *Food Packaging and Shelf Life* **2020**, *25*, 100515. [CrossRef]
33. Boonruang, K.; Chinsirikul, W.; Hararak, B.; Kerddonfag, N.; Chonhenchob, V. Antifungal Poly(lactic acid) Films Containing Thymol and Carvone. *MATEC Web Conf.* **2016**, *67*, 6107. [CrossRef]
34. Yang, W.; Owczarek, J.; Fortunati, E.; Kozanecki, M.; Mazzaglia, A.; Balestra, G.; Kenny, J.; Torre, L.; Puglia, D. Antioxidant and antibacterial lignin nanoparticles in polyvinyl alcohol/chitosan films for active packaging. *Ind. Crop. Prod.* **2016**, *94*, 800–811. [CrossRef]
35. Platnieks, O.; Gaidukovs, S.; Barkane, A.; Sereda, A.; Gaidukova, G.; Grase, L.; Thakur, V.; Filipova, I.; Fridrihsone, V.; Skute, M.; et al. Bio-Based Poly(butylene succinate)/Microcrystalline Cellulose/Nanofibrillated Cellulose-Based Sustainable Polymer Composites: Thermo-Mechanical and Biodegradation Studies. *Polymers* **2020**, *12*, 1472. [CrossRef]
36. Khan, M.; Chonhenchob, V.; Huang, C.; Suwanamornlert, P. Antifungal Activity of Propyl Disulfide from Neem (*Azadirachta indica*) in Vapor and Agar Diffusion Assays against Anthracnose Pathogens (*Colletotrichum gloeosporioides* and *Colletotrichum acutatum*) in Mango Fruit. *Microorganisms* **2021**, *9*, 839. [CrossRef] [PubMed]
37. Balouiri, M.; Sadiki, M.; Ibnsouda, S.K. Methods for in vitro evaluating antimicrobial activity: A review. *J. Pharm. Anal.* **2015**, *6*, 71–79. [CrossRef] [PubMed]
38. Balaguer, M.P.; Lopez-Carballo, G.; Catala, R.; Gavara, R.; Hernandez-Munoz, P. Antifungal properties of gliadin films incorporating cinnamaldehyde and application in active food packaging of bread and cheese spread foodstuffs. *Int. J. Food Microbiol.* **2013**, *166*, 369–377. [CrossRef] [PubMed]
39. Phua, Y.J. Reactive processing of maleic anhydride-grafted poly(butylene succinate) and the compatibilizing effect on poly(butylene succinate) nanocomposites. *Express Polym. Lett.* **2013**, *7*, 340–354. [CrossRef]
40. Saffian, H.; Yamaguchi, M.; Ariffin, H.; Abdan, K.; Kassim, N.; Lee, S.; Lee, C.; Shafi, A.; Alias, A.H. Thermal, Physical and Mechanical Properties of Poly(Butylene Succinate)/Kenaf Core Fibers Composites Reinforced with Esterified Lignin. *Polymers* **2021**, *13*, 2359. [CrossRef]
41. Valderrama, A.C.S.; Rojas De, G.C. Traceability of Active Compounds of Essential Oils in Antimicrobial Food Packaging Using a Chemometric Method by ATR-FTIR. *Am. J. Anal. Chem.* **2017**, *8*, 726–741. [CrossRef]
42. Mohamad, N.; Mazlan, M.M.; Tawakkal, I.S.M.A.; Talib, R.A.; Kian, L.K.; Jawaid, M. Characterization of Active Polybutylene Succinate Films Filled Essential Oils for Food Packaging Application. *J. Polym. Environ.* **2021**, *30*, 585–596. [CrossRef]
43. Zarei, M.; El Fray, M. Synthesis of Hydrophilic Poly(butylene succinate-butylene dilinoleate) (PBS-DLS) Copolymers Containing Poly(Ethylene Glycol) (PEG) of Variable Molecular Weights. *Polymers* **2021**, *13*, 3177. [CrossRef]
44. Rajkumar, P.; Selvaraj, S.; Suganya, R.; Velmurugan, D.; Gunasekaran, S.; Kumaresan, S. Vibrational and electronic spectral analysis of thymol an isomer of carvacrol isolated from Trachyspermum ammi seed: A combined experimental and theoretical study. *Chem. Data Collect.* **2018**, *15–16*, 10–31. [CrossRef]
45. Sahoo, S.; Misra, M.; Mohanty, A.K. Enhanced properties of lignin-based biodegradable polymer composites using injection moulding process. *Compos. Part A-Appl. Sci. Manuf.* **2011**, *42*, 1710–1718. [CrossRef]
46. Sen, S.; Patil, S.; Argyropoulos, D. Thermal properties of lignin in copolymers, blends, and composites: A review. *Green Chem.* **2015**, *17*, 4862–4887. [CrossRef]
47. Pereira, L.A.S.; Silva, P.D.C.E.; Pagnossa, J.P.; Miranda, K.W.E.; Medeiros, E.S.; Piccoli, R.H.; de Oliveira, J.E. Antimicrobial zein coatings plasticized with garlic and thyme essential oils. *Braz. J. Food Technol.* **2019**, *22*. [CrossRef]
48. Javidi, Z.; Hosseini, S.F.; Rezaei, M. Development of flexible bactericidal films based on poly(lactic acid) and essential oil and its effectiveness to reduce microbial growth of refrigerated rainbow trout. *Lwt* **2016**, *72*, 251–260. [CrossRef]
49. Balani, K.; Verma, V.; Agarwal, A.; Narayan, R. Physical, thermal, and mechanical properties of polymers. In *Biosurfaces: A Materials Science and Engineering Perspective*; Balani, K., Verma, V., Agarwal, A., Narayan, R., Eds.; John Wiley & Sons, Inc: Hoboken, NJ, USA, 2015; pp. 329–344.
50. Celebi, H.; Gunes, E. Combined effect of a plasticizer and carvacrol and thymol on the mechanical, thermal, morphological properties of poly(lactic acid). *J. Appl. Polym. Sci.* **2017**, *135*, 45895. [CrossRef]
51. Berlin, H. Investigation of Polymers with Differential Scanning Calorimetry. Available online: https://polymerscience.physik.hu-berlin.de/docs/manuals/DSC.pdf (accessed on 27 November 2022).
52. Basu, P. Biomass gasification, pyrolysis and torrefaction: Practical design and theory. In *Biomass Gasification, Pyrolysis and Torrefaction*, 3rd ed.; Basu, P., Ed.; Academic press: Cambridge, MA, USA, 2018; pp. 479–495.
53. Król-Morkisz, K.; Pielichowska, K. Thermal decomposition of polymer nanocomposites with functionalized nanoparticles. In *Polymer composites with functionalized nanoparticles*; Pielichowski, K., Majka, T.M., Eds.; Elsevier: Amsterdam, The Netherlands, 2019; pp. 405–435.

54. Mariën, H.; Peeters, L.; Harumashi, T.; Rubens, M.; Vendamme, R.; Vleeschouwers, R.; Vanbroekhoven, K. Improving the Thermal Stability of MS Polymers with Lignin Fractions. *Adhes. Adhes.* **2022**, *19*, 30–33. [CrossRef]
55. Mousavioun, P.; Doherty, W.O.S.; George, G. Thermal stability and miscibility of poly(hydroxybutyrate) and soda lignin blends. *Ind. Crops Prod.* **2010**, *32*, 656–661. [CrossRef]
56. Crawford, C.B.; Quinn, B. Physiochemical properties and degradation. In *Microplastic Pollutants*; Elsevier: Amsterdam, The Netherlands, 2017; pp. 57–100. [CrossRef]
57. Weihua, K.; He, Y.; Asakawa, N.; Inoue, Y. Effect of lignin particles as a nucleating agent on crystallization of poly(3-hydroxybutyrate). *J. Appl. Polym. Sci.* **2004**, *94*, 2466–2474. [CrossRef]
58. Purnama, P.; Kim, S.H. Biodegradable blends of stereocomplex polylactide and lignin by supercritical carbon dioxide-solvent system. *Macromol. Res.* **2013**, *22*, 74–78. [CrossRef]
59. Petchwattana, N.; Naknaen, P. Utilization of thymol as an antimicrobial agent for biodegradable poly(butylene succinate). *Mater. Chem. Phys.* **2015**, *163*, 369–375. [CrossRef]
60. Salamon, D. Advanced ceramics. In *Advanced ceramics for dentistry*; Shen, J.Z., Kosmač, T., Eds.; Elsevier: Amsterdam, The Netherlands, 2014; pp. 103–122.
61. Mariana, M.; Alfatah, T.; Abdul Khalil, H.P.S.; Yahya, E.B.; Olaiya, N.; Nuryawan, A.; Mistar, E.; Abdullah, C.; Abdulmadjid, S.; Ismail, H. A current advancement on the role of lignin as sustainable reinforcement material in biopolymeric blends. *J. Mater. Res. Technol.* **2021**, *15*, 2287–2316. [CrossRef]
62. Chantapet, P.; Kunanopparat, T.; Menut, P.; Siriwattanayotin, S. Extrusion Processing of Wheat Gluten Bioplastic: Effect of the Addition of Kraft Lignin. *J. Polym. Environ.* **2012**, *21*, 864–873. [CrossRef]
63. Sakunkittiyut, Y.; Kunanopparat, T.; Menut, P.; Siriwattanayotin, S. Effect of kraft lignin on protein aggregation, functional, and rheological properties of fish protein-based material. *J. Appl. Polym. Sci.* **2012**, *127*, 1703–1710. [CrossRef]
64. Bhat, R.; Abdullah, N.; Din, R.H.; Tay, G.-S. Producing novel sago starch based food packaging films by incorporating lignin isolated from oil palm black liquor waste. *J. Food Eng.* **2013**, *119*, 707–713. [CrossRef]
65. Shankar, S.; Reddy, J.P.; Rhim, J.-W. Effect of lignin on water vapor barrier, mechanical, and structural properties of agar/lignin composite films. *Int. J. Biol. Macromol.* **2015**, *81*, 267–273. [CrossRef] [PubMed]
66. Gomide, R.A.C.; De Oliveira, A.C.S.; Luvizaro, L.B.; Yoshida, M.I.; De Oliveira, C.R.; Borges, S.V. Biopolymeric films based on whey protein isolate/lignin microparticles for waste recovery. *J. Food Process. Eng.* **2020**, *44*. [CrossRef]
67. Lian, H.; Wei, W.; Wang, D.; Jia, L.; Yang, X. Effect of thymol on physical properties, antimicrobial properties and fresh-keeping application of cherry tomato of starch/PBAT extrusion blowing films. *Food Sci. Technol.* **2022**, *42*. [CrossRef]
68. Aadil, K.R.; Prajapati, D.; Jha, H. Improvement of physcio-chemical and functional properties of alginate film by Acacia lignin. *Food Packag. Shelf Life* **2016**, *10*, 25–33. [CrossRef]
69. Izaguirre, N.; Gordobil, O.; Robles, E.; Labidi, J. Enhancement of UV absorbance and mechanical properties of chitosan films by the incorporation of solvolytically fractionated lignins. *Int. J. Biol. Macromol.* **2020**, *155*, 447–455. [CrossRef]
70. Zhou, S.-J.; Wang, H.-M.; Xiong, S.-J.; Sun, J.-M.; Wang, Y.-Y.; Yu, S.; Sun, Z.; Wen, J.-L.; Yuan, T.-Q. Technical Lignin Valorization in Biodegradable Polyester-Based Plastics (BPPs). *ACS Sustain. Chem. Eng.* **2021**, *9*, 12017–12042. [CrossRef]
71. Müller, K.; Bugnicourt, E.; Latorre, M.; Jorda, M.; Echegoyen Sanz, Y.E.; Lagaron, J.M.; Miesbauer, O.; Bianchin, A.; Hankin, S.; Bölz, U.; et al. Review on the Processing and Properties of Polymer Nanocomposites and Nanocoatings and Their Applications in the Packaging, Automotive and Solar Energy Fields. *Nanomaterials* **2017**, *7*, 74. [CrossRef]
72. Abdan, K.B.; Yong, S.C.; Chiang, E.C.W.; Talib, R.A.; Hui, T.C.; Hao, L.C. Barrier properties, antimicrobial and antifungal activities of chitin and chitosan-based IPNs, gels, blends, composites, and nanocomposites. In *Handbook of Chitin and Chitosan*; Elsevier: Amsterdam, The Netherlands, 2020; pp. 175–227. [CrossRef]
73. Kovalcik, A.; Machovsky, M.; Kozakova, Z.; Koller, M. Designing packaging materials with viscoelastic and gas barrier properties by optimized processing of poly(3-hydroxybutyrate-co-3-hydroxyvalerate) with lignin. *React. Funct. Polym.* **2015**, *94*, 25–34. [CrossRef]
74. Othman, S.H.; Nordin, N.; Azman, N.A.A.; Tawakkal, I.S.M.A.; Basha, R.K. Effects of nanocellulose fiber and thymol on mechanical, thermal, and barrier properties of corn starch films. *Int. J. Biol. Macromol.* **2021**, *183*, 1352–1361. [CrossRef] [PubMed]
75. Zhang, C.-W.; Nair, S.S.; Chen, H.; Yan, N.; Farnood, R.; Li, F.-Y. Thermally stable, enhanced water barrier, high strength starch bio-composite reinforced with lignin containing cellulose nanofibrils. *Carbohydr. Polym.* **2019**, *230*, 115626. [CrossRef] [PubMed]
76. Zhang, Z.; Terrasson, V.; Guénin, E. Lignin Nanoparticles and Their Nanocomposites. *Nanomaterials* **2021**, *11*, 1336. [CrossRef] [PubMed]
77. Zhang, J. Lasiodiplodia theobromae in Citrus Fruit (Diplodia Stem-End Rot). In *Postharvest Decay: Control Strategies*; Bautista-Baños, S., Ed.; Academic Press: Cambridge, MA, USA, 2014; pp. 309–335.
78. Chen, K.; Qiu, X.; Yang, D.; Qian, Y. Amino acid-functionalized polyampholytes as natural broad-spectrum antimicrobial agents for high-efficient personal protection. *Green Chem.* **2020**, *22*, 6357–6371. [CrossRef]
79. Spasojevic, D.; Zmejkoski, D.; Glamoclija, J.; Nikolic, M.; Sokovic, M.; Milosevic, V.; Jaric, I.; Stojanovic, M.; Marinkovic, E.; Barisani-Asenbauer, T.; et al. Lignin model compound in alginate hydrogel: A strong antimicrobial agent with high potential in wound treatment. *Int. J. Antimicrob. Agents* **2016**, *48*, 732–735. [CrossRef] [PubMed]
80. Abu Bakar, N.; Karsani, S.A.; Alias, S.A. Fungal survival under temperature stress: A proteomic perspective. *Peerj* **2020**, *8*, e10423. [CrossRef]

81. Mensah-Attipoe, J.; Toyinbo, O. Fungal growth and aerosolization from various conditions and materials. In *Fungal infection*; de Loreto, É.S., Tondolo, J.S.M., Eds.; IntechOpen: London, UK, 2019; pp. 1–10.
82. Zikeli, F.; Vinciguerra, V.; Sennato, S.; Mugnozza, G.S.; Romagnoli, M. Preparation of Lignin Nanoparticles with Entrapped Essential Oil as a Bio-Based Biocide Delivery System. *ACS Omega* **2019**, *5*, 358–368. [CrossRef]
83. Kamle, M.; Kumar, P. Colletotrichum gloeosporioides: Pathogen of Anthracnose Disease in Mango (Mangifera indica L.). In *Current Trends in Plant Disease Diagnostics and Management Practices*; Fungal Biology: Berlin/Heidelberg, Germany, 2016; pp. 207–219.
84. Galsurker, O.; Diskin, S.; Duanis-Assaf, D.; Doron-Faigenboim, A.; Maurer, D.; Feygenberg, O.; Alkan, N. Harvesting Mango Fruit with a Short Stem-End Altered Endophytic Microbiome and Reduce Stem-End Rot. *Microorganisms* **2020**, *8*, 558. [CrossRef]
85. Dannemiller, K.C.; Weschler, C.J.; Peccia, J. Fungal and bacterial growth in floor dust at elevated relative humidity levels. *Indoor Air* **2016**, *27*, 354–363. [CrossRef]

Disclaimer/Publisher's Note: The statements, opinions and data contained in all publications are solely those of the individual author(s) and contributor(s) and not of MDPI and/or the editor(s). MDPI and/or the editor(s) disclaim responsibility for any injury to people or property resulting from any ideas, methods, instructions or products referred to in the content.

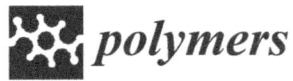

Article

Evaluation of Active LDPE Films for Packaging of Fresh Orange Juice

Pedro V. Rodrigues [1], Dalila M. Vieira [1], Paola Chaves Martins [2], Vilásia Guimarães Martins [2], M. Cidália R. Castro [1,*] and Ana V. Machado [1]

[1] Instituto de Polímeros e Compósitos, Universidade do Minho, Campus de Azurém, 4800-058 Guimarães, Portugal
[2] Laboratory of Food Technology, School of Chemistry and Food, Federal University of Rio Grande (FURG), Rio Grande 96203-900, Brazil
* Correspondence: cidaliacastro@dep.uminho.pt; Tel.: +351-253510320

Abstract: Microbial development, enzymatic action, and chemical reactions influence the quality of untreated natural orange juice, compromising its organoleptic characteristics and causing nutritional value loss. Active low-density polyethylene (LDPE) films containing green tea extract (GTE) were previously prepared by a blown film extrusion process. Small bags were prepared from the produced films, which were then filled with fresh orange juice and stored at 4 °C. Ascorbic acid (AA) content, sugar content, browning index, color parameters, pH, total acidity (TA) and microbial stability were evaluated after 3, 7, and 14 days of storage. The packaging containing GTE maintained the microbial load of fresh juice beneath the limit of microbial shelf-life (6 log CFU/mL) for the bacterial growth, with a more prominent effect for LDPE with 3%GTE. Regarding yeasts and molds, only the CO_LDPE_3GTE package maintained the microbial load of fresh juice below the limit for up to 14 days. At 14 days, the lowest levels of AA degradation (32.60 mg/100 mL of juice) and development of brown pigments (browning index = 0.139) were observed for the packages containing 3% of GTE, which had a pH of 3.87 and sugar content of 11.4 g/100 mL of juice at this time. Therefore, active LDPE films containing 3% of GTE increase the shelf-life of fresh juice and can be a promising option for storage of this food product while increasing sustainability.

Keywords: active packaging; orange juice; physicochemical characteristics; microbial stability; shelf-life

1. Introduction

Fruit juices are known for being a source of vitamins, soluble/insoluble fiber and minerals, and their characteristic flavor makes them a product of high consumption [1]. In fact, processed products, specifically juice, are very popular because they are easily consumed [2]. The processing and storage conditions, packaging and raw material are very important factors for the stability of citrus juice as these factors determine the microbiological, enzymatic, chemical, and physical changes that can spoil the juice's sensory and nutritional characteristics [1,3,4].

The high content of vitamin C (ascorbic acid), an essential nutrient for humans, in orange juice and its pleasant taste makes it the most appreciated and consumed citrus juice [1,2,5–7]. However, due to its nature, vitamin C can oxidize and be lost during the juice storage period. Its degradation rate is highly dependent on storage conditions such as temperature, dissolved oxygen, and the oxygen permeability of the packaging material [2,8].

Fresh orange juice is extremely susceptible to microbial growth, which results in fast deterioration. The deterioration of the organoleptic and physicochemical characteristics is a major reason for the rejection of juice by customers [9]. Lactic and acetic acid bacteria have been isolated from fruit juices, and many microorganisms found in juices are acid-tolerant bacteria and fungi [10,11]. Usually, the most reported bacteria genera include

Acetobacter, Alicyclobacillus, Bacillus, Gluconobacter, Lactobacillus, Leuconostoc, Zymomonas, and *Zymobacter* [12,13].

The availability of nutrients, presence of antimicrobial compounds, oxidation-reduction potential, water activity, and pH are the critical factors that influence the spoilage of juices, with the last two factors being of crucial importance. The spoilage in juices results in degradation of the product, which induces changes in appearance, color, texture, CO_2 production, cloud loss and the development of off-flavors [3,11,14,15]. The acidic properties of fruit juices (pH < 4.5) act as a vital barrier to microbial growth. However, foodborne pathogens such as, *E. coli* and *Salmonella* can persist at an acidic pH level due to the acid stress response. Thus, in the last twenty years, several foodborne outbreaks related to unpasteurized fruit juices have been reported in several countries [11,13,16,17].

The shelf-life of a food product is commonly recognized as the length of time during which it is still suitable for consumption and sale with acceptable characteristics under specified storage conditions, and is determined by its sensory characteristics (color, aroma and taste). On a food label, shelf-life can be indicated by either a "best before" date that indicates the quality of the food or a "use by" date that is linked to food safety. The accuracy of the shelf-life prediction displayed on the package is important for both food industries and consumers [18]. Active compounds can provide several functions when incorporated into the packaging materials, which are an alternative to conventional packaging systems [19,20]. There are several natural antioxidants, sweeteners, coloring and antimicrobial agents originating from animals, plants, or even microorganisms, although they have not been defined as a specific category for natural additives [21]. Several natural substances can have an active function in the package. These include essential oils (EOs) or extracts of plants that are known as Generally Recognized as Safe (GRAS). They can be used as food additives not only to extend the shelf-life but also to preserve the food's quality for a long period of time [22–28]. The EOs/extracts have the potential function of inhibiting microorganisms and reducing lipid oxidation due to their high content in phenolic compounds and volatile terpenoids [27–29]. These compounds have several biological properties, such as antioxidant and antimicrobial activities [26–28,30–40]. Through disruption of the cytoplasmic membrane, the active components of plant EOs/extracts inhibit microorganism proliferation. In fact, these components disrupt the electron flow, active transport and proton motive forces and inhibit protein synthesis [28,29]. The use of EOs/extracts in food could decrease or substitute the dependence on synthetic antioxidants and antimicrobial compounds, thereby meeting the consumer's demand for more natural products [26,27].

Since consumers demand more natural products, much research has been conducted aiming to replace synthetic compounds with natural ones, such as plant extracts and EOs, due to their benefits to human health. GTE is rich in flavonols and gallic acid derivatives, namely, (+)-catechin, (-)-epicatechin, (+)-gallocatechin, (-)-epicatechin gallate, (-)-epigallocatechin, and (-)-epigallocatechin gallate. GTE is described as a powerful source of polyphenol antioxidants and already has the status of food additive [20,39–43]. Studies on GTE have revealed its excellent antioxidant properties and nontoxicity in various food model systems, which encourages its incorporation into polymer matrixes for the development of active packaging films to prevent food oxidation [42–46].

This active packaging could help the food packaging industry to eliminate or reduce spoilage and foodborne pathogens on the surface of products, and thereby, increase a product's shelf life. Thus, active packaging incorporating antimicrobial activity is one the most promising methods to extend shelf-life while sustaining the nutritional and sensory quality of food [47,48]. Moreover, in recent years, researchers have made efforts to develop alternatives to multi-material packages in order to improve their recyclability. Thus, LDPE packaging systems have been produced with the incorporation of nanoparticles or natural compounds to increase juice preservation [7,49–51]. Therefore, the purpose of this study is to evaluate the potential of GTE in a LDPE matrix as an innovative packaging to preserve and extend orange juice shelf-life.

2. Materials and Methods

2.1. Materials

Low-density polyethylene (LDPE) was kindly provided by Vizelpas—Flexible Films, S.A. (Portugal) and GTE was supplied from ESSÊNCIAD'UMSEGREDO, LDA (Portugal). Distilled water, phenolphthalein (indicator ACS, Merck, Darmstadt, Germany), and sodium hydroxide (NaOH $\geq$ 98%, Merck, Darmstadt, Germany) were used in total acidity (TA) tests. Ethanol (absolute $\geq$ 99%, Thermo Fisher Scientific, Waltham, MA, USA) was used in the browning index assay. Sulfuric acid ($H_2SO_4 \geq$ 99%, Merck, Darmstadt, Germany), starch, iodine solution (Merck, Darmstadt, Germany), and sodium thiosulfate ($Na_2S_2O_3 \geq$ 99.99%, Merck, Darmstadt, Germany) were used for the determination of ascorbic acid (AA) content. Peptone water, plate count agar (PCA), and Dichloran Rose-Bengal Chloramphenicol agar (DRBC) were acquired from Merck (Germany) and were used in microbiological tests.

2.2. Preparation of LDPE Active Films

LDPE active films were prepared according to the methodology already reported by our group [46]. Briefly, a co-rotating twin-screw extruder (Leistritz AG LSM 34 6L) was used to prepare a masterbatch of LDPE/10 wt.% GTE, at 170 °C. Then, this was diluted into neat LDPE to produce monolayer and coextruded films of LDPE containing 1.5 and 3 wt.% of GTE by blown film extrusion.

Descriptions of the different active films produced for juice packaging based on LDPE containing GTE are shown in Table 1.

Table 1. Packaging films based on LDPE/GTE.

Film	Description
LDPE	Monolayer film of LDPE
LDPE_1.5GTE	Monolayer film of LDPE with 1.5 wt.% GTE
LDPE_3GTE	Monolayer film of LDPE with 3 wt.% GTE
CO_LDPE_1.5GTE	LDPE film coextruded with LDPE_1.5GTE
CO_LDPE_3GTE	LDPE film coextruded with LDPE_3GTE

2.3. Packaging Orange Juice

To evaluate the packaging potential, an orange juice was selected which was produced under ideal hygienic-sanitary conditions and without the addition of any preservatives or preservation processes. Before the production of juice, the oranges were sanitized with water and soap and all the equipment used in the process was previously sterilized in an autoclave (Technal, AV-18, São Paulo, Brazil). The oranges were peeled and the juice was extracted using a juicer. The juice produced containing pieces of orange pulp was standardized with a 1 mm mesh filter and stored in a sterile glass container. The films used for application were previously decontaminated using a laminar flow cabinet with UV light (15 W) (Solab, SLH-656/4, New York, NY, USA) for 15 min of exposure on each side. Then, small bags (14 × 13 cm) made from the produced films were sealed at the bottom, filled with 300 mL of juice and closed aseptically.

2.4. Storage

The orange juice packages were stored in the dark and in cold conditions (4 °C). The samples from a single package for each treatment were assessed with a total of 5 tests for physicochemical properties including color, pH, sugar content, TA, browning index, AA content and microbiological growth. The tests were conducted immediately after packaging and after 3, 7, and 14 storage days.

2.5. Measurement of Color, pH and Sugar Content

The juices were evaluated for color variations using a colorimeter (Minolta Chroma Meter, CR-400, Konica Minolta, NJ, USA) with triplicate measurements. The equipment uses the CIELab measurement system that measures the L* parameter (lightness index scale) in a range from 0 (black) to 100 (white), the a* parameter that indicates the degree of red (+a) or green (−a*) color and the b* parameter which measures the degree of yellow (+b) or blue (−b*) color.

The pH measurement was performed using a pH meter (Even, PHS-3E, USA) at room temperature. The sugar concentration was measured using a refractometer (Hanna, HI 96801, Judetul Cluj, Romania) which provides values referring to the amount in mg of sugar in 100 mL of juice.

2.6. Total Acidity (TA)

To determine the TA, 5 mL of previously filtered juice was used, and homogenized with 25 mL of distilled water with 2 drops of 1% phenolphthalein solution. The mixture was titrated with 0.1 M NaOH until a pink color appeared. TA was calculated using Equation (1) [52]:

$$TA = \frac{(V \times M \times 100)}{p} \quad (1)$$

where V (mL) is the volume of NaOH spent in the titration of the juice; M is the molarity of the standardized NaOH solution, and p (mL) is the amount of juice used.

2.7. Browning Index Measurement

A 10 mL sample of juice was collected from the package and centrifuged at 2000 rpm for 20 min. The supernatant was homogenized in a 1:1 ratio with ethanol and filtered with a 0.45-mm filter paper to obtain a clarified extract. The extract absorbance was read on a UV-Vis. spectrophotometer at 420 nm [53].

2.8. Determination of Ascorbic Acid (AA)

The AA content was calculated according to Zambiazi [52], where 20 ml of juice was mixed with 3 mL of H_2SO_4 (12 M) and 3 mL of starch (0.5% m/v). After homogenization, the mixture was titrated with a standardized 0.01 M iodine solution until a dark color appeared. Afterward, the solution was titrated again using 0.01 M sodium thiosulfate until the dark color disappeared; finally, the solution was titrated one more time with 0.01 M iodine until the reappearance of the dark color. The amount of AA present was calculated by applying Equation (2):

$$AA = [(V_i \times F_i) - (V_t \times F_t)] \times 0.88 \quad (2)$$

where AA is the content of ascorbic acid present in the juice expressed in mg of ascorbic acid/mL of juice, Vi (mL) is the total volume of iodine used in the titrations, Fi is the correction factor obtained in the standardization of the iodine solution, Vt (mL) is the volume of sodium thiosulfate used in the titration, and Ft is the correction factor for the standardized sodium thiosulfate solution.

2.9. Microbiological Growth Tests

The microbiological growth of the bacteria, molds and yeasts in the juice were evaluated. A sample of 1 mL of juice was aseptically collected for each different treatment and diluted in tubes containing 9 mL of 0.1% (w/v; peptone/water) sterile peptone water. The tubes containing juice and peptone water were manually shaken for approximately 1 min at room temperature (25 °C). Then, the serial dilutions of the homogenates were prepared for each treatment (10^0 to 10^{-6}). For the growth of bacteria, 1 mL of each dilution was added to petri plates and placed on plate count agar for in-depth homogenization. The plates were incubated at 37 °C for 48 h, and subsequently counted for the determination of colony forming units (CFU). For the determination of the molds and yeasts, 100 µL of each dilution was taken and inoculated on the surface of dicloran rose bengal agar. The plates were incubated in an oven incubator at 25 °C for 5 days, and after that, the colonies were counted as described above. All the microbiological tests were performed in triplicate [54].

2.10. Statistical Analysis

The resulting data was evaluated with Microsoft Windows Excel 365 and Origin-Pro (Version 17) software. At least three replicates were used to express the results as mean ± standard deviation. For the color analysis, an analysis of variance (ANOVA) was applied as well as the Tukey's test to determine significant differences with a 95% significance interval. The software used was Statistics 5.0.

3. Results and Discussion

3.1. Ascorbic Acid (AA) and Browning Index

The evolution of AA content in orange juice packed in active LDPE and in LDPE film, stored at 4 °C for 14 days, is shown in Figure 1A. The film that presented the best retention of AA was LDPE_3GTE followed by CO_LDPE_3GTE, while the LDPE without GTE had poor retention of AA, as expected. Since oxygen is one of the main components that contribute to AA degradation and considering that the headspace was the same for all packages, the only factor that can explain these variations in AA retention is oxygen permeability [2]. In fact, the results indicate that the LDPE_3GTE film had the lowest permeability followed by CO_LDPE_3GTE, LDPE_1.5GTE, and CO_LDPE_1.5GTE film, respectively. Considering the limit of 20 mg/100 mL of AA value for shelf-life estimation [55], all LDPE active films presented a higher value than the AA limit after storage for 14 days, whereas the lowest value, 25.83 mg/100 mL, was obtained for LDPE_1.5GTE and CO_LDPE_1.5GTE film. During the early stage of storage, the results indicate a fast degradation of AA, which was followed by a gradual loss. This agrees with results obtained by other researchers [56,57], and can be attributed mostly to the oxygen dissolved in the juice and in the headspace of the package at the beginning of storage [53]. Indeed, the dissolved oxygen concentration has a great impact on the AA oxidation rate. Solomon et al. [58] and Wilson et al. [59] demonstrate that the rate of oxidation of AA is significantly associated with the level of dissolved oxygen and with the duration of storage time. Throughout the storage period, the oxygen permeation across the packaging contributes significantly to the extension of the aerobic mechanism of AA oxidation in the active LDPE films [53,60]. Parameters such as light, heat, oxygen, enzymes, and peroxides stimulated the oxidative process of AA [7,8].

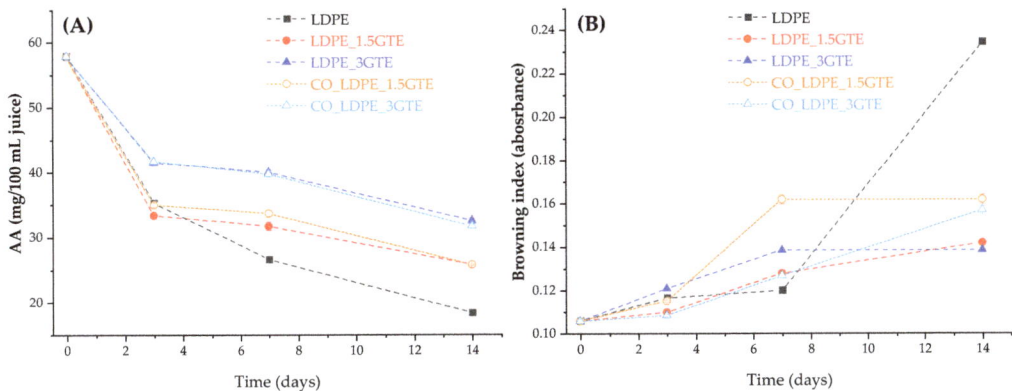

Figure 1. The content of AA (**A**), and browning index (**B**) in orange juice packed with neat LDPE films and active LDPE films containing GTE stored at 4 °C for 14 days.

Overall, after 14 days of storage, the final AA content in the juice varied from 18.36 to 32.60 mg/100 mL. Comparing these data with the minimum values recommended for processed orange juice, it can be seen that they were lower than the value indicated as minimum for industrialized juice, 40 mg/100 mL [61].

During the storage of citrus products, non-enzymatic processes are one of the most critical chemical phenomena responsible for quality and color variations. Moreover, ascorbic acid degradation into dehydroascorbic acid (DHA) is known as the main chemical reaction that occurs during the storage of all kinds of juice. More specifically, the resulting DHA is converted to 2,3-diketogulonic acid (DKG), forming xylosone through the aerobic pathway, which degrades to form reductones or ethylglyoxal. Then, these compounds react with amino acids, yielding brown compounds. Therefore, there is a high-level of correlation between the percentage loss of ascorbic acid and the increase in the browning index [6,53,62]. This relationship is noticeable in Figure 1, where a significant decrease in the AA content is observed for the packages during the storage at 4 °C, while browning index, only increased slightly. The values of the browning index in fresh juice measured immediately after packaging were 0.106 (Figure 1B). Leizerson and Shimoni [63] reported values of the browning index up to 0.367, which leads to the conclusion that is still undetected in our case. It can also be seen in Figure 1B that the package with only LDPE is the one that exhibits the highest browning index, as expected, following by a pronounced increase from day 7 onwards. The package that demonstrated the lowest browning index was LDPE_3GTE, which proves the influence of GTE as an antioxidant agent. In fact, Roig et al. [57] and Bharate et al. found a relationship between the browning index and the oxidative loss of L-ascorbic acid in citrus juices [62]. The results obtained for the browning index are similar to the Zerdin et al. [53] study that determined the extent of AA loss due to oxygen and temperature for orange juice packed in oxygen scavenging film and oxygen barrier film. For the browning index, Cortés et al. [64], obtained lower values (0.093) than those obtained in the present study; however, the temperature used in their study (2 °C) was lower. According to the results of Emamifar et al. [7], the browning index increased significantly for all packaging tested stored at 4 °C, agreeing with the values reached in this study.

3.2. Color

The color of an orange juice is a crucial characteristic for the consumer's initial purchasing decision and for consumer perception about the food quality. Carotenoid pigments are responsible for orange juice color and can be affected by product ripening, processing treatments, storage conditions and browning reactions [7,8]. Table 2 depicts the evolution of color parameters of orange juice packed in active LDPE films with and without GTE, stored at 4 °C. It can be noticed that color parameters did not show significant variations until 3 days of storage. After this time, L^* values start to increase, suggesting an increase in the brightness and light; a^* and b^* parameters did not present significant variations until 14 days of storage. Thus, there was no considerable variation in juice color. The L^* parameter increased in LDPE_1.5GTE after 7 days but at a lower rate than for CO_LDPE_1.5GTE. Moreover, it was observed that LDPE_3GTE and CO_LDPE_1.5GTE were the samples with higher brightness after 14 days. Concerning the a^* parameter (variation between red and green color), an increase in a^* value was verified with a higher amount of GTE (LDPE_3GTE = -0.13 ± 0.03) when compared with only LDPE (-1.26 ± 0.05), after 14 days storage. Parameter b^* (variation between yellow and blue color) showed an increase for all packaging after 14 days of storage, with the largest increase for packaging with 3% GTE. Thus, a color shift toward positive b^* and negative a^* directions indicate greater values of yellow and green colors in the orange juice. These changes show the progressive deterioration of the juice due to changes in the color spectrum.

Table 2. The color parameters obtained for orange juice packaged in active LDPE films containing GTE at different days of storage.

Orange Juice in Film:		LDPE	LDPE 1.5GTE	LDPE 3GTE	CO_LDPE 1.5GTE	CO_LDPE 3GTE
Day 0	L*	41.47 [a]	40.95 [a]	39.04 [b]	40.96 [a]	40.29 [a,b]
	a*	−1.67 [d]	−1.10 [c]	0.30 [a]	−1.25 [c]	−0.87 [b]
	b*	20.10 [a]	20.04 [a]	18.32 [b]	18.80 [b]	18.78 [b]
Day 3	L*	40.67 [c]	41.59 [b,c]	42.98 [a]	42.94 [a]	42.57 [a,b]
	a*	−0.50 [b,c]	−0.37 [b]	0.87 [a]	−1.25 [d]	−0.77 [c]
	b*	20.17 [c]	22.77 [b]	25.96 [a]	22.89 [b]	23.42 [b]
Day 7	L*	43.05 [b]	42.56 [b]	46.15 [a]	47.10 [a]	43.67 [b]
	a*	−0.64 [c]	−0.91 [d]	0.79 [a]	−0.09 [b]	−0.94 [d]
	b*	23.36 [b,c]	22.46 [c]	29.07 [a]	28.79 [a]	24.26 [b]
Day 14	L*	46.65 [a]	43.33 [c]	45.65 [a,b]	45.51 [b]	43.41 [c]
	a*	−1.26 [c]	−0.59 [b]	−0.13 [a]	−1.10 [c]	−0.96 [c]
	b*	28.03 [a]	24.27 [c,d]	26.40 [b]	25.40 [b,c]	23.68 [d]

Different superscript letters in the same line indicate a statistically significant difference ($p < 0.05$). Values are given as mean, and standard deviation values are under 5% for all samples.

These color changes have a good correlation with the reduction of AA content and the development of brown pigments during storage. A bleaching effect may be due to the oxidation of carotenoids; consequently, the free radicals formed might be responsible for the changes in the orange juice color [7,8]. Bull et al. reported an increase in the total color variation with time during storage in fresh orange juice, regardless of treatment [65]. Esteve et al. obtained a slight decrease in L* at 4 °C for different commercial orange juices [61]. Lee and Coates studied pasteurized orange juice and reported a small increase in L* value from 40.22 to 41.22 [66]. Rivas et al. describe a decrease in the parameter L* for pasteurized orange-carrot juice during refrigerated storage, and Cortés et al. observed that L* values increased substantially after one week of refrigerated storage, which is also in agreement with the findings of this study [64,67].

3.3. pH and Total Acidity (TA)

After 14 days of storage at 4 °C, the pH values of the juice studied in five different packages were within the normal range (3–4), however, with significant differences among them, as presented in Figure 2A. After 7 days of storage at 4 °C, the pH values of the juice in different packages decrease from 4.73 (initial) to close to 3.85 (day 7), where the CO_LDPE_3GTE and LDPE films had a larger decrease in pH values. In fact, the results obtained are in agreement with the study by Touati et al. [68], which found that pH values become significantly lower with storage, independent of the temperature. In addition, Bull et al. observed a significant variation in pH in studies of pasteurized and high pressurized orange juices stored for 12 weeks [65]. In contrast, Esteve et al. [61] did not detected significant changes in pH values of various pasteurized orange and carrot-orange juices refrigerated at 4 °C and 10 °C, and in the study by Cortés et al. [64], there was a statistically significant increase in pH values for all the juices analyzed. This increase can be related to a microbiological deterioration of juice, as described by Del Caro et al. [69], who studied the changes in pH in citrus segments and juices during storage at 4 °C.

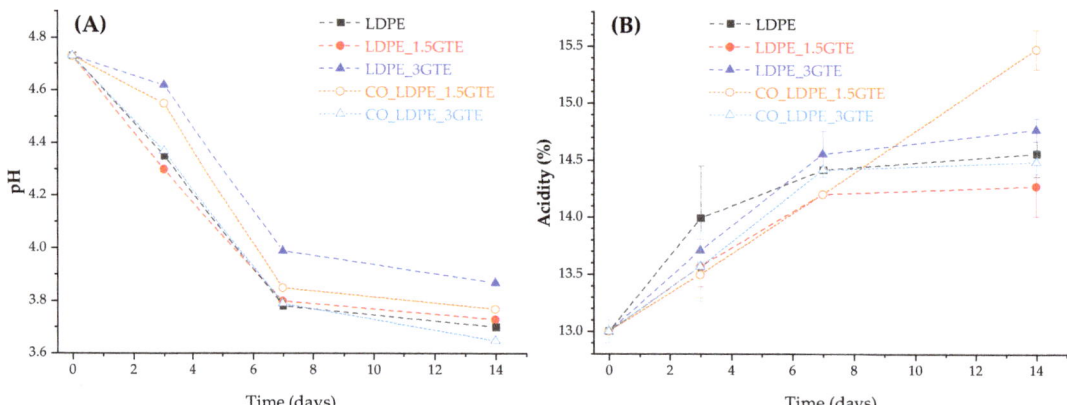

Figure 2. pH indexes (**A**), and TA (**B**) of orange juice packed in active LDPE films containing GTE and stored at 4 °C for 14 days.

When fermentation of orange juice occurs, the organic acids (produced from the biochemical process due to the development of spoilage microorganisms) lead to pH reductions which results in a specific flavor and palatability of the juice. In general, acid environments protect against the growth of pathogens [9]. Citric acid is the most abundant free acid in orange juice, followed by malic acid, and although in limited quantities, they also appear as citrates or malates giving a buffer effect to orange juice. Non-volatile free acids, such as oxalic, galacturonic and quinic acids and many others are found in smaller quantities [61]. As would be expected, higher acidity corresponds to lower pH value. Owing to the presence of this natural buffer medium in orange juice (based on mainly potassium citrates and malates), pH variations are slightly more pronounced than acidity variations. The TA in the five studied packages present similar behavior (Figure 2B), except for the package made of CO_LDPE_1.5GTE film. During storage, acidity increased in all juices until day 7, reaching a plateau. However, the CO_LDPE_1.5GTE film exhibited a linear increase in total acidity with time, indicating the start of spoilage or fermentation of the sample. These results are in agreement with those reported by Esteve et al. [58] and by Supraditareporn and Pinthong [70] where a significant increase in acidity with storage time was observed. The low pH values of orange juices (3–4) significantly limit the number and types of bacteria that can survive or grow, especially the lactic acid bacteria, which are spoilage microorganisms that cause the development of slime, gas, off-flavors, turbidity, and changes in acidity [70].

3.4. Microbiological Analysis and Sugar Content

At the moment of packaging, the initial population of microorganisms inside the orange juice was 1.79×10^3 CFU/mL for yeast and molds and 4.57×10^2 CFU/mL for bacteria (Figure 3). These results indicate that despite the large population of bacteria, the yeast and molds increased during storage, meaning that yeast and molds are better adapted than bacteria to orange juice under refrigeration, as reported by Sadler et al. [71] and Emamifar et al. [7]. Figure 3 shows that the population of yeast and molds, and bacterial growth increased to 1.55×10^6 CFU/mL and 6.15×10^4 CFU/mL, respectively, after 14 days of storage inside the LDPE_3GTE package. Nevertheless, a significant deceleration was observed in the growth rate and in the total count of bacteria population after 7 days of storage, especially for LDPE_3GTE films.

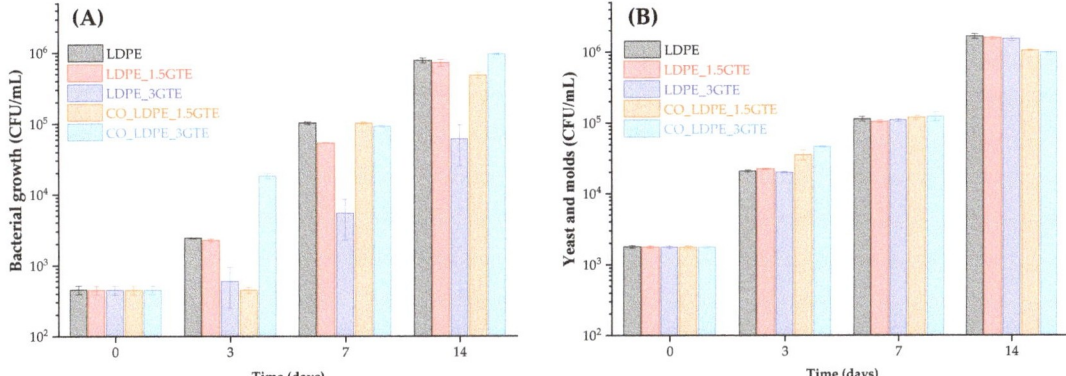

Figure 3. Count of bacterial (**A**) and yeast and molds (**B**) growth in orange juice packed from active LDPE films containing GTE stored at 4 °C for 14 days.

For fresh orange juice, the shelf-life is defined as the required time need to reach a microbial population of 6 log CFU/mL [72]. Moreover, previous studies have shown a shelf life of 14 days for refrigerated orange juice (4 °C) [5,65,73,74]. The average population of bacterial growth remained below 6 log CFU/mL until 7 days in all the packages, yet in the case of yeast and molds, only the CO_LDPE_3GTE package remained below 6 log films at 7 days. It is noteworthy that as the GTE concentration increases, the antimicrobial activity is enhanced, yet for the same GTE concentrations, the LDPE_3GTE package exhibited a higher antibacterial activity compared with CO_LDPE_3GTE, even after 14 days of storage. Considering the results obtained, it is possible to verify that the increased GTE concentration in the packages has a more prominent effect on antibacterial activity than on antifungal activity after a week of storage, and the LDPE_3GTE packages maintained the same pattern over time, always having higher antimicrobial activity than the other active packages. Thus, yeast, molds, and bacteria exhibit different levels of sensitivity to the GTE incorporated in active LDPE films. Published studies demonstrate that the yeast growth during storage is the principal parameter that affects the shelf life of natural orange juice [1,74].

Muriel-Galet et al. characterized the antimicrobial efficiency of polypropylene/ethylene-vinyl alcohol (EVOH) films with oregano essential oil and citral and verified that antimicrobial activity reduced spoilage flora on salad and was more effective against Gram-negative bacteria [75]. Another study assessed the antimicrobial effect of GTE and oregano essential oil incorporated in EVOH films, which showed strong antimicrobial activity against the tested microorganisms, and films containing GTE also inhibited the growth of *L. monocytogenes* and *E. coli* in liquid media; however, a synergistic antimicrobial effect was not detected [39]. The study of Dong et al. [28] based on bilayer LDPE active packaging with the incorporation of rosemary and cinnamon essential oils revealed an effective retardation of the growth of the total viable count in Pacific white shrimps, showing that the cinnamon essential oil exhibited stronger antimicrobial effects than rosemary essential oil.

The initial value of sugar concentration in orange juice for the different packages was 12.7 g/100 mL. After 3 days of storage, orange juice showed a decrease in sugar concentration by about 9.45% in all packages, where juice in the LDPE_1.5GTE packaging film showed the higher reduction (≈11%). From the day 3 to the end of storage (14 days), the sugar concentration remained practically constant, as can be seen in Figure 4. The reduction in sugars concentration is correlated with the increase in microorganism growth in the juice, as can be seen in Figures 3 and 4. As yeast and mold populations increase, there is a consumption of sugars that are transformed into carbon dioxide through a fermentation process and, which consequently, contributes to a decrease in the sugar concentrations.

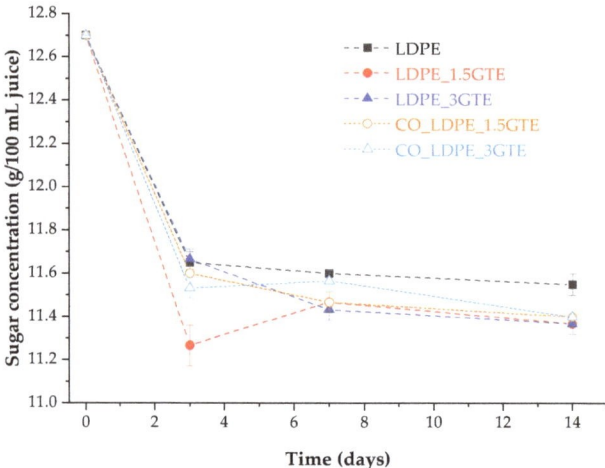

Figure 4. Sugar content in orange juice packed in active LDPE films stored at 4 °C for 14 days.

4. Conclusions

The results show that final content of AA after 14 days of storage varied from 18.36 to 32.60 mg/100 mL, which is lower than the reference value for industrialized juice (40 mg/100 mL). The decrease in the amount of AA is correlated with an increase in browning values; thus, the LDPE packages had the greatest decrease in AA content and also the highest browning index. The juices with higher AA content and lowest browning index were the ones packed in film containing 3% GTE (both monolayer and co-extruded). Other parameters, such as pH and TA, showed different patterns. Although pH decreased by approximately 21%, the TA increased. The most promising packaging for increasing the juice shelf life was verified through microbiological analysis. This analysis showed that the increased GTE concentration in the LDPE films had a more pronounced effect on the bacteria than the fungi after a week of storage. Therefore, it can be concluded that GTE is more effective as an inhibitor of bacterial growth in orange juice.

It is important to mention that the orange juice used in this study is a natural juice without any additional preservatives and usually has a short shelf life between 3 and 4 days. Overall, the microbiological activity of the produced active LDPE films demonstrates that, at least 14 days are necessary for the growth of bacteria, yeasts and molds to reach the limit value of 6 log. The addition of GTE had a positive effect on the inhibition of bacterial growth, being most effective for the monolayer film with 3% GTE.

Based on the results from this investigation, we conclude that the LDPE_3GTE package is the most suitable for storage of orange juice for 14 days at 4 °C. Juice stored in this package maintained a higher concentration of AA, had a lower browning index and had the most resistance to bacterial growth. Thus, active LDPE films containing GTE are effective as a new approach to preserve and extend the shelf-life of fresh orange juice at 4 °C.

As a general conclusion, to achieve their desired properties as a gas/light barrier or for mechanical stability, conventional food packaging systems are made of multi-material products. For example, Tetra Pak@ packages have paperboard, aluminum, and LDPE layers, which have a complex manufacturing process and are difficult to recycle. Therefore, a package made of a single polymer will have lower production costs, a smaller carbon footprint as well as increased shelf-life and sustainability.

Author Contributions: Conceptualization, V.G.M. and A.V.M.; Methodology, V.G.M. and A.V.M.; Validation, P.V.R., M.C.R.C., V.G.M. and A.V.M.; Formal Analysis, D.M.V., P.C.M., P.V.R. and M.C.R.C.; Investigation, D.M.V., P.C.M. and P.V.R.; Resources, V.G.M. and A.V.M.; Data Curation, D.M.V., P.C.M., P.V.R. and M.C.R.C.; Writing—Original Draft Preparation, D.M.V. and P.C.M.; Writing—Review and Editing, P.V.R., M.C.R.C., V.G.M. and A.V.M.; Visualization, D.M.V.; Supervision, V.G.M. and A.V.M.; Project Administration, A.V.M.; Funding Acquisition, A.V.M. All authors have read and agreed to the published version of the manuscript.

Funding: This work was founded by Portugal 2020, and Fundo Social Europeu (FSE) through Programa Operacional Regional do NORTE (NORTE-08-5369–FSE-000034), developed under the program "IMPULSE—Polímeros e Compósitos: Drivers da inovação tecnológica e da competitividade industrial".

Institutional Review Board Statement: Not applicable.

Informed Consent Statement: Not applicable.

Data Availability Statement: The data presented in this study is available in the article.

Conflicts of Interest: The authors declare no conflict of interest.

References

1. De Souza, M.C.C.; Benassi, M.d.T.; Meneghel, R.F.d.A.; da Silva, R.S.d.S.F. Stability of unpasteurized and refrigerated orange juice. *Braz. Arch. Biol. Technol.* **2004**, *47*, 391–397. [CrossRef]
2. Ros-Chumillas, M.; Belissario, Y.; Iguaz, A.; López, A. Quality and shelf life of orange juice aseptically packaged in PET bottles. *J. Food Eng.* **2007**, *79*, 234–242. [CrossRef]
3. Jia, X.; Ren, J.; Fan, G.; Reineccius, G.A.; Li, X.; Zhang, N.; An, Q.; Wang, Q.; Pan, S. Citrus juice off-flavor during different processing and storage: Review of odorants, formation pathways, and analytical techniques. *Crit. Rev. Food Sci. Nutr.* **2022**, 1–26. [CrossRef] [PubMed]
4. Perez-Cacho, P.R.; Rouseff, R. Processing and storage effects on orange juice aroma: A review. *J. Agric. Food Chem.* **2008**, *56*, 9785–9796. [CrossRef]
5. López-Gómez, A.; Ros-Chumillas, M. Packaging and the Shelf Life of Orange Juice. In *Food Packaging and Shelf Life: A Practical Guide*; Robertson, G.L., Ed.; CRC Press: Boca Raton, FL, USA, 2009; pp. 179–198.
6. Akyildiz, A.; Mertoglu, T.S.; Agcam, E. Kinetic study for ascorbic acid degradation, hydroxymethylfurfural and furfural formations in orange juice. *J. Food Compos. Anal.* **2021**, *102*, 103996. [CrossRef]
7. Emamifar, A.; Kadivar, M.; Shahedi, M.; Soleimanian-Zad, S. Evaluation of nanocomposite packaging containing Ag and ZnO on shelf life of fresh orange juice. *Innov. Food Sci. Emerg. Technol.* **2010**, *11*, 742–748. [CrossRef]
8. Polat, S. Color quality, ascorbic acid, total carotenoid, and volatile compounds of dried orange slices as influenced by packaging methods and storage conditions. *J. Food Process. Preserv.* **2022**, *46*, e15898. [CrossRef]
9. Kaddumukasa, P.P.; Imathiu, S.M.; Mathara, J.M.; Nakavuma, J.L. Influence of physicochemical parameters on storage stability: Microbiological quality of fresh unpasteurized fruit juices. *Food Sci. Nutr.* **2017**, *5*, 1098–1105. [CrossRef]
10. Roberts, T.A.; Cordier, J.-L.; Gram, L.; Tompkin, R.B.; Pitt, J.I.; Gorris, L.G.M.; Swanson, K.M.J. (Eds.) Soft drinks, fruit juices, concentrates, and fruit preserves. In *Microorganisms in Foods 6: Microbial Ecology of Food Commodity*; Kluwer Academic: Boston, MA, USA, 2005; pp. 544–573. [CrossRef]
11. Aneja, K.R.; Dhiman, R.; Aggarwal, N.K.; Kumar, V.; Kaur, M. Microbes Associated with Freshly Prepared Juices of Citrus and Carrots. *Int. J. Food Sci.* **2014**, *2014*, 408085. [CrossRef]
12. Bevilacqua, A.; Corbo, M.R.; Campaniello, D.; D'Amato, D.; Gallo, M.; Speranza, B. Shelf life prolongation of fruit juices through essential oils and homogenization: A review. In *Science against Microbial Pathogens: Communicating Current Research and Technological Advances*; FORMATEX: Badajoz, Spain, 2011; pp. 1157–1166.
13. Pala, Ç.U.; Toklucu, A.K. Microbial, physicochemical and sensory properties of UV-C processed orange juice and its microbial stability during refrigerated storage. *LWT—Food Sci. Technol.* **2013**, *50*, 426–431. [CrossRef]
14. Lawlor, K.A.; Schuman, J.D.; Simpson, P.G.; Taormina, P.J. Microbiological Spoilage of Beverages BT. In *Compendium of the Microbiological Spoilage of Foods and Beverages*; Sperber, W.H., Doyle, M.P., Eds.; Springer: New York, NY, USA, 2009; pp. 245–284. [CrossRef]
15. Sospedra, I.; Rubert, J.; Soriano, J.M.; Mañes, J. Incidence of microorganisms from fresh orange juice processed by squeezing machines. *Food Control* **2012**, *23*, 282–285. [CrossRef]
16. Ghenghesh, K.S.; Belhaj, K.; El-Amin, W.B.; El-Nefathi, S.E.; Zalmum, A. Microbiological quality of fruit juices sold in Tripoli–Libya. *Food Control* **2005**, *16*, 855–858. [CrossRef]
17. Raybaudi-Massilia, R.M.; Mosqueda-Melgar, J.; Soliva-Fortuny, R.; Martín-Belloso, O. Control of Pathogenic and Spoilage Microorganisms in Fresh-cut Fruits and Fruit Juices by Traditional and Alternative Natural Antimicrobials. *Compr. Rev. Food Sci. Food Saf.* **2009**, *8*, 157–180. [CrossRef]

18. Wibowo, S.; Buvé, C.; Hendrickx, M.; Van Loey, A.; Grauwet, T. Integrated science-based approach to study quality changes of shelf-stable food products during storage: A proof of concept on orange and mango juices. *Trends Food Sci. Technol.* **2018**, *73*, 76–86. [CrossRef]
19. Nerín, C.; Tovar, L.; Djenane, D.; Camo, J.; Salafranca, J.; Beltrán, J.A.; Roncalés, P. Stabilization of Beef Meat by a New Active Packaging Containing Natural Antioxidants. *J. Agric. Food Chem.* **2006**, *54*, 7840–7846. [CrossRef] [PubMed]
20. Carrizo, D.; Taborda, G.; Nerín, C.; Bosetti, O. Extension of shelf life of two fatty foods using a new antioxidant multilayer packaging containing green tea extract. *Innov. Food Sci. Emerg. Technol.* **2016**, *33*, 534–541. [CrossRef]
21. Khorshidian, N.; Yousefi, M.; Khanniri, E.; Mortazavian, A.M. Potential application of essential oils as antimicrobial preservatives in cheese. *Innov. Food Sci. Emerg. Technol.* **2018**, *45*, 62–72. [CrossRef]
22. Sanches-Silva, A.; Costa, D.; Albuquerque, T.G.; Buonocore, G.G.; Ramos, F.; Castilho, M.C.; Machado, A.V.; Costa, H.S. Trends in the use of natural antioxidants in active food packaging: A review. *Food Addit. Contam. Part A* **2014**, *31*, 374–395. [CrossRef] [PubMed]
23. Sung, S.-Y.; Sin, L.T.; Tee, T.-T.; Bee, S.-T.; Rahmat, A.; Rahman, W.; Tan, A.-C.; Vikhraman, M. Antimicrobial agents for food packaging applications. *Trends Food Sci. Technol.* **2013**, *33*, 110–123. [CrossRef]
24. Ahmed, I.; Lin, H.; Zou, L.; Brody, A.L.; Li, Z.; Qazi, I.M.; Pavase, T.R.; Lv, L. A comprehensive review on the application of active packaging technologies to muscle foods. *Food Control* **2017**, *82*, 163–178. [CrossRef]
25. Singh, S.; Chaurasia, P.K.; Bharati, S.L. Functional roles of Essential oils as an effective alternative of synthetic food preservatives: A review. *J. Food Process. Preserv.* **2022**, *46*, e16804. [CrossRef]
26. Kumar, S.; Basumatary, I.B.; Sudhani, H.P.; Bajpai, V.K.; Chen, L.; Shukla, S.; Mukherjee, A. Plant extract mediated silver nanoparticles and their applications as antimicrobials and in sustainable food packaging: A state-of-the-art review. *Trends Food Sci. Technol.* **2021**, *112*, 651–666. [CrossRef]
27. Ribeiro-Santos, R.; De Melo, N.R.; Andrade, M.; Azevedo, G.; Machado, A.; Carvalho-Costa, D.; Sanches-Silva, A. Whey protein active films incorporated with a blend of essential oils: Characterization and effectiveness. *Packag. Technol. Sci.* **2018**, *31*, 27–40. [CrossRef]
28. Dong, Z.; Xu, F.; Ahmed, I.; Li, Z.; Lin, H. Characterization and preservation performance of active polyethylene films containing rosemary and cinnamon essential oils for Pacific white shrimp packaging. *Food Control* **2018**, *92*, 37–46. [CrossRef]
29. Ribeiro-Santos, R.; Andrade, M.; de Melo, N.R.; Sanches-Silva, A. Use of essential oils in active food packaging: Recent advances and future trends. *Trends Food Sci. Technol.* **2017**, *61*, 132–140. [CrossRef]
30. Estevez-Areco, S.; Guz, L.; Candal, R.; Goyanes, S. Release kinetics of rosemary (*Rosmarinus officinalis*) polyphenols from polyvinyl alcohol (PVA) electrospun nanofibers in several food simulants. *Food Packag. Shelf Life* **2018**, *18*, 42–50. [CrossRef]
31. Andrade, M.A.; Ribeiro-Santos, R.; Costa Bonito, M.C.; Saraiva, M.; Sanches-Silva, A. Characterization of rosemary and thyme extracts for incorporation into a whey protein based film. *LWT* **2018**, *92*, 497–508. [CrossRef]
32. Feng, K.; Wen, P.; Yang, H.; Li, N.; Lou, W.Y.; Zong, M.H.; Wu, H. Enhancement of the antimicrobial activity of cinnamon essential oil-loaded electrospun nanofilm by the incorporation of lysozyme. *RSC Adv.* **2017**, *7*, 1572–1580. [CrossRef]
33. Pola, C.C.; Medeiros, E.A.; Pereira, O.L.; Souza, V.G.; Otoni, C.G.; Camilloto, G.P.; Soares, N.F. Cellulose acetate active films incorporated with oregano (*Origanum vulgare*) essential oil and organophilic montmorillonite clay control the growth of phytopathogenic fungi. *Food Packag. Shelf Life* **2016**, *9*, 69–78. [CrossRef]
34. Lorenzo, J.M.; Batlle, R.; Gómez, M. Extension of the shelf-life of foal meat with two antioxidant active packaging systems. *LWT—Food Sci. Technol.* **2014**, *59*, 181–188. [CrossRef]
35. Wrona, M.; Nerín, C.; Alfonso, M.J.; Caballero, M.Á. Antioxidant packaging with encapsulated green tea for fresh minced meat. *Innov. Food Sci. Emerg. Technol.* **2017**, *41*, 307–313. [CrossRef]
36. Delgado-Adámez, J.; Bote, E.; Parra-Testal, V.; Martín, M.J.; Ramírez, R. Effect of the Olive Leaf Extracts In Vitro and in Active Packaging of Sliced Iberian Pork Loin. *Packag. Technol. Sci.* **2016**, *29*, 649–660. [CrossRef]
37. Song, N.-B.; Lee, J.-H.; Al Mijan, M.; Bin Song, K. Development of a chicken feather protein film containing clove oil and its application in smoked salmon packaging. *LWT—Food Sci. Technol.* **2014**, *57*, 453–460. [CrossRef]
38. Panrong, T.; Karbowiak, T.; Harnkarnsujarit, N. Thermoplastic starch and green tea blends with LLDPE films for active packaging of meat and oil-based products. *Food Packag. Shelf Life* **2019**, *21*, 100331. [CrossRef]
39. Muriel-Galet, V.; Cran, M.J.; Bigger, S.W.; Hernández-Muñoz, P.; Gavara, R. Antioxidant and antimicrobial properties of ethylene vinyl alcohol copolymer films based on the release of oregano essential oil and green tea extract components. *J. Food Eng.* **2015**, *149*, 9–16. [CrossRef]
40. Martins, C.; Vilarinho, F.; Silva, A.S.; Andrade, M.; Machado, A.V.; Castilho, M.C.; Sá, A.; Cunha, A.; Vaz, M.F.; Ramos, F. Active polylactic acid film incorporated with green tea extract: Development, characterization and effectiveness. *Ind. Crops Prod.* **2018**, *123*, 100–110. [CrossRef]
41. Vieira, D.M.; Pereira, C.; Calhelha, R.C.; Barros, L.; Petrovic, J.; Sokovic, M.; Barreiro, M.F.; Ferreira, I.C.; Castro, M.C.R.; Rodrigues, P.V.; et al. Evaluation of plant extracts as an efficient source of additives for active food packaging. *Food Front.* **2022**, *3*, 480–488. [CrossRef]
42. Hu, J.; Zhou, D.; Chen, Y. Preparation and Antioxidant Activity of Green Tea Extract Enriched in Epigallocatechin (EGC) and Epigallocatechin Gallate (EGCG). *J. Agric. Food Chem.* **2009**, *57*, 1349–1353. [CrossRef]

43. Chen, C.-W.; Xie, J.; Yang, F.-X.; Zhang, H.-L.; Xu, Z.-W.; Liu, J.-L.; Chen, Y.-J. Development of moisture-absorbing and antioxidant active packaging film based on poly(vinyl alcohol) incorporated with green tea extract and its effect on the quality of dried eel. *J. Food Process. Preserv.* **2018**, *42*, e13374. [CrossRef]
44. Lee, L.-S.; Kim, S.-H.; Kim, Y.-B.; Kim, Y.-C. Quantitative Analysis of Major Constituents in Green Tea with Different Plucking Periods and Their Antioxidant Activity. *Molecules* **2014**, *19*, 9173–9186. [CrossRef]
45. Xing, L.; Zhang, H.; Qi, R.; Tsao, R.; Mine, Y. Recent Advances in the Understanding of the Health Benefits and Molecular Mechanisms Associated with Green Tea Polyphenols. *J. Agric. Food Chem.* **2019**, *67*, 1029–1043. [CrossRef]
46. Vieira, D.M.; Andrade, M.A.; Vilarinho, F.; Silva, A.S.; Rodrigues, P.V.; Castro, M.C.R.; Machado, A.V. Mono and multilayer active films containing green tea to extend food shelf life. *Food Packag. Shelf Life* **2022**, *33*, 100918. [CrossRef]
47. Wen, P.; Zhu, D.-H.; Wu, H.; Zong, M.-H.; Jing, Y.-R.; Han, S.-Y. Encapsulation of cinnamon essential oil in electrospun nanofibrous film for active food packaging. *Food Control* **2016**, *59*, 366–376. [CrossRef]
48. Soltani Firouz, M.; Mohi-Alden, K.; Omid, M. A critical review on intelligent and active packaging in the food industry: Research and development. *Food Res. Int.* **2021**, *141*, 110113. [CrossRef] [PubMed]
49. Manikantan, M.R.; Pandiselvam, R.; Arumuganathan, T.; Indurani, C.; Varadharaju, N. Low-density polyethylene based nanocomposite packaging films for the preservation of sugarcane juice. *J. Food Sci. Technol.* **2022**, *59*, 1629–1636. [CrossRef]
50. Azevedo, A.G.; Barros, C.; Miranda, S.; Machado, A.V.; Castro, O.; Silva, B.; Saraiva, M.; Silva, A.S.; Pastrana, L.; Carneiro, O.S.; et al. Active Flexible Films for Food Packaging: A Review. *Polymers* **2022**, *14*, 2442. [CrossRef]
51. Wrona, M.; Silva, F.; Salafranca, J.; Nerín, C.; Alfonso, M.J.; Caballero, M. Design of new natural antioxidant active packaging: Screening flowsheet from pure essential oils and vegetable oils to ex vivo testing in meat samples. *Food Control* **2021**, *120*, 107536. [CrossRef]
52. Zambiazi, R.C. *Análise Físico-Química de Alimentos*, 1st ed.; UFPel: Pelotas, Brazil, 2010.
53. Zerdin, K.; Rooney, M.L.; Vermuë, J. The vitamin C content of orange juice packed in an oxygen scavenger material. *Food Chem.* **2003**, *82*, 387–395. [CrossRef]
54. American Public Health Association. *Compendium of Methods for the Microbiological Examination of Foods*, 4th ed.; APHA: Washington, DC, USA, 2001.
55. Polydera, A.C.; Stoforos, N.G.; Taoukis, P.S. Comparative shelf life study and vitamin C loss kinetics in pasteurised and high pressure processed reconstituted orange juice. *J. Food Eng.* **2003**, *60*, 21–29. [CrossRef]
56. Soares, N.F.F.; Hotchkiss, J.H. Comparative effects of de-aeration and package permeability on ascorbic acid loss in refrigerated orange juice. *Packag. Technol. Sci.* **1999**, *12*, 111–118. [CrossRef]
57. Roig, M.G.; Bello, J.F.; Rivera, Z.S.; Kennedy, J.F. Studies on the occurrence of non-enzymatic browning during storage of citrus juice. *Food Res. Int.* **1999**, *32*, 609–619. [CrossRef]
58. Solomon, O.; Svanberg, U.; Sahlström, A. Effect of oxygen and fluorescent light on the quality of orange juice during storage at 8 °C. *Food Chem.* **1995**, *53*, 363–368. [CrossRef]
59. Wilson, R.J.; Beezer, A.E.; Mitchell, J.C. A kinetic study of the oxidation of L-ascorbic acid (vitamin C) in solution using an isothermal microcalorimeter. *Thermochim. Acta* **1995**, *264*, 27–40. [CrossRef]
60. Viana Batista, R.; Gonçalves Wanzeller, W.; Lim, L.-T.; Quast, E.; Zanella Pinto, V.; Machado de Menezes, V. Food packaging and its oxygen transfer models in active multilayer structures: A theoretical review. *J. Plast. Film Sheeting* **2022**, *38*, 458–488. [CrossRef]
61. Esteve, M.J.; Frígola, A.; Rodrigo, C.; Rodrigo, D. Effect of storage period under variable conditions on the chemical and physical composition and colour of Spanish refrigerated orange juices. *Food Chem. Toxicol.* **2005**, *43*, 1413–1422. [CrossRef]
62. Bharate, S.S.; Bharate, S.B. Non-enzymatic browning in citrus juice: Chemical markers, their detection and ways to improve product quality. *J. Food Sci. Technol.* **2014**, *51*, 2271–2288. [CrossRef]
63. Leizerson, S.; Shimoni, E. Stability and Sensory Shelf Life of Orange Juice Pasteurized by Continuous Ohmic Heating. *J. Agric. Food Chem.* **2005**, *53*, 4012–4018. [CrossRef]
64. Cortés, C.; Esteve, M.J.; Frígola, A. Color of orange juice treated by High Intensity Pulsed Electric Fields during refrigerated storage and comparison with pasteurized juice. *Food Control* **2008**, *19*, 151–158. [CrossRef]
65. Bull, M.K.; Zerdin, K.; Howe, E.; Goicoechea, D.; Paramanandhan, P.; Stockman, R.; Sellahewa, J.; A Szabo, E.; Johnson, R.L.; Stewart, C.M. The effect of high pressure processing on the microbial, physical and chemical properties of Valencia and Navel orange juice. *Innov. Food Sci. Emerg. Technol.* **2004**, *5*, 135–149. [CrossRef]
66. Lee, H.S.; Coates, G.A. Effect of thermal pasteurization on Valencia orange juice color and pigments. *LWT—Food Sci. Technol.* **2003**, *36*, 153–156. [CrossRef]
67. Rivas, A.; Rodrigo, D.; Martínez, A.; Barbosa-Cánovas, G.V.; Rodrigo, M. Effect of PEF and heat pasteurization on the physical–chemical characteristics of blended orange and carrot juice. *LWT—Food Sci. Technol.* **2006**, *39*, 1163–1170. [CrossRef]
68. Touati, N.; Barba, F.J.; Louaileche, H.; Frigola, A.; Esteve, M.J. Effect of Storage Time and Temperature on the Quality of Fruit Nectars: Determination of Nutritional Loss Indicators. *J. Food Qual.* **2016**, *39*, 209–217. [CrossRef]
69. Del Caro, A.; Piga, A.; Vacca, V.; Agabbio, M. Changes of flavonoids, vitamin C and antioxidant capacity in minimally processed citrus segments and juices during storage. *Food Chem.* **2004**, *84*, 99–105. [CrossRef]
70. Supraditareporn, W.; Pinthong, R. Physical, Chemical and Microbiological Changes during Storage of Orange Juices cv. Sai Nam Pung and cv. Khieo Waan in Northern Thailand. *Int. J. Agric. Biol.* **2007**, *9*, 726–730.

71. Sadler, G.D.; Parish, M.E.; Wicker, L. Microbial, Enzymatic, and Chemical Changes During Storage of Fresh and Processed Orange Juice. *J. Food Sci.* **1992**, *57*, 1187–1197. [CrossRef]
72. Raccach, M.; Mellatdoust, M. The Effect of Temperature on Microbial Growth in Orange Juice. *J. Food Process. Preserv.* **2007**, *31*, 129–142. [CrossRef]
73. Fellers, P.J. Shelf Life and Quality of Freshly Squeezed, Unpasteurized, Polyethylene-Bottled Citrus Juice. *J. Food Sci.* **1988**, *53*, 1699–1702. [CrossRef]
74. Zanoni, B.; Pagliarini, E.; Galli, A.; Laureati, M. Shelf-life prediction of fresh blood orange juice. *J. Food Eng.* **2005**, *70*, 512–517. [CrossRef]
75. Muriel-Galet, V.; Cerisuelo, J.P.; López-Carballo, G.; Aucejo, S.; Gavara, R.; Hernández-Muñoz, P. Evaluation of EVOH-coated PP films with oregano essential oil and citral to improve the shelf-life of packaged salad. *Food Control* **2013**, *30*, 137–143. [CrossRef]

Disclaimer/Publisher's Note: The statements, opinions and data contained in all publications are solely those of the individual author(s) and contributor(s) and not of MDPI and/or the editor(s). MDPI and/or the editor(s) disclaim responsibility for any injury to people or property resulting from any ideas, methods, instructions or products referred to in the content.

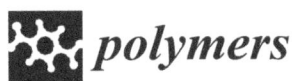

Article

Designing an Oxygen Scavenger Multilayer System Including Volatile Organic Compound (VOC) Adsorbents for Potential Use in Food Packaging

Carol López-de-Dicastillo [1], Gracia López-Carballo [1], Pedro Vázquez [2], Florian Schwager [2], Alejandro Aragón-Gutiérrez [3], José M. Alonso [3], Pilar Hernández-Muñoz [1] and Rafael Gavara [1,*]

[1] Packaging Group, Institute of Agrochemistry and Food Technology IATA-CSIC, Av. Agustín Escardino 7, 46980 Paterna, Spain; clopezdedicastillo@iata.csic.es (C.L.-d.-D.)
[2] Evonik Operations GmbH, Germany
[3] Grupo de Tecnología de Envases y Embalajes, Instituto Tecnológico del Embalaje, Transporte y Logística, ITENE, Unidad Asociada al CSIC, calle de Albert Einstein 1, 46980 Paterna, Spain
* Correspondence: rgavara@iata.csic.es

Abstract: Oxygen scavengers are valuable active packaging systems because several types of food deterioration processes are initiated by oxygen. Although the incorporation of oxygen scavenger agents into the polymeric matrices has been the trend in recent years, the release of volatile organic compounds (VOC) as a result of the reaction between oxygen and oxygen scavenger substances is an issue to take into account. This is the case of an oxygen scavenger based on a trans-polyoctenamer rubber (TOR). In this work, the design of an oxygen scavenger multilayer system was carried out considering the selection of appropriate adsorbents of VOCs to the proposed layer structure. Firstly, the retention of some representative organic compounds by several adsorbent substances, such as zeolites, silicas, cyclodextrins and polymers, was studied in order to select those with the best performances. A hydrophilic silica and an odor-adsorbing agent based on zinc ricinoleate were the selected adsorbing agents. The principal VOCs released from TOR-containing films were carefully identified, and their retention first by the pure adsorbents, and then by polyethylene incorporated with the selected compounds was quantified. Detected concentrations decreased by 10- to 100-fold, depending on the VOC.

Keywords: oxygen scavenger; active packaging; food packaging; multilayer film; volatile organic compound; VOC

1. Introduction

Active packaging is an emerging food technology based on a deliberate interaction of the packaging with the product and/or its headspace to improve food quality and safety. Active packaging refers to those materials intended to interact with the internal gas environment and/or directly with the product with a beneficial outcome. Some of these newly employed technologies modify the gas environment by removing gases from or adding gases to the package headspace [1,2]. Oxygen scavengers are one of the most interesting active packaging systems since the presence of oxygen is known to trigger many food deterioration reactions, such as lipid oxidation, color changes, nutrient losses and microbial growth [1–3]. In comparison with other packaging technologies, such as Modified Atmosphere Packaging (MAP) and vacuum packaging, oxygen scavengers may greatly reduce the oxygen level in the headspace to less than 0.1 vol%, leading to an extended shelf life. In most commercial applications, oxygen-adsorbing substances have been included in sachets that are inserted into the package or as adhesive labels bonded to the inner wall of the package [4–7]. Nevertheless, this system presents some disadvantages, such as a possible accidental ingestion or misuse by the consumer, consumer rejection, and the need of an additional packaging operation to insert the sachet in each package. Therefore, although this technology is well established, these inconveniences have been the driving force of more recent developments, such as closure liners containing the oxygen scavenger, dissolution or dispersion of the active substances in the plastic material [8]. In this regard, designing functional polymeric materials that include the active agents in their structure and ensuring that these active substances do not cause relevant effects on packaging functional properties are both key factors. Metal and metal derivatives, such as finely divided palladium, cobalt or iron, and organic and fatty acids have been broadly used as active substances in the development of oxygen scavengers [3,4,6,9]. Regarding the polymeric material for use in an oxygen scavenging composition, this should exhibit good processing characteristics, be able to be formed directly into useful packaging materials, or have high compatibility with those polymers commonly used in food packaging designs.

Several research works and patent applications have disclosed that ethylenic-unsaturated hydrocarbons, such as squalene, fatty acids, or polybutadiene, present sufficient commercial oxygen scavenging capacity to extend the shelf life of oxygen-sensitive products [9–11]. These unsaturated hydrocarbons, after being functionally terminated with a chemical group to make them compatible with the packaging materials, can be added during conventional melt–mixing processes to thermoplastics, such as polyesters, low-density polyethylene (LDPE) and polypropylene (PP), and the films can be obtained using most conventional plastic-processing techniques, such as co-injection or co-extrusion. 1,2-polybutadiene is especially preferred thanks to its transparency, mechanical properties and processing characteristics similar to those of polyethylene [9,10,12]. The main problem of this technology is that during the reaction between these polyunsaturated macromolecules and oxygen, byproducts such as organic acids, aldehydes, or ketones can be generated, which are volatile organic compounds (VOCs) that can affect the sensory quality of the food or raise food regulatory issues [4,9]. This problem can be minimized through two strategies: (i) the use of functional barriers located between the food product and the scavenger layer that impede the migration of undesirable oxidation products, but allow oxygen transfer [13]; and (ii) the use of VOC adsorbent materials, either polymers with inherent organic compound-scavenging properties or the incorporation of adsorbers within the polymer structure (i.e., silica gel, zeolites, etc.).

In the present work, the oxygen scavenger component is based on trans-polyoctenamer rubber (TOR) in combination with a catalyst with several characteristics that made this product attractive for food packaging applications. Because the release of VOCs, such as organic acids, aldehydes and ketones, can become a drawback, the design of an oxygen scavenger multilayer system was carried out from the selection of appropriate adsorbents of VOCs to the proposal of the multilayer structure.

Therefore, the aim of this study was the reduction of VOCs released by TOR-containing films into the package headspace through the incorporation of a functional barrier with VOC adsorbents in the packaging system. Different adsorbing substances, such as zeolites, cyclodextrines and polymers, were tested in order to select and later develop the multilayer packaging system. The necessary milestones for the development of an efficient commercial prototype were detailed, from the selection of efficient adsorbents to the processing of the resulting multilayer system.

2. Materials and Methods

2.1. Materials

Films from TOR masterbatch (TM) were prepared and supplied by Evonik as experimental test specimens. These films, 50 μm low-density polyethylene (LDPE) film containing 10%wt. of TOR and tri-layer 10/10/10 μm (passive/active/passive) LDPE film containing 10%wt. of TOR in the active layer were supplied in high barrier Aluminum/LDPE bags under vacuum until use.

Adsorbents (supplier): Sylysia (Syl) SY350 and Dumacil (Duma) 100 FG K (Fuji Sylysia, Bussi, Italy), Tego Sorb PY 88 TQ (Tego) (zinc ricinoleate) (Evonik, Essen, Germany), β-cyclodextrin (βCD) and hydroxypropyl-β-CD (HPβCD), commercially named as Cavasol W7 and Cavamax W7, respectively (Wacker Chemie AG, Munich, Germany), Selvol Ultiloc 5003 (polyvinyl alcohol/polyvinyl amine) (Sekisui, La Canonja, Spain) and Tenax TA 60–80 mesh (modified polyphenylene oxide) (Merck Life Science S.L.U., Madrid, Spain).

Formic acid, acetic acid, hexanoic acid, acetaldehyde, propanal, valeraldehyde, hexanal, acetone, amyl formate and cyclooctane were purchased from Merck Life Science S.L.U. (Madrid, Spain).

AGILITY EC 7000 Performance LDPE, 0.919 g/cm^3, melt index 3.9 g/10 min, and maleic anhydride grafted polyethylene used as tie layer Amplify TY-1057-H were purchased from Dow Chemical (Barcelona, Spain). Ethylene vinyl alcohol copolymer (EVOH) Eval F171B, metallocene linear low-density polyetlylene (mPE) Supeer 8115 and polypropylene (PP) Moplen RP 310 were supplied by Kuraray-Eval Europe (Zwijndrecht, Belgium), Sabic (Cartagena, Spain) and Lyondell Basell (Vilaseca, Spain), respectively.

2.2. Selection of Adsorbents

2.2.1. Analysis of Retention Capacity of Adsorbents

Formic acid, acetic acid, hexanoic acid, acetaldehyde, propanal, valeraldehyde, hexanal, acetone, amyl formate and cyclooctane were selected as representative volatile organic compounds (VOCs) to carry out the evaluation of the adsorption capacity of adsorbents.

Fifty mg samples of each specific adsorbent were placed in a 22 mL vial, and the vial was hermetically sealed with an aluminum cap and butyl rubber/polytetrafluoroethylene (PTFE) septum. In parallel and for each organic compound, a known volume of the VOC was injected in a 120 mL glass vial closed with a butyl rubber/PTFE septum previously heated at 150 °C. The substance was allowed to evaporate for 30 min and then a calculated amount of the headspace was withdrawn and injected in the 22 mL vial with the adsorbent, to put quantities in the range 20–100 μg/vial in gas phase. These samples were allowed to equilibrate during 24 h at 23 °C. Then, 1 mL of the headspace was withdrawn and injected in the injection port of an Agilent 7890 gas chromatograph (GC) equipped with flame-ionization detection (FID) (Agilent, Las Rozas de Madrid, Spain). The GC conditions were as follows: 200 and 300 °C were the injector and detector temperatures, respectively; Agilent HP5 column (30 m, 0.32 mm diameter, 0.25 μm) with a 15 mL/min constant flow of He and 5:1 split; oven at 40 °C for 3 min, 10 °C/min to 80 °C, 40 °C/min to 200 °C and 4 isothermal min at 200 °C. The response of the GC was previously calibrated for each VOC by injecting known amounts. Vials without adsorbent were also included as controls. The amount of VOC retained by the adsorbents was indirectly calculated through the difference between the concentrations of VOCs in samples with adsorbents and controls. The results were calculated following Equation (1) and expressed as partition coefficient (K) values:

$$[VOC]_{adsorbent} \left(\frac{g}{g}\right) = \frac{\left[\left([VOC]_{headspace} \left(\frac{g}{mL}\right)^{control} - [VOC]_{headspace} \left(\frac{g}{mL}\right)^{sample}\right) \cdot 22 \text{ mL}\right]}{50 \cdot 10^{-3} \text{ g}};$$

$$K = \frac{[VOC]_{adsorbent} \left(\frac{g}{g}\right)}{[VOC]_{headspace} \left(\frac{g}{mL}\right)}$$

(1)

2.2.2. Analysis of Retention Capacity of Adsorbents

The three substances with best performances as VOC adsorbents were selected and tested via their exposure to 100 μg of a single VOC in vapor state. The retention of VOCs by increasing amounts of pure adsorbents was analyzed (approximately from 30–50 mg). The sorption was analyzed by GC, following the same procedure described in Section 2.2.1.

2.3. Assessment of the Oxygen Scavenging Activity of the TM Product

The oxygen scavenging activity of these two types of films, a 50 μm thick LDPE-based monolayer containing the TM oxygen scavenger at a 10%wt. concentration and a 30 μm 3-layer (10/10/10) LDPE film with the center layer containing 10%wt. scavenger were analyzed in order to confirm their efficiency. The oxygen concentration in the headspace was measured over time by a non-invasive method based on phosphorescence quenching of a sensor dye. An Oxy-4 multichannel oxygen meter equipped with 4 polymer optical fibers and the Measurement Studio 2 software (PreSens, Regensburg, Germany) were used. In 250 mL wide-mouth glass bottles with glass stoppers, calibrated oxygen sensor spots PSt3 (Presens) were adhered on the inner wall of the bottle, and a film sample (1.2 g and 3.6 g of monolayer and 3-layer films, respectively) was introduced in the bottle and was immediately closed with vacuum paste. Adapters for round containers were set on the outside of the bottles fixing the optical fibers for sequential readings.

2.4. Analysis of the Capacity of Retention of VOCs Released by TM-Containing Films by Selected Adsorbents

The principal released VOCs by these films were identified and their retention capacity by the adsorbents selected in Section 2.2 was tested alone and in combination in order to study their synergistic/antagonistic effect. Adsorbents were exposed in a closed container to the VOCs delivered to the headspace by the oxygen scavenging 3-layer 10/10/10 µm film and their adsorption capacities were tested. The 3-layer film was selected because the potential application on food packaging would be more similar to this system because the oxygen scavenger components should not be in direct contact with the food. The amount of film was ca. 250 mg, and approx. 10 mg of every adsorbent was used on every test.

In a 22 mL vial, weighted amounts of the specific adsorbent or adsorbents were introduced. Then, pieces of the active film were weighed and included in the vial, which was hermetically sealed with an aluminum cap and rubber/PTFE septum. The samples were allowed to equilibrate during 72 h. Then, a carboxen/polydimethylsiloxane CAR/PDMS Solid Phase Microextraction (SPME) fiber was introduced in the vial and allowed to adsorb for 20 min. The SPME fiber was introduced in the injector of a gas chromatograph for the desorption of all VOCs. The GC was equipped with a mass detector and the compounds were identified using the NIST (National Institute of Standards and Technology) library. Samples without adsorbent were also included as controls.

The GC conditions were as follows: 250 °C the injector temperature; Agilent HP-5MS (5% phenyl methylpolysiloxane) column (30 m, 0.25 mm diameter, 0.25 µm) with a 0.92 mL/min constant flow of He and 5:1 split; oven at 40 °C for 4 min, 10 °C/min to 220 °C and 10 isothermal min at 220 °C. The conditions of the MS detector were: 1000 gain factor, voltage of 1692 V, frequency at 3.1 scans/s, step size at 0.1 m/z, MS source at 230 °C and MS quad at 150 °C. For each VOC found, a characteristic ion of the substance's MS spectra was selected, and the number of those characteristic ions was monitored to determine the adsorbent efficiency. No calibration was carried out. The percentage of VOCs retained by the adsorbent was indirectly calculated by the difference between the samples with adsorbents and the controls.

2.5. Development of Films including Adsorbents

LDPE formulation compounded with 5 wt.% of Sylysia and 5 wt.% of TegoSorb was prepared by melt extrusion processing using a twin-screw extruder Coperion ZSK 26 Mc (Coperion, Stuttgart, Germany). LDPE pellets and Sylysia were fed through the main hopper of the extruder while Tegosorb was introduced inside the extruder through a side feeder. The temperature profile along the 10-barrel zones from hopper (zone 1) to die (zone 10) was 185–195–200–205–200–200–200–205–205–205 °C, the applied screw rotation speed was set at 550 rpm, and the throughput was 12 kg/h. Then, the compounded sample in pellet form was transformed into a film of around 35 µm using a single-screw extruder Brabender Stand-alone KE 30/32 (Brabender® GmbH & Co. KG, Duisburg, Germany) equipped with an extrusion roller calender line. The temperature profile along the 6-barrel zones from hopper (zone 1) to die (zone 6) was 185–185–190–190–200–200 °C, screw speed 50 rpm and a roll speed of 14 m/min.

Their activities to scavenge released VOCs by TM-including films were analyzed as follows. In a 22 mL vial, a weighted mass of TM-monolayer film (Section 2.3) was included in the vial and then the same weight ("one side"), twice the weight ("two sides"), and four times the weight ("double two sides") of the VOC-scavenger film were included in vials and closed. After 5 days, the headspace of the vials was analyzed by GC-MS. The sample named as "one side" represented a potential 3-layer film containing one pristine PE layer, the TM layer and the VOC scavenging layer. The "two sides" sample represented a potential 3-layer film containing a VOC scavenging layer, the TM layer and VOC scavenging layer. The "Double two sides" sample represented a potential 3-layer film containing VOC scavenging layer, the TOR layer and the VOC scavenging layer but with double the content of Sylysia and Tegosorb adsorbents. The GC conditions of analysis were analogous to Section 2.3.

2.6. Development and Efficiency of Oxygen Scavenger Packaging Systems including TM-Active Agent and VOC Adsorbers

The 10-layer structure was developed in three different steps by employing a coextrusion line equipped with three single-screw extruders (Dr Collin E30P, 25 L/D, COLLIN Lab & Pilot Solutions GmbH, Maitenbeth, Germany). The equipment has a feed-block that can provide up to five layers (ABCBA), and different multilayer rearrangements (e.g., ABC, BCB, ACA), bilayer (AC, BC) or monolayer structures. The first step consisted of the preparation of a 5-layer LDPE/tie/EVOH/tie/LDPE coextruded structure with an overall thickness of 33–36 µm. The processing parameters to obtain the symmetrical five layer "ABCBA" structure are displayed in Table 1. The chill roll temperature and the line speed were set at 60 °C and 20 m/min, respectively.

Table 1. Parameters and scheme of the first step of the multilayer development: a LDPE/tie/EVOH/tie/LDPE film structure.

	Screw Speed (rpm)	T1 [1] (°C)	T2 [1] (°C)	T3 [1] (°C)	T4 [1] (°C)	T5 [1] (°C)	COEX (°C)	DIE (°C)
Extruder A	120	165	180	180	180	180	210	200
Extruder B	25	165	175	175	180	185		
Extruder C	55	205	220	220	220	220		
Scheme	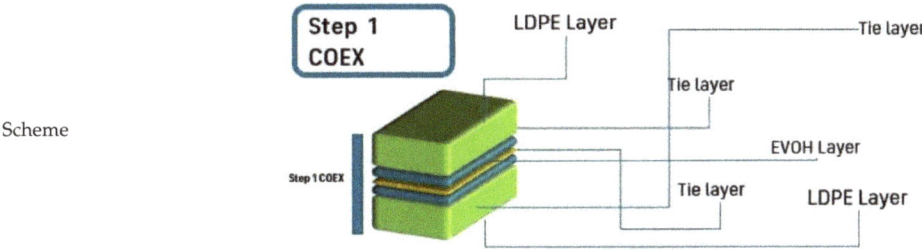							

[1] T1 to T5 are the five heating zones of the extruder screw.

Thereafter, a coextrusion lamination of a PP/mPE bilayer structure over the 5-layer system was carried out, leading to a final overall thickness of 53–60 µm after this stage. Thus, a 7-layer structure as described in Table 2 was obtained. Finally, the third stage consisted of the coextrusion lamination of three-layer structure LDPE-VOC/LDPE + 33% TOR/LDPE-VOC over the 7-layer structure obtained previously to obtain a final multilayer system of 10 layers. The processing parameters of this stage to obtain the three-layer structure "ACA" extrusion-laminated over the 7-layer substrate are described in Table 3. The chill roll temperature and the line speed were set at 90 °C and 22 m/min, respectively, and the final dimensions of the 10-layer (PE-VOC/PE-TM/PE-VOC/LDPE/tie/EVOH/tie/LDPE/mPE/PP) structure were 300 mm width and 85 µm thickness, approximately. A similar structure was prepared by extruding LDPE without adsorbents in A to prepare a control with TOR but without VOC adsorbents (PE/PE-TM/PE/LDPE/tie/EVOH/tie/LDPE/mPE/PP).

Table 2. Parameters and scheme of the second step of the multilayer development: PP/mPE coextrusion lamination onto LDPE/tie/EVOH/tie/LDPE.

	Screw Speed (rpm)	T1 [1] (°C)	T2 [1] (°C)	T3 [1] (°C)	T4 [1] (°C)	T5 [1] (°C)	COEX (°C)	DIE (°C)
Extruder A								
Extruder B	5	190	215	225	240	245	210	200
Extruder C	80	210	240	240	240	240		
Scheme	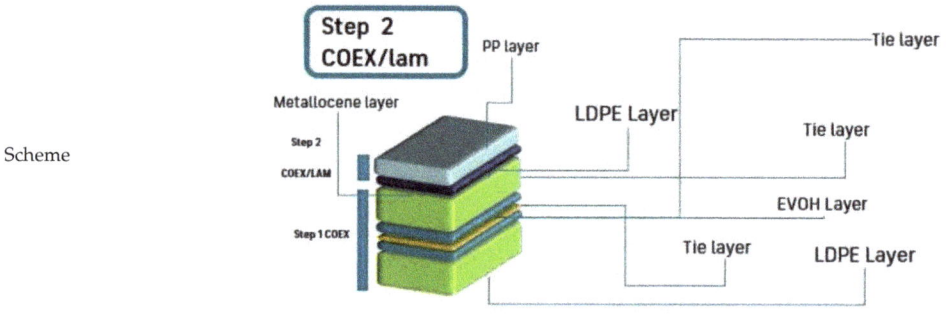							

[1] T1 to T5 are the five heating zones of the extruder screw.

Table 3. Parameters and scheme of the third step of the multilayer development: PE-VOC/PE-TM/PE-VOC coextrusion lamination onto LDPE/tie/EVOH/tie/LDPE/mPE/PP.

	Screw Speed (rpm)	T1 [1] (°C)	T2 [1] (°C)	T3 [1] (°C)	T4 [1] (°C)	T5 [1] (°C)	COEX (°C)	DIE (°C)
Extruder A	120	165	190	190	195	200		
Extruder B	-	-	-	-	-	-	205	205
Extruder C	100	190	190	190	195	200		
Scheme	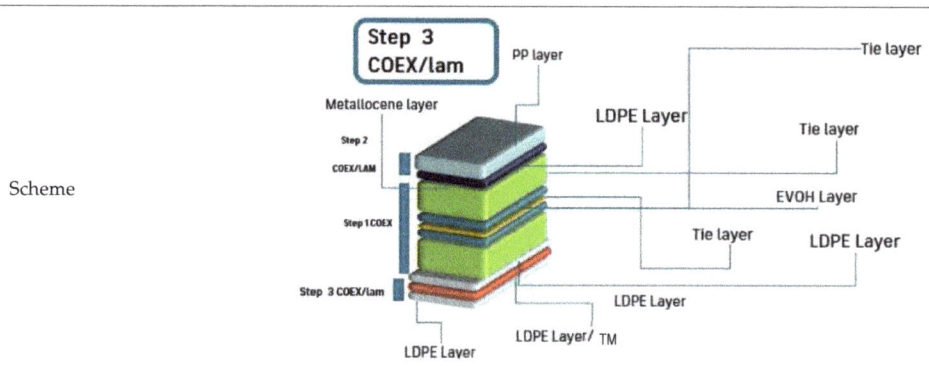							

[1] T1 to T5 are the five heating zones of the extruder screw.

The final thickness distribution of the 10-layer structure was evaluated by optical microscopy in transmittance mode using a Leica DM/LM optical microscope.

An amount of 1 g of the 10-layer film was hermetically closed in a 22 mL vial with air. After 5 days, an analysis of the vial headspace was conducted to compare the amount of volatile compounds released from films. A similar set-up was prepared with the control 10-layer film. The percentage of VOCs retained by adsorbents was evaluated.

2.7. Statistical Analysis

Analysis of variance (ANOVA) and Fisher's multiple range test of the DSC and mechanical parameters, permeability, and overall migration data were performed using Statgraphics Plus 5.1. The number of replicates of the ANOVA analysis was stated in the experimental procedure for each test. Statistical comparisons were made from a randomized experimental design with a confidence level of 95%. Results were reported as the mean and standard deviation. A p-value less than 0.05 indicated that the mean values were significantly different between the samples.

3. Results and Discussion

3.1. Adsorption Capacity of VOC Adsorbents

As mentioned in the introduction, unsaturated polyolefins are excellent oxygen scavenger materials but, during the oxygen reaction process, chain scission occurs releasing low-molecular-weight oxidized organic compounds including organic acids, aldehydes and ketones. This is what occurs when TOR containing films are exposed to atmospheres containing oxygen. Thus, in order to reduce the release of VOCs into package headspace, several adsorbent materials were exposed to individual volatile compounds to check their adsorption capacity. The selected volatiles were mainly short chain organic acids and aldehydes. Also, acetone, amyl formate and cyclooctane were selected as representative compounds of the ketone, esters and hydrocarbon families.

Figure 1 shows the partition coefficients between adsorbents and headspace for all tested volatile organic compounds. Silica is a solid form of silicon dioxide and is specially synthesized to be highly porous. Although SiO_2 is a polar compound, the surfaces of functionalized silica are usually covered with a layer of hydroxyl groups which can be largely substituted through treatment with suitable reagents to reduce polarity [14]. Adsorption results of Sylysia SY350 (Syl) in this first adsorption analysis evidenced this hydrophilic silica as a potential adsorbent to scavenge VOCs from TOR byproducts. This silica presented K values above 1000 for carboxylic acids, aldehydes, ketones and esters, although K values were lower for cyclooctane due to the increased apolar character of this alkane. On the contrary, the hydrophobic silica Dumacil 100 FG (Duma) did not present relevant adsorption values for any of the tested organic substances. Previous works have already shown the effects of physical structure and chemical modification of adsorbents on the adsorption performance [15,16].

Tego active agent, a product from Evonik, is a masterbatch of LDPE-containing zinc ricinoleate, the zinc salt of the major fatty acid found in castor oil often used as an odor-adsorbing agent [17,18]. The most relevant K values presented by this product were towards organic acids, but relevantly towards more apolar compounds and those with higher molecular weights, such as cyclooctane.

Cyclodextrins (CDs) are cyclic oligosaccharides which have been widely used as scavengers of organic compounds of low polarity. They present a truncated cone shape with a polar outer surface and an apolar inner surface [19,20]. Among the diverse CDs, βCD has been reported to complex several types of organic compounds. This CD has also been functionalized to improve their compatibility with polymers or water solubility. This is the case of hydroxypropyl-βCD (HPβCD), obtained through the functionalization of hydroxyl by hydroxypropyl groups, and commonly used for the complexation of poorly soluble drugs [21,22]. In general, both CDs presented acceptable K values, with the retention capacity of both tested CDs being very similar.

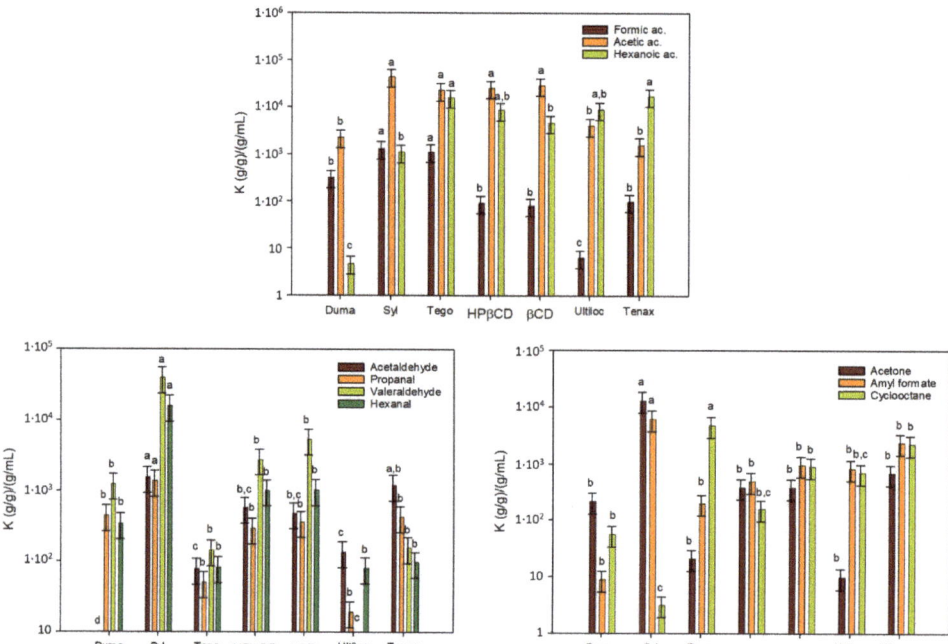

Figure 1. Partition coefficients (K) for the selected volatile organic compounds between the tested adsorbents named in the x-axis and the headspace. "a, b, c and d" indicate significant differences among the K values of each organic compound between adsorbents.

Ultiloc 5003 (Ultiloc) is a novel copolymer of vinyl alcohol and vinyl amine that was selected as a potential aldehyde scavenger since aldehydes are known to react with primary amines to produce Schiff bases [23,24], and these aldehydes, especially C5 to C8, are known to promote rancid off-flavor which can affect sensory properties in food packaging applications. Nevertheless, a low adsorption capacity was observed for polyvinyl alcohol/polyvinyl amine Ultiloc.

Finally, Tenax, a modified polyphenylene oxide and powerful adsorbent commonly used for the trapping of VOCs in water samples, or as food simulant of solid foods in migration tests, did not present the expected results for most VOCs, although its adsorption capacity towards molecules with higher molecular size, such as hexanoic acid, amylformate and cyclooctane, was quite high.

From this first analysis, Syl, Tego and βCD were the adsorbents with better performances, and particularly, Syl was the material that presented the greatest adsorption capacity on most VOCs; Tego seemed to provide good adsorption properties to organic acids and alkanes but showed a lower capacity with aldehydes, and βCD presented an average adsorption capacity to all compounds and their incorporation in polymers can be performed easily through extrusion rather than HPβCD. Tenax (60–80 mesh), due to its high cost, low availability and large particle size, was discontinued for further tests.

The effect of increasing the amount of adsorbents with the best performances on VOC retention was also analyzed to assess whether the retention capacity measured is a consequence of a partition or is an exhaustion of the adsorption capacity. Figure 2 shows two examples of the results obtained for the capacity of increasing amounts of Syl, Tego and βCD in retaining acetic acid, and valeraldehyde. As can be seen, K values for the tested VOCs and the three adsorbents did not present any trend with respect to the amount of the solid sample included in the analyses. Although some variations were observed, in general, it can be concluded that partition equilibria of the VOCs between the adsorbent and the headspace were achieved. These results also confirmed that Syl retained relevant amounts of most polar VOCs thanks to its hydrophilic character but failed to retain alkanes and alkenes. Tego was good for acids but performed greatly with hydrocarbons. βCD always presented intermediate K values, being, therefore, discarded for the following trials.

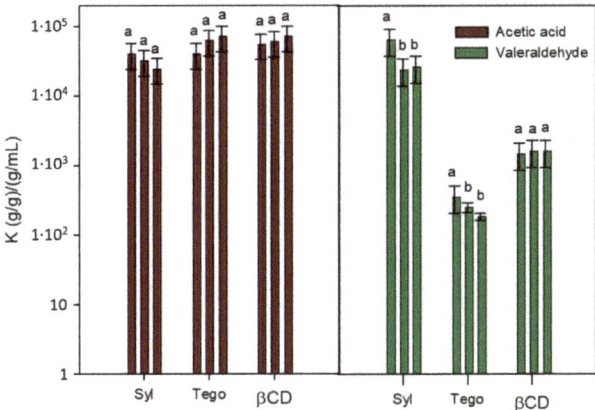

Figure 2. Partition coefficients (K) for acetic acid and valeraldehyde between Syl, Tego and βCD adsorbents and the headspace (the amount of adsorbent increases from left to right for each adsorbent and VOC). "a and b" indicate significant differences among the K values of each organic compound between adsorbents.

3.2. Oxygen Scavenging of TOR-Containing Films

Two LDPE films containing oxygen scavenger TM were successfully developed through melting extrusion: a 50 μm thick LDPE based monolayer containing TM at 10%wt. and a 30 μm 3-layer (10/10/10) LDPE film with the center layer containing 10%wt. scavenger. Both films were supplied by Evonik in aluminum/PE pouches under vacuum.

First, the oxygen scavenging activities of these active films were analyzed using Oxi-4 Presens equipment. Oxygen scavenging activities were periodically measured for 5 days and their resulting kinetics are graphed in Figure 3.

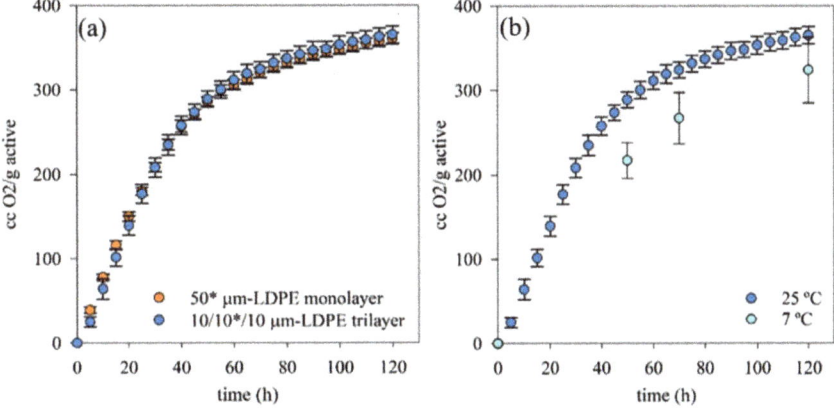

Figure 3. Kinetic oxygen scavenging activities of (**a**) both mono (50 μm) and trilayer (10/10*/10 μm) active films at room temperature; (**b**) and trilayer films at room temperature and 7 °C. Note: (*) indicates the layer containing TM.

As Figure 3a shows, both oxygen scavenger films presented a similar kinetic profile and final oxygen scavenging activities. At first, the monolayer showed slightly faster scavenging activity probably due to the easier accessibility of TM scavenger to oxygen, but, in general, oxygen scavenging activities of both films were not relevantly different, probably due to the highly oxygen-permeable characteristic of LDPE. The final data of the oxygen scavenging capacities were 351 ± 9 cm^3 and 349 ± 17 cm^3 per gram of active agent incorporated into the film for the monolayer and the 3-layer, respectively, values obtained after 5 days of exposition to air. Assuming the 3-layer 10/10*/10 μm structure with the scavenger at 10%wt. load, the scavenging capacity will be 316 ± 8 cm^3/m^2. These values are excellent when compared to other potential scavengers: polybutadiene with cobalt salt scavenge of 15 mg/100 mg of polymer (0.2 mL/g) [12,25], iron-based sachets, ca. 45 mL/g, 2.5 mL/g of an iron–kaolinite composite in LDPE [26], or polymer composites including TiO$_2$ cs. 30 mL/g [27].

Figure 3b showed that the scavenging activity of the 3-layer active film was slightly slower and lower at lower temperature (7 °C), and higher deviation values between samples were found. The increase in oxygen scavenging activity with the temperature has also been observed when using antioxidants as oxygen scavenger, such as gallic acid and tea polyphenols, because their activities increased at higher temperatures [28,29].

3.3. VOC Retention of Selected Adsorbents in Real Condition Testing

The use of polyalkenes for oxygen scavenging activity results in the generation of oxidation products from allylic carbon hydrogen bonds oxidative degradation that can turn out a problem due to their pungent odor of these byproducts, referred in this work as Volatile Organic Compounds (VOCs) [9,30]. Therefore, the capacity of adsorbents selected in Section 3.1 to retain VOCs released by the active 3-layer films containing TM was analyzed in order to diminish this secondary effect.

First, the main volatile compounds released from TM were identified by gas chromatography mass spectrometry using the NIST library. The corresponding compounds were quantitatively measured attending to their most relevant ions. Table 4 presents the identified compounds, their retention time, and the ion mass utilized for quantification.

Table 4. Compounds tentatively identified by GC-MS using NIST library, their retention time, and the ion mass (uma) utilized for quantification.

Peak	Retention Time (min)	Volatile Compound	Ion Mass	Peak	Retention Time (min)	Volatile Compound	Ion Mass
1	1.63	pentane	43	17	6.98	2-pentenol	68
2	1.73	formic acid	46	18	7.63	2-heptanone	43
3	2.05	hexane	57.1	19	7.69	cyclohexanone	55
4	2.24	acetic acid	60	20	8.1	pentanoic acid	60
5	2.65	cyclohexane	84.1	21	8.23	cyclooctane	56
6	2.69	butanol	56	22	8.44	hexyl formate	56.1
7	3.11	3-methylbutanal	57	23	9.13	cyclohexyl formate	67
8	3.4	propanoic acid	74	24	9.73	2-octanone	43
9	3.62	Butyl formate	56.1	25	9.96	hexanoic acid	60
10	4.12	trimethylpentane	71.1	26	10.41	heptyl formate	70.1
11	4.72	pentanol	55	27	10.53	cyclopentyl carboxilic acid	73
12	4.94	trimethylhexane	57.1	28	11.48	heptanoic acid	60
13	5.1	3-hexanone	57	29	12.16	cyclohexane carboxilic acid	55.1
	5.19	2-hexanone	43	30	12.92	octanoic acid	60
14	5.44	hexanal	56	31	16.18	tetradecanal	82.1
15	5.75	butanoic acid	60	32	16.93	1-phenyl 1-hexanone	105
16	6.13	Pentyl formate	70.1				

They all were quantified and their adsorptions in both pure and mixed adsorbents were analyzed after 72 h of exposure. Due to their amount and their sensory impact, carboxylic acids and esters of formic acid are of paramount importance. Gas chromatograms of samples with Sylysia, Tegosorb adsorbents and their mixture are compared to the chromatogram of the control sample without adsorbent substances in Figure 4.

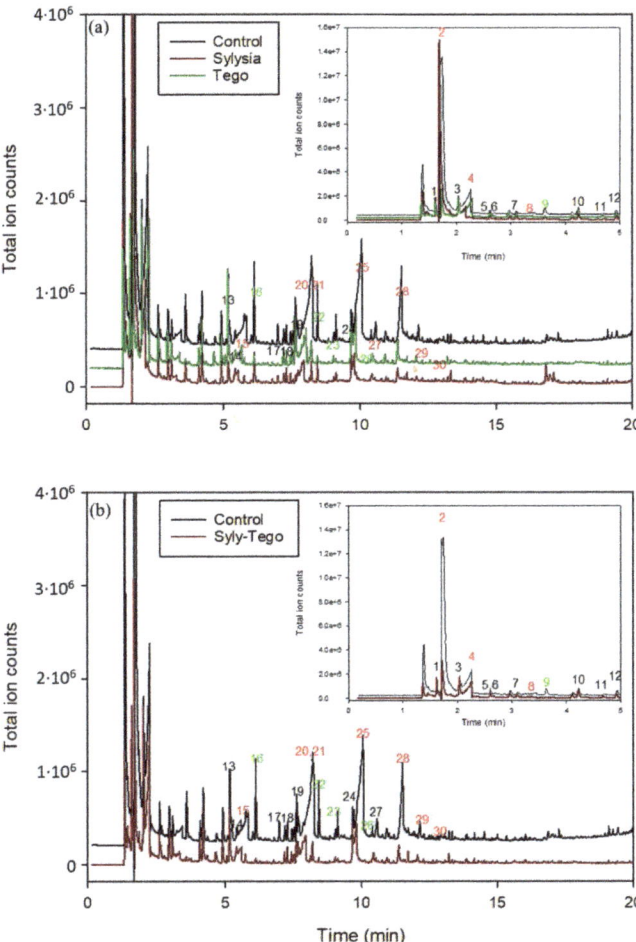

Figure 4. Representative GC-MS chromatograms from headspace of vial containing (**a**) black line, the active trilayer film (control); red line, the vial containing the active film and Sylysia; and green line, the vial containing the active film and Tegosorb (Tego); (**b**) black line, the active 3-layer film (control); red line, the vial containing the active film and mixture of Sylysia and Tegosorb. Numbers in figures indicate the identified compound according to Table 4.

The adsorbent retention percentages of most important VOCs released by the 3-layer film are presented in Figure 5. In general, most identified VOCs were relevantly retained by the adsorbents (Figure 5). VOC adsorption depends on various factors, such as VOC type and concentration, and the interactions between the adsorbent substance and VOCs that are mainly governed by their corresponding polarities [31,32].

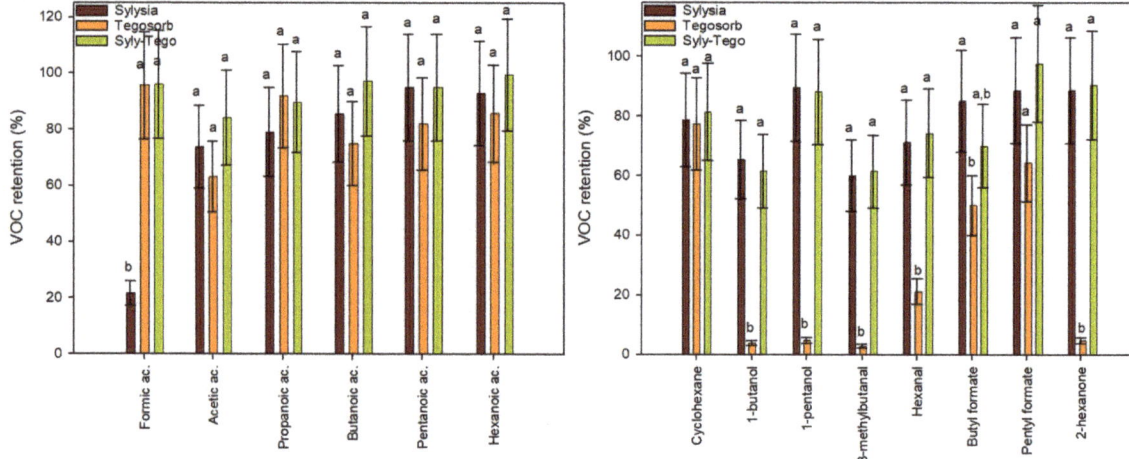

Figure 5. Kinetic retention of VOCs by adsorbents Syl, Tego, and their mixture after 72 h in air at room temperature of exposure to 3-layer films of organic acids (**left**); other VOCs (**right**). "a and b" indicate significant differences among VOC retention values of each organic compound between adsorbents.

As can be seen in Figure 5, the adsorbents, individually or mixed, retained a large part of the carboxylic acids released by the scavenger. Sylysia retained above 80% of them with the exception of formic acid. Tegosorb performed above 80% for these compounds and above 95% for formic acid. Vials with both adsorbents performed in general better than individually. Also, the scavengers performed well with esters of formic acid, especially Sylysia or the combination of both adsorbents.

Several alkanes from C5 to C9 were identified in low amounts and were scarcely retained by adsorbents either individually or mixed, with the exception of cyclohexane. Three alcohols were present in the vial headspace at low concentrations, according to the small peaks observed in the chromatograms. The alcohol retention by Tegosorb was very limited. On the contrary, Sylysia or the mixture retained ca. 60% of butanol, and ca. 80% for C5-alcohols. Other works have also demonstrated various silica as efficient adsorbents for various VOCs' removal thanks to their uniform and open-pore structures [33,34]. Similar comments are appropriate for retention capacity of adsorbents on aldehydes and ketones Their retention, especially in vials containing Sylysia was found to be between 60 and 80% (Figure 5). Considering the large retention observed for these compounds when tested as single contaminants (Figure 1), the preferential retention of other compounds, present in large quantities (acids and esters) could have reduced the scavenging of aldehydes and alcohols.

Some studies have evidenced carbon sorbents quantitively trapped in a wide range of VOCs from C3 to C12 whereas mesoporous silica trapped considerably larger molecules from C8 to C12 with the potential to go beyond C12 [32]. This was confirmed by the great retention of Syl and their mixtures on VOCs with high molecular weight (Figure 5).

After these results, a mixture of Syl and Tego at equal concentrations was selected for incorporation in a polyethylene film and confirmed their capacity to retain VOCs when included in a polymer matrix.

3.4. VOC Retention of Developed TM Adsorbents Prototypes

Figure 6 shows the results obtained comparing the content of 15 selected compounds in control samples (only the TM film) with the other samples including films containing the adsorbents at a 5%wt. concentration. The sample named "One side" included the same weight of film than the oxygen scavenger monolayer to be simulated as a trilayer in which one of the external side layers contains the VOC adsorbents. The "Two sides" sample included twice the weight of the oxygen scavenger film to simulate as a trilayer in which the oxygen adsorber layer is sandwiched between two layers of VOC scavenger LDPE. Finally, "Double two sides" included double quantity of VOC adsorbents film to simulate a trilayer like the "Two sides" but with twice the concentration of Sylysia and Tegosorb. As can be seen, the VOC scavenging film systems retained most of the released compounds, achieving percentages of VOC retained values between 85 and 100%. As was expected, the designed system improved the retention as the amount of adsorbents increased, although no significant differences were observed. The "Two sides" samples presented slightly higher VOC retention than "One side" samples. Therefore, the design of a system including a trilayer of LDPE containing 10%wt. TM in the central layer and two identical external layers containing 5%wt. of Sylysia and Tegosorb appeared to be sufficient to reduce the risk of sensorial damage on packaged products caused by the release of low-molecular-weight compounds produce by oxidative reactions of TOR.

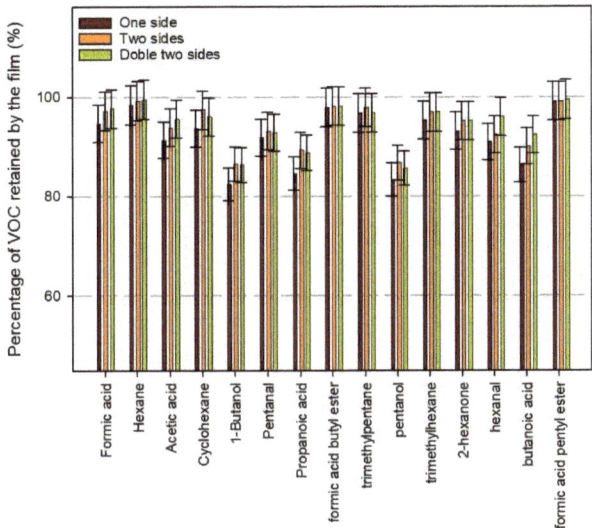

Figure 6. Retention of some volatile compounds released from TM-containing films containing VOC scavengers.

3.5. VOC Retention of Developed Multilayer Active Systems

As described in Section 3.3, it was evidenced that the best performance of VOC retention occurred at both sides, and therefore, a multilayer system was designed in which the layer containing TM was sandwiched between two layers of PE containing the adsorbent. The multilayer system PP/tie/EVOH/tie/PP/PE-VOC/PE-TM/PE-VOC was successfully developed. PE-VOC corresponded to PE layer including the adsorbents at 5%wt. An EVOH layer was included in order to protect the PE-TM layer from outer oxygen. Other oxygen scavenger developments were also based on the design of multilayer systems including one or more layers which are permeable to oxygen between the food product and the oxygen scavenger layer and an oxygen barrier layer towards the outside of the package [10]. The final 10-layer films were evaluated by optical microscopy in transmittance mode, to determine the thickness of each layer, with an accuracy of 3 µm. Figure 7 shows an image of the film with the thickness of the different layers, the right layers correspond to the PE-VOC/PE-TM/PE-VOC.

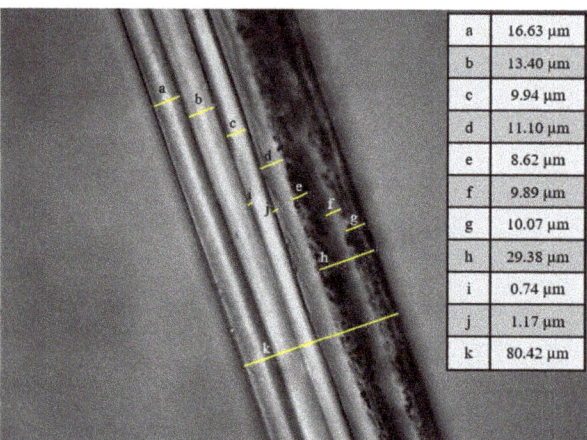

Figure 7. Thicknesses of the diverse layers of the multilayer structure developed. Left, PP layer; right, the PE-VOC layer.

As can be seen in Figure 8, most of the volatile compounds were reduced by the inclusion of the VOC scavengers, especially alkanes, acids, ketones and esters. Alkanes, as cyclohexane and trimethylpentane, and ketones, such as 3-hexanone and 2-hexanone, displayed retention values higher than 95%. Organic acids, such as formic, acetic and propanoic acids, evinced intermediate retention values between 50–60% with respect to the control released amount. On the contrary, alcohols, such as butanol and pentanol, and hexanal were apparently released with the same intensity from the film with VOC scavengers. Nevertheless, these compounds were present as a trace, and were difficult to identify, and therefore, scarcely relevant as compounds with potential sensory effects on packaged products.

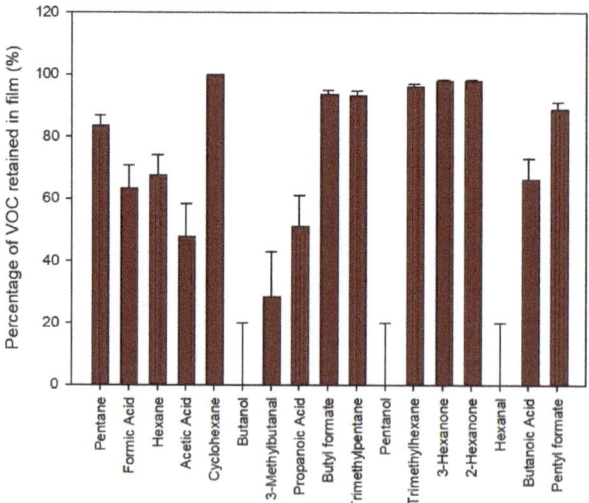

Figure 8. Percentage of diverse VOCs retained by active film with oxygen scavenger and VOC scavenger (PP/tie/EVOH/tie/PP/PE-VOC/PE-TM/PE-VOC) related to the concentrations measured in vials with film containing only oxygen scavenger (PP/tie/EVOH/tie/PP/PE/PE-TM/PE) after 5 days of storage at room temperature.

4. Conclusions

This work has presented the scientific and technological steps for the development of an active packaging with oxygen scavenger capacity. This study has included the study of the efficiency of the TOR oxygen scavenger (TM) films, the formation of volatile organic compounds (VOCs) generated by the TM active compound in the presence of oxygen, the search for VOC adsorbers, and the development of a multilayer film that included both the oxygen scavenger and VOCs adsorbents. First, an optimal product including TOR was prepared as a masterbatch blended with LDPE (among other polyolefins), that was further diluted to a preferred used concentration of ca. 10%wt. Subsequently, the scavenging material was incorporated in a multilayer system that included a high barrier layer to avoid the reaction with the oxygen from the external atmosphere. Also, a functional layer was required to impede direct contact of the scavenger material with the food product.

Among the advantages of the TOR masterbatch, their extraordinarily high scavenging capacity above 350 mL of oxygen per gram of material, their fast kinetics (80% of its capacity in 5 days), and the independency of this activity with humidity were highlighted. Sylysia, a hydrophilic silica, and Tegosorb 88, a product based on zinc ricinoleate, have demonstrated a good retention of VOCs. TOR product presented efficient oxygen scavenger values at low and room temperatures. Several VOCs were generated as byproducts of the oxodegradation of this oxygen scavenger in the presence of oxygen. The incorporation of layers including absorbent substances have demonstrated the efficiency on VOC retention. The design and the incorporation of this layer was successfully carried out in a multilayer system that can be shaped into sheets, containers or other packaging materials through thermal forming technologies, such as melting extrusion or compression/blow molding. The main advantage of this kind of oxygen scavenging system lies with the fact that it can be used to package dry food products sensitive to oxidation, from nuts to fish rich in unsaturated fatty acids such as salmon.

Author Contributions: Conceptualization, P.V., P.H.-M. and R.G.; methodology, G.L.-C., F.S., A.A.-G. and J.M.A.; formal analysis, C.L.-d.-D., P.V., P.H.-M. and R.G.; writing—original draft preparation, C.L.-d.-D. and R.G.; writing—review and editing, C.L.-d.-D. and R.G. All authors have read and agreed to the published version of the manuscript.

Funding: This research was funded by Evonik Resources Efficiency GmbH, Contract 20196079.

Data Availability Statement: Data will be available upon request.

Acknowledgments: C.L.-d.-D. acknowledges the "Ramon & Cajal" Fellowship RYC2020-029874-I funded by MCIN/AEI/10.13039/501100011033 and by "European Union NextGenerationEU/PRTR".

Conflicts of Interest: Authors Pedro Vazquez and Florian Schwager were employed by the company Evonik Operations GmbH. The remaining authors declare that the research was conducted in the absence of any commercial or financial relationships that could be construed as a potential conflict of interest.

References

1. Firouz, M.S.; Mohi-Alden, K.; Omid, M. A critical review on intelligent and active packaging in the food industry: Research and development. *Food Res. Int.* **2021**, *141*, 110113. [CrossRef] [PubMed]
2. Ahari, H.; Soufiani, S.P. Smart and Active Food Packaging: Insights in Novel Food Packaging. *Front. Microbiol.* **2021**, *12*, 1366. [CrossRef] [PubMed]
3. Rüegg, N.; Röcker, B.; Yildirim, S. Application of palladium-based oxygen scavenger to extend the mould free shelf life of bakery products. *Food Packag. Shelf Life* **2022**, *31*, 100771. [CrossRef]
4. Pourshahbazi, H.; Javanmard dakheli, M.; Salehirad, A.S. farhadi, Novel oxygen scavenger screw-cap for shelf-life improvement in virgin olive oil packaging during storage. *J. Food Meas. Charact.* **2022**, *2022*, 1–7. [CrossRef]
5. Ramakanth, D.; Akhila, K.; Gaikwad, K.K.; Maji, P.K. UV-activated oxygen scavenging system based on natural rubber latex from Hevea brasiliensis for active packaging applications. *Ind. Crops Prod.* **2022**, *178*, 114658. [CrossRef]
6. di Giuseppe, F.; Coffigniez, F.; Aouf, C.; Guillard, V.; Torrieri, E. Activated gallic acid as radical and oxygen scavenger in biodegradable packaging film. *Food Packag. Shelf Life* **2022**, *31*, 100811. [CrossRef]

7. Martín-Mateos, M.J.; Amaro-Blanco, G.; Manzano, R.; Andrés, A.I.; Ramírez, R. Efficacy of modified active packaging with oxygen scavengers for the preservation of sliced Iberian dry-cured shoulder. *Food Sci. Technol. Int.* **2022**, *29*, 318–330. [CrossRef]
8. Kruijf, D.D.; Beest, V.V.; Rijk, R.; Sipiläinen-Malm, T.; Losada, P.P.; Meulenaer, D.D. Active and intelligent packaging: Applications and regulatory aspects. *Food Addit. Contam.* **2002**, *19*, 144–162. [CrossRef]
9. Dey, A.; Neogi, S. Oxygen scavengers for food packaging applications: A review. *Trends Food Sci. Technol.* **2019**, *90*, 26–34. [CrossRef]
10. US5211875A—Methods and Compositions for Oxygen Scavenging—Google Patents. Available online: https://patents.google.com/patent/US5211875A/en (accessed on 27 May 2022).
11. WO1996040799 Compositions Having Ethylenic Backbone and Benzylic, Allylic, or Ether-Containing Side-Chains, Oxygen Scavenging Compositions Containing Same, and Process for Making These Compositions by Esterification or Transesterification of a Polymer Melt. Available online: https://patentscope.wipo.int/search/en/detail.jsf?docId=WO1996040799 (accessed on 27 May 2022).
12. Li, H.; Tung, K.K.; Paul, D.R.; Freeman, B.D.; Stewart, M.E.; Jenkins, J.C. Characterization of Oxygen Scavenging Films Based on 1,4-Polybutadiene. *Ind. Eng. Chem. Res.* **2012**, *51*, 7138–7145. [CrossRef]
13. US6908652B1—Poly(lactic acid) in Oxygen Scavenging Article—Google Patents. Available online: https://patents.google.com/patent/US6908652B1/en (accessed on 24 July 2023).
14. Peri, J.B.; Hensley, A.L. The surface structure of silica gel. *J. Phys. Chem.* **1968**, *72*, 2926–2933. [CrossRef]
15. Shen, X.; Du, X.; Yang, D.; Ran, J.; Yang, Z.; Chen, Y. Influence of physical structures and chemical modification on VOCs adsorption characteristics of molecular sieves. *J. Environ. Chem. Eng.* **2021**, *9*, 106729. [CrossRef]
16. Hou, S.; Tang, Y.; Zhu, T.; Huang, Z.H.; Liu, Y.; Sun, Y.; Li, X.; Shen, F. The molecular simulation and experimental investigation of toluene and naphthalene adsorption on ordered porous silica. *Chem. Eng. J.* **2022**, *435*, 134844. [CrossRef]
17. WO2007053790A1—Fabric Softening Dryer Sheet with Odor Control—Google Patents. Available online: https://patents.google.com/patent/WO2007053790A1/en (accessed on 20 June 2022).
18. WO2007057043 Absorbent Articles Comprising Acidic Superabsorber and an Organic Zinc Salt. Available online: https://patentscope.wipo.int/search/en/detail.jsf?docId=WO2007057043 (accessed on 20 June 2022).
19. Mura, P. Advantages of the combined use of cyclodextrins and nanocarriers in drug delivery: A review. *Int. J. Pharm.* **2020**, *579*. [CrossRef] [PubMed]
20. Cid-Samamed, A.; Rakmai, J.; Mejuto, J.C.; Simal-Gandara, J.; Astray, G. Cyclodextrins inclusion complex: Preparation methods, analytical techniques and food industry applications. *Food Chem.* **2022**, *384*, 132467. [CrossRef] [PubMed]
21. Zhang, G.; Yuan, C.; Sun, Y. Effect of Selective Encapsulation of Hydroxypropyl-β-cyclodextrin on Components and Antibacterial Properties of Star Anise Essential Oil. *Mol. A J. Synth. Chem. Nat. Prod. Chem.* **2018**, *23*, 1126. [CrossRef]
22. Alopaeus, J.F.; Göbel, A.; Breitkreutz, J.; Sande, S.A.; Tho, I. Investigation of hydroxypropyl-β-cyclodextrin inclusion complexation of two poorly soluble model drugs and their taste-sensation—Effect of electrolytes, freeze-drying and incorporation into oral film formulations. *J. Drug Deliv. Sci. Technol.* **2021**, *61*, 102245. [CrossRef]
23. Higueras, L.; López-Carballo, G.; Gavara, R.; Hernández-Muñoz, P. Reversible Covalent Immobilization of Cinnamaldehyde on Chitosan Films via Schiff Base Formation and Their Application in Active Food Packaging. *Food Bioprocess Technol.* **2015**, *8*, 526–538. [CrossRef]
24. Balaguer, M.P.; Gómez-Estaca, J.; Gavara, R.; Hernandez-Munoz, P. Biochemical properties of bioplastics made from wheat gliadins cross-linked with cinnamaldehyde. *J. Agric. Food Chem.* **2011**, *59*, 13212–13220. [CrossRef]
25. US20160311949A1—Epoxy-Terminated Polybutadiene as Oxygen Scavenger—Google Patents. Available online: https://patents.google.com/patent/US20160311949A1/en (accessed on 25 July 2023).
26. Kordjazi, Z.; Ajji, A.; Tabatabaei, H. Development of Oxygen Scavenging Polymeric Systems for Food Packaging Applications. Ph.D. Thesis, Polytechnic University Montreal, Montreal, QC, Canada, 2019.
27. Kordjazi, Z.; Ajji, A. Development of TiO_2 catalyzed HTPB based oxygen scavenging films for food packaging applications. *Food Control* **2021**, *121*, 107639. [CrossRef]
28. Pan, L.; Zhang, M.; Lu, L.; Ou, B.; Chen, X. A polyethylene base moisture activating oxygen scavenging film co-extruded with tea polyphenols-β-cyclodextrin inclusion complex. *Materials* **2020**, *13*, 3857. [CrossRef] [PubMed]
29. US4968438A—Gallic Acid as an Oxygen Scavenger—Google Patents. Available online: https://patents.google.com/patent/US4968438A/en (accessed on 27 June 2022).
30. Coquillat, M.; Verdu, J.; Colin, X.; Audouin, L.; Nevière, R. Thermal oxidation of polybutadiene. Part 1: Effect of temperature, oxygen pressure and sample thickness on the thermal oxidation of hydroxyl-terminated polybutadiene. *Polym. Degrad. Stab.* **2007**, *92*, 1326–1333. [CrossRef]
31. Senthilnathan, J.; Kim, K.H.; Kim, J.C.; Lee, J.H.; Song, H.N. Indoor air pollution; sorbent selection, and analytical techniques for volatile organic compounds. *Asian J. Atmos. Environ.* **2018**, *12*, 289–310. [CrossRef]
32. Idris, S.A.; Robertson, C.; Morris, M.A.; Gibson, L.T. A comparative study of selected sorbents for sampling of aromatic VOCs from indoor air. *Anal. Methods* **2010**, *2*, 1803–1809. [CrossRef]

33. Gibson, L.T. Mesosilica materials and organic pollutant adsorption: Part A removal from air. *Chem. Soc. Rev.* **2014**, *43*, 5163. [CrossRef]
34. Ewlad-Ahmed, A.M.; Morris, M.; Holmes, J.; Belton, D.J.; Patwardhan, S.V.; Gibson, L.T. Green Nanosilicas for Monoaromatic Hydrocarbons Removal from Air. *Silicon* **2022**, *14*, 1447–1454. [CrossRef]

Disclaimer/Publisher's Note: The statements, opinions and data contained in all publications are solely those of the individual author(s) and contributor(s) and not of MDPI and/or the editor(s). MDPI and/or the editor(s) disclaim responsibility for any injury to people or property resulting from any ideas, methods, instructions or products referred to in the content.

Article

Study of Ethylene-Removing Materials Based on Eco-Friendly Composites with Nano-TiO$_2$

Alba Maldonado [1,2,*], Paulina Cheuquepan [1,2], Sofía Gutiérrez [1,2], Nayareth Gallegos [1,2], Makarena Donoso [1,2], Carolin Hauser [3], Marina P. Arrieta [4], Alejandra Torres [1,2,5], Julio Bruna [1,2,5], Ximena Valenzuela [1,2,5], Abel Guarda [1,2,5], María Galotto [1,2,5] and Francisco Rodríguez-Mercado [1,2,5,*]

[1] Packaging Innovation Center (LABEN–Chile), Universidad de Santiago de Chile, Obispo Umaña 050, Santiago 9170201, Chile; paulina.cheuquepan@usach.cl (P.C.); sofia.gutierrez.b@usach.cl (S.G.); nayareth.gallegos@usach.cl (N.G.); makarena.donoso@usach.cl (M.D.); alejandra.torresm@usach.cl (A.T.); julio.bruna@usach.cl (J.B.); ximena.valenzuela@usach.cl (X.V.); abel.guarda@usach.cl (A.G.); maria.galotto@usach.cl (M.G.)

[2] Center for the Development of Nanoscience and Nanotechnology (CEDENNA), Universidad de Santiago de Chile, Alameda 3363, Santiago 9170022, Chile

[3] Department of Applied Chemistry, Nuremberg Institute of Technology Georg Simon Ohm, Keßlerplatz 12, 90489 Nuremberg, Germany; carolin.hauser@th-nuernberg.de

[4] Departamento Ingeniería Química Industrial y del Medio Ambiente, Escuela Técnica Superior de Ingenieros Industriales, Universidad Politécnica de Madrid, (ETSII-UPM), Calle José Gutiérrez Abascal 2, 28006 Madrid, Spain; m.arrieta@upm.es

[5] Department of Food Science and Technology, Faculty of Technology, Universidad de Santiago de Chile, Avenida Víctor Jara 3769, Santiago 9170124, Chile

* Correspondence: alba.maldonado@usach.cl (A.M.); francisco.rodriguezm@usach.cl (F.R.-M.)

Citation: Maldonado, A.; Cheuquepan, P.; Gutiérrez, S.; Gallegos, N.; Donoso, M.; Hauser, C.; Arrieta, M.P.; Torres, A.; Bruna, J.; Valenzuela, X.; et al. Study of Ethylene-Removing Materials Based on Eco-Friendly Composites with Nano-TiO$_2$. *Polymers* **2023**, *15*, 3369. https://doi.org/10.3390/polym15163369

Academic Editors: Ana Luisa Fernando, Lorenzo M. Pastrana and Victor G. L. Souza

Received: 7 July 2023
Revised: 4 August 2023
Accepted: 8 August 2023
Published: 11 August 2023

Copyright: © 2023 by the authors. Licensee MDPI, Basel, Switzerland. This article is an open access article distributed under the terms and conditions of the Creative Commons Attribution (CC BY) license (https://creativecommons.org/licenses/by/4.0/).

Abstract: Ethylene is a phytohormone that is responsible of fruit and vegetable ripening. TiO$_2$ has been studied as a possible solution to slowing down unwanted ripening processes, due to its photocatalytic capacity which enables it to remove ethylene. Thus, the objective of this study was to develop nanocomposites based on two types of eco-friendly materials: Mater-Bi® (MB) and poly(lactic acid) (PLA) combined with nano-TiO$_2$ for ethylene removal and to determine their ethylene-removal capacity. First, a physical–chemical characterization of nano-TiO$_2$ of different particle sizes (15, 21, 40 and 100 nm) was done through structural and morphological analysis (DRX, FTIR and TEM). Then, its photocatalytic activity and the ethylene-removal capacity were determined, evaluating the effects of time and the type of light irradiation. With respect to the analysis of TiO$_2$ nanoparticles, the whole samples had an anatase structure. According to the photocatalytic activity, nanoparticles of 21 nm showed the highest activity against ethylene (~73%). The results also showed significant differences in ethylene-removal activity when comparing particle size and type and radiation time. Thus, 21 nm nano-TiO$_2$ was used to produce nanocomposites through the melt-extrusion process to simulate industrial processing conditions. With respect to the nanocomposites' ethylene-removing properties, there were significant differences between TiO$_2$ concentrations, with samples with 5% of active showed the highest activity (~57%). The results obtained are promising and new studies are needed to focus on changes in material format and the evaluation in ethylene-sensitive fruits.

Keywords: titanium dioxide; photocatalytic activity; ethylene-removing; active packaging; nanocomposites; eco-friendly materials

1. Introduction

Ethylene (C$_2$H$_4$) is an unsaturated hydrocarbon and a gaseous plant hormone that is involved in almost all phases of fruit and vegetable (F&V) growth and development [1]. Ethylene can regulate ripening and senescence of F&V at molecular, biochemical and physiological levels, due to ethylene-stimulated gene expression for the synthesis of enzymes that promote these processes [2]. Although it is not yet clear whether ethylene is responsible for the initiation of ripening in F&V, or is an accelerator of ripening, it has been demonstrated that its elimination from storage chambers causes a slowdown in ripening [3]. In addition, it is well-known that ripening is dependent on whether the fruit is climacteric or non-climacteric. On the other hand, the relationship between F&V and ethylene has a higher consequence with the worldwide problem of food being lost or wasted. In this regard, the FAO has reported that approximately 45% of fruits and vegetables (F&V) are lost or wasted worldwide [4]. Due to their high metabolic activity, F&V are highly perishable products and the application of different methods to control their ripening in postharvest storage is key to adequately preserving these types of food.

Different technologies to minimize the negative effects of ethylene have been studied. These include modified and controlled atmospheres and cool storage; however, the use of chemical products has shown the best effects on F&V shelf-life [5]. Therefore, photocatalysis is a mechanism that has received a lot of attention in both scientific and industrial fields. Titanium dioxide (TiO_2) is among the most well-researched substances. This oxide has been widely studied because it is harmless to both the environment and humans, as well as having applications in various productive areas, such as pharmaceutical, cosmetic and packaging industries, as well as being used as a pigment in paints and coatings, as a photocatalysts, and as an antimicrobial [6]. It is also a biologically and chemically stable substance with low cost, low toxicity, large surface area and high photocatalytic activity [7]. TiO_2 can produce photocatalytic oxidation of C_2H_4 under ultraviolet (UV) radiation. The oxidation process begins on the TiO_2 surface where UV radiation produces reactive oxygen species (ROS) which oxidize C_2H_4 into carbon dioxide (CO_2) and water (H_2O) [8,9]. Considering this property, TiO_2 can be used as an active substance in the development of ethylene-removal plastic materials oriented to F&V. Moreover, application of titanium dioxide nanoparticles (nano-TiO_2) has shown interesting results in ethylene control [10].

Considering the environmental problems generated by the inappropriate management of plastic materials from fossil resources, the generation of new eco-friendly plastic materials with specific functionality becomes relevant for industry and society. In this field, the use of eco-friendly polymers has been widely studied and it is possible to find several commercial alternatives that are available for the development of active materials. Poly(lactic acid) (PLA) and Mater-Bi® are bioplastics which are produced from biological resources and are recognized for being biodegradable and compostable [11,12]. Due to its properties, PLA is recognized as the most suitable bio-based polymer material for food-packaging applications [13]. Indeed, it is versatile, compostable, recyclable, and moreover, it has high transparency, high molecular weight, high water-solubility resistance and good processability [14]. On the other hand, Mater-Bi® is a commercial biodegradable material based on modified starch and bio-based polyester blends. Mater-Bi® has important commercial applications because it shows interesting mechanical properties, thermal stability, processability and biodegradability [11].

Although, to date, it has been possible to find some studies where the antimicrobial and ethylene-removing activity of TiO_2-based eco-friendly materials is reported [3,15,16], there have been few studies where this type of material was obtained through an extrusion process. In this sense, the aim of this study was to develop nanocomposites based on Mater-Bi® and PLA with nano-TiO_2 for ethylene removal and to evaluate the photocatalytic activity of different sizes of nano-TiO_2 in removing ethylene. We used a melt-extrusion process which is potentially scalable to the industrial sector, for the development of active packaging for fruit and vegetables.

2. Materials and Methods
2.1. Materials

Nanoparticles of titanium dioxide (nano-TiO_2) anatase were supplied by Sigma Aldrich (21 nm, 99.5%) and US Research Nanomaterials Inc. (Houston, TX, USA) (15 nm—99.5%, 40 nm—99.5% and 100 nm 99.9%). Its crystalline structure was anatase, 99.5% purity, with four different particle sizes: 15, 21, 40 and 100 nm. To determine photocatalytic activity methylene blue (MeB) (Sigma Aldrich, Darmstadt, Germany) was used. To elaborate nanocomposites, Mater-Bi® EF51L (Novamont SpA, Milan, Italy) and PLA Ingeo Biopolymer 2003D (NatureWork, Plymouth, MN, USA) were provided used by Novamont SpA and NatureWork, respectively.

2.2. X-ray Diffraction (XRD)

To analyze the crystalline structure of nanoparticles, an analysis of X-ray diffraction was carried out on a powder X-ray diffractometer (Bruker, D8-Advance, Urbana, IL, USA) using CuKα (λ = 1.5406 Å) radiation at room temperature at 40 kV and 35 mA. All scans were performed at 2θ angles from 2° to 80° at 0.02° every 0.2 s. The lattice parameters were calculated according to Equation (1), which is used for tetragonal crystallites:

$$\frac{1}{d^2} = \frac{h^2 + k^2}{a^2} + \frac{l^2}{c^2} \qquad (1)$$

where h, k and l are the Miller indices obtained from the crystallographic parameters of anatase (JCPDS 84-1286); a and c are the lattice parameters, and d value was obtained by the Bragg's law Equation (2):

$$n\lambda = 2d\sin\theta \qquad (2)$$

where λ is wavelength of incidence X ray (1.5406 Å), θ is peak position in radians, n is the order of diffraction (1) and d is the interplanar spacing or d-spacing.

2.3. Fourier Transform Spectroscopy (FTIR)

Spectroscopic analysis was carried out in an IR-ATR spectrometer (Bruker, Alpha, Germany). Absorbance was measured with a wavelength range from 4000 cm^{-1} to 400 cm^{-1} using 24 scans, and a resolution of 2 cm^{-1}.

2.4. Transmission Electron Microscopy (TEM) and Scanning Electron Microscopy (SEM)

For nanoparticle samples, TEM analysis was performed on a transmission electron microscope (Hitachi, HT-7700, Tokyo, Japan) with a magnification of 50×. Before this analysis, the samples were diluted and dispersed in ethanol (95%) and placed on a copper grid. For SEM analysis the samples were coated with a gold–palladium film on a Hummer 6.2 metallizer. Images were obtained on a microscope (Tescan, Vega, Brno, Czech Republic) from 3 to 20 kV.

The proper incorporation and dispersion of nano-TiO_2 into the polymeric matrix were evaluated by TEM. The nanocomposite was embedded in a low-viscosity TEM-grade epoxy resin (Taab Laboratories, Aldermaston, UK) and cured for 3 days at 40 °C. To observe the cross-section of the nanocomposite, the resins embedded with the films were cut with a Reichert Ultracut S Leica ultra-microtome (Leica Biosystems, Wetzlar, Germany) mounted on copper grids and the TEM images were obtained with a JEOL JEM-1010 transmission electron microscope. Only the samples of PLA could be measured because the samples of nanocomposites of MB could not be prepared by the method described above due to the soft composition of the MB matrix. The TiO_2 particles' sizes within the polymeric matrix were measured through ImageJ software (version 1.51 k, Java 1.6.0_24, Wayen Rasband, US National Institutes of Health, Bethesda, MD, USA).

2.5. Particle Sizes (PS)

The particle size of different nanoparticles was determined by dynamic light scattering (DLS) at 90° using a Zetasizer (NanoS90, Malvern Instruments, Worcestershire, UK). A refractive index of TiO_2 equal to 2.49 and of continuous phase (ethanol) equal to 1.36 was used for measurements. The reported values corresponded to an average of 10 measurements, where each sample was measured in triplicate.

2.6. Thermogravimetric Analysis (TGA)

Thermograms were obtained with a TGA/DSC 1 analyzer (Mettler Toledo, STARe System, Bern, Switzerland) with STARe V.12.0 software (Bern, Switzerland). Approximately 10 mg of sample was placed into an alumina crucible (70 µL). The experiment was performed in a high-purity dynamic nitrogen flow of 50 mL/min at a heating rate of 10 °C/min. The analysis was performed in a temperature range of 25 to 900 °C.

2.7. Photocatalytic Activity of TiO_2

The photocatalytic activity of nano-TiO_2 was determined against the degradation of MeB. This experiment considered the application of ultraviolet (UV) and visible (VIS) light. For this, irradiation chambers were used for UV light (Philips Actinic BL TL-D 15W/10, 365 nm, Hamburg, Germany) and VIS light (General Electric R7s 150 W/118 mm, 630 nm, CA, USA). For this purpose, 100 mL of a solution of MeB (17 ppm) was prepared, and 100 mg of TiO_2 was added to this solution. The mixture was stirred in the dark for 30 min. The solution was then placed inside UV or VIS light chambers for 270 min, and 2 mL aliquots were taken every 30 min. The aliquots were centrifuged at 7000 rpm for 5 min in a centrifuge (Hettich, Universal 32R, Tuttlingen, Germany). Then, the supernatant was taken and placed into a cuvette (2 mL) and the absorbance was measured at 665 nm in a UV-VIS spectrophotometer (Shimadzu, UV-1900i, Tokyo, Japan). A calibration curve was prepared to determine the MeB concentration, where the absorbance of different MeB solutions was measured (1–20 ppm).

2.8. Ethylene-Removal Study

The study of the ethylene-removal kinetics of nano-TiO_2 and nanocomposites was carried out according to the methodology described below. First, 0.15 g of TiO_2 (1 g for nanocomposite samples) was introduced into a 22 mL vial; then, a gaseous ethylene–nitrogen mixture was injected into each of the vials at a concentration of 50,000 ppm. The vials were placed 15 cm from the lamp and were irradiated inside a special chamber with UV or VIS light, for 5 or 15 min at 20 °C. Then, samples from the head space were analyzed by gas chromatography at 0, 1, 2, 3, 4 and 5 days. For this, ethylene quantification was done in a Perkin Elmer Clarus 580 gas chromatograph (Waltham, MA, USA) with a Head Space Turbo Matrix 40-Perkin Elmer sampler (Waltham, MA, USA) with an RtTM-alumina PLOT column of 50 m length and 0.53 mm diameter. The gas chromatograph conditions consisted of an injector temperature of 180 °C, oven temperature of 180 °C and flame ionization detector (FID) temperature of 250 °C. Head space autosampler conditions included thermostatization temperature of 45 °C and needle and transfer-line temperature of 70 °C; N_2 was used as carrier gas at a flow rate of 10 mL min^{-1}.

2.9. Fabrication of Nanocomposites

Poly(lactic acid) (PLA) and Mater-Bi® (MB) were used as polymeric matrices to produce nanocomposites with different contents of nano-TiO_2 (0, 5 and 10% wt.). The nanocomposites were produced by extrusion, using a twin-screw micro-extruder (Thermo Scientific, Thermo Process 11, Erlangen, Germany). The extrusion conditions included a screw feed speed of 15 rpm, a twin-screw speed of 30 rpm, a die temperature of 190 °C and temperature profiles of 120–180 °C and 160–185 °C for PLA and MB, respectively.

2.10. Colorimetric Properties

Color analysis was carried out on an UltraScan VIS spectrophotometer (HunterLab, VA, USA). Prior to the measurement, the instrument was calibrated with the black and white cells. For this purpose, the sample was placed in a glass optics cell and inserted into the instrument, and the L*, a*, b* and ΔE_{00} were measured. In addition, ΔE parameter was determined from [17].

3. Results and Discussion

3.1. Characterization of Nano-TiO$_2$

X-ray diffractogram analysis allows us to identify the crystalline structures of a chemical compound. Figure 1A shows the diffraction patterns of different samples of nano-TiO$_2$. All samples showed the characteristic typical anatase diffraction 2θ peaks at 25.41°, 37.94°, 48.05°, 54.13°, 55.39°, 70.24° and 75.14° which correspond to planes (101), (004), (200), (105), (211), (220) and (215) of its crystalline structure (JCPDS 84-1286) [18]. Anatase has a tetragonal structure of crystallite, and the lattice parameters a, b and c were calculated, considering a = b due its structure. The values were (1) a: 3.7889, 3.7787, 3.7787 and 37824 Å and (2) c: 9.4983, 9.4597, 9.4983 and 9.5047 Å for 15, 21, 40 and 100 nm, respectively. These values are in agreement with those reported for an anatase standard (a = 3.7848 and c = 9.5124 Å) [19].

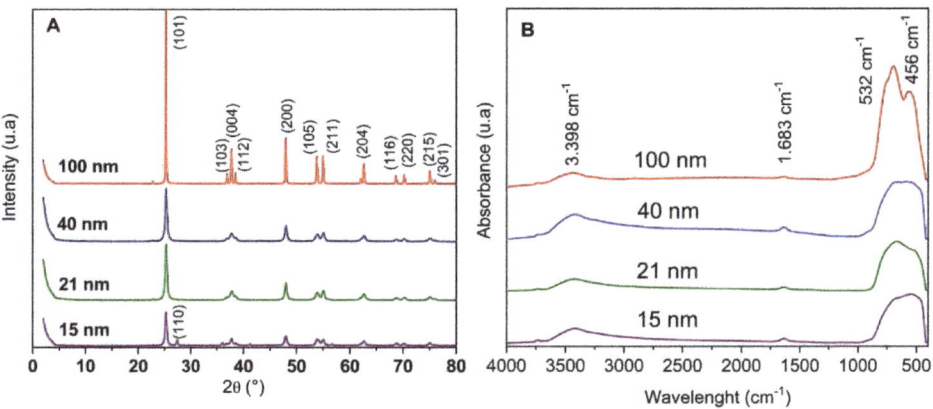

Figure 1. XRD (**A**) and FTIR (**B**) spectra of nano-TiO$_2$ with different particle sizes.

It should be noted that the peak intensities increased with increasing particle size. This could be observed in nano-TiO$_2$ (100 nm), where the peak intensity was higher than for samples of 40, 21 and 15 nm. This was because nano-TiO$_2$, with lower particle size is composed of irregular and amorphous structures crystals which produce a reduction of the peak intensities [20]. These amorphous zones revealed a broad pattern of low intensity in the shape of the peaks; however, the effect of the amorphous materials on the broadening of the XRD patterns of the nano-TiO$_2$ samples was negligible.

The FTIR spectra of different nano-TiO$_2$ samples are shown in Figure 1B. According to this analysis, all samples of nano-TiO$_2$ had similar spectra. A small band was observed around 3400 between 3500 and 3000 cm^{-1} and there was a small peak that could be assigned to the stretching of surface hydroxyl groups of TiO$_2$. In addition, there was a band around 1680 and between 1500 and 1600 cm^{-1}. This small band could evidence the presence of -OH bonds due to of moisture content in the TiO$_2$ [21]. Finally, the broad band located at 750–500 cm^{-1} was assigned to the bending vibration of Ti-O-Ti bonds in the TiO$_2$ lattice. An important band close to 1000 cm^{-1} to 400 cm^{-1} was also observed. This band reflected the vibration of O-Ti-O bonds in the molecules [20].

Figure 2 shows the TEM images of different nano-TiO$_2$ samples. The morphology was determined through TEM microscopy. The nanoparticles were characterized by a rounded shape and the formation of agglomerates (confirmed by PS analysis). From this analysis it was possible to determine the sizes and shapes of the nanoparticles. The nano particles in samples were sphere-shaped, while the sizes, determined through DLS equipment, were heterogeneous: 272.2 ± 22.6; 335.5 ± 44.5; 597.4 ± 55.2 and 779.9 ± 28.5 nm for samples of 15, 21, 40 and 100 nm, respectively. These results show that the agglomeration of the nanoparticles overestimated the real particle size, resulting in sizes above those reported in the data sheets. The particle sizes increased due to all samples being agglomerated and was caused by the characteristics of the samples because these were powdered nanoparticles.

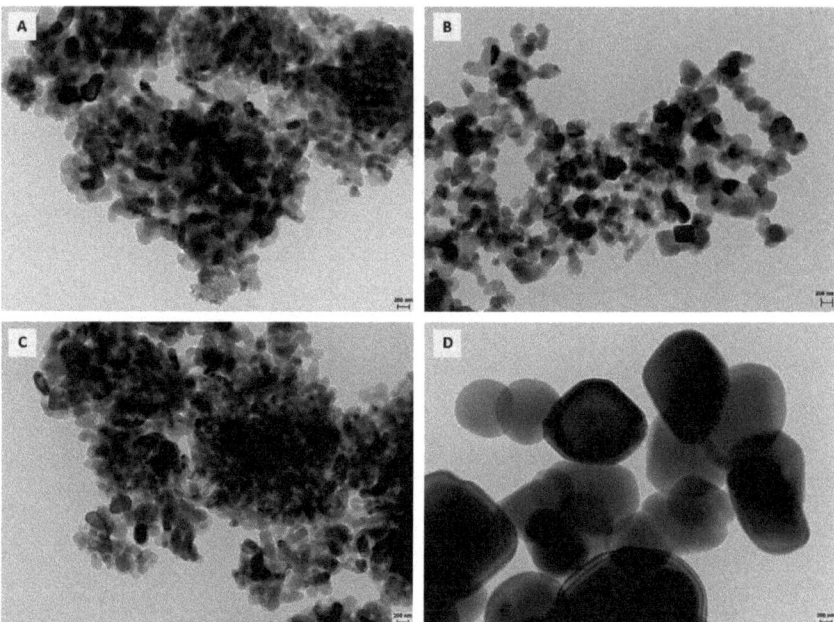

Figure 2. TEM micrographs of the different nano-TiO$_2$ samples: 15 nm (**A**), 21 nm (**B**), 40 nm (**C**) and 100 nm (**D**). Magnification 50 Kx.

As the nanoparticles would be used in the elaboration of nanocomposites through an extrusion process, it was important to determine their thermal stability. For this reason, the nanoparticles were analyzed by TGA. All samples exhibited a single mass loss of about 5% at 100 °C which was associated with the evaporation of water (previously evidenced in the FTIR analysis). At higher temperatures, no changes in nano-TiO$_2$ mass were observed, which confirms the high thermal stability of this compound [22]. With respect to thermal properties, these were determined through thermogravimetric analysis (TGA). All samples had a difference in weight between 30 and 100 °C, due to evaporation of water; this could be attributed to a certain humidity present in the samples. Therefore, it can be said that there was a weight loss of approximately 5% for all samples of nano-TiO$_2$. The presence of water could be observed through the FTIR test, which is explained above. However, after this initial degradation, the mass of active compound (TiO$_2$) was constant until 900 °C, so it is possible to conclude that nano-TiO$_2$ is highly temperature-stable, and this characteristic is desirable when producing nanocomposites by the extrusion process. These results were expected since TiO$_2$ is an inorganic chemical compound, which is highly stable at high temperatures [22].

3.2. Photocatalytic Activity

In order to assess the photocatalytic activity of TiO$_2$ nanoparticles to oxidize organic molecules, their activity against methylene blue (MeB) was determined. It is well-known that the use of nanoparticles provides a larger contact area during photocatalytic reactions and furthermore that it is favored when anatase structure is used [21]. This assay considered the monitoring of MeB concentration for 270 min when this compound was exposed to both nano-TiO$_2$ of different sizes and UV or VIS light (Figure 3). The results showed that the type of light had a clear effect on the degradation of MeB. Thus, samples irradiated with UV light were able to significantly degrade the MeB, while samples irradiated with VIS light were not able to affect this compound.

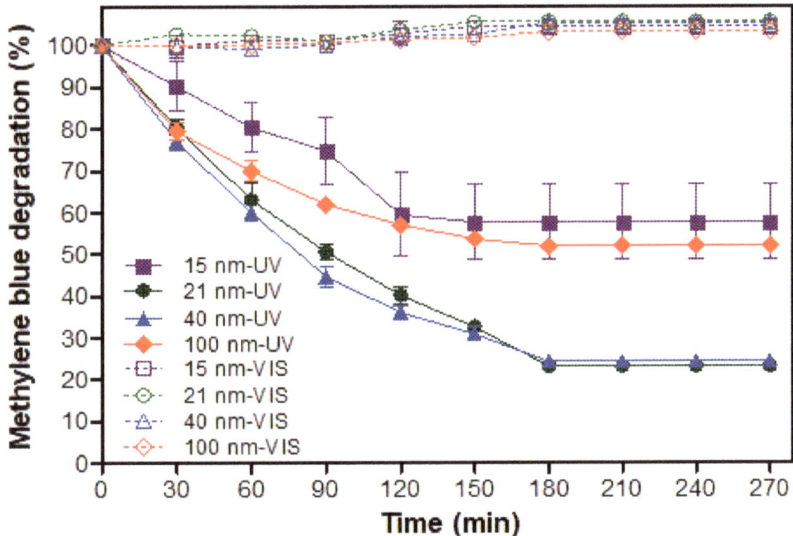

Figure 3. Degradation kinetics of methylene blue under UV and VIS irradiation.

This can be explained due to the characteristic of the crystalline structure that was used. It is known that anatase does not have photocatalytic activity under VIS light due to its wide band gap (3.2 eV) [23]. Hence, only UV radiation has sufficient energy to generate reactive oxygen species (ROS), which can act as oxidants against organic molecules [24]. ROS are formed from photoexcitation of electrons (e−) in TiO_2. The electrons move from the valence band to the conduction band leaving holes (h+) in the first band [7]. Here pairs of e− and h+, called excitons, are generated, causing oxidation/reduction reactions with H_2O and O_2 which are present in the reaction medium. In this way, ROS are produced. Within these species, the hydroxyl radicals (OH) are recognized as being responsible for the degradation of MeB and other organic compounds into CO_2 and H_2O [24]. On the other hand, despite sunlight having a low amount of UV light (~5%), this is not enough to activate the nanoparticles. Consequently, it is essential to use artificial UV light to activate TiO_2 anatase [25].

Regarding particle size, it was observed that nano-TiO_2 samples of 21 and 40 nm were more active in degrading MeB than nanoparticles of 15 and 100 nm. These results were expected, especially for the 100 nm particles since it has been reported that the photocatalytic activity of TiO_2 decreases with increasing particle size due to decreasing surface area [24]. However, in the case of the 15 nm particles the explanation would be different. Once e− and h+ are generated by the interaction between TiO_2 and UV light, two processes can occur: (i) recombination of e− and h+ and the dissipation of their energy in the form of electromagnetic radiation or heat and (ii) migration to the TiO_2 surface to react by absorbing molecules [7]. Thus, the e− can move to the particle surface through electron traps on the crystalline net, which can be of two types—superficial or deep. Superficial traps are beneficial to the electron migration process due to the photocatalyser surface, which allows the reaction, whereas the deeper traps constitute recombination centers. Therefore, when particle size reduces under an optimal size in nanocrystalline TiO_2, most electrons and holes generated close to the surface recombine faster than the interfacial charge transfer processes. Similar results were obtained in a study where different particle sizes (8, 18 and 27 nm) of the crystalline structure of titanium dioxide were evaluated according to their photocatalytic properties against MeB. In this study, the authors were able to evidence that smaller-sized nanoparticles (8 nm) showed lower photocatalytic activity against organic molecules [26]. Additionally, the agglomeration of nanoparticles could be another factor to consider. As observed in the TEM analysis (Section 3.1), agglomeration was more important in smaller particles. Thus, agglomeration would allow the interaction of light only with the most exposed particles in the agglomerate, thus affecting the photocatalytic process.

3.3. Evaluation of Ethylene Degradation by Nano-TiO$_2$

The ethylene-removal capacity, which is expressed as the reduction percentage of the ethylene concentration of different nanoparticles, was determined over a 5-day period under two types of light (UV or VIS) and two irradiation times (5 or 15 min) (Figure 4).

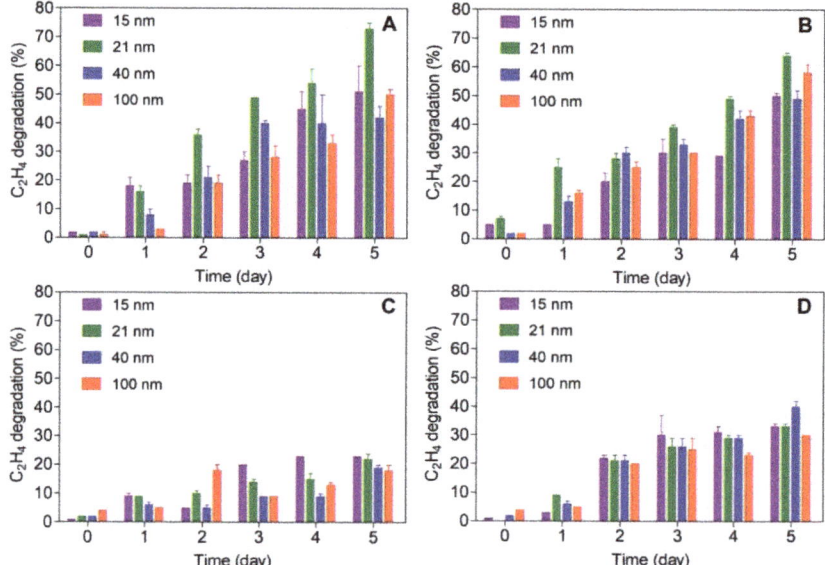

Figure 4. Ethylene degradation of TiO$_2$ nanoparticles under different types of light and irradiation times: 5 min—UV (**A**), 15 min—UV (**B**), 5 min—VIS (**C**) and 15 min—VIS (**D**).

In general, as time increased, an increase in ethylene removal was evidenced for all the nanoparticles studied; however, significant differences were observed when comparing the type of light used. Thus, the nanoparticles irradiated with UV light showed higher ethylene-removal capacity than those treated with VIS light (Tables 1 and 2, respectively). However, there were significant variations in this activity for the different nanoparticle sizes at the two irradiation times. As a result, the 21 nm samples were the most active, reducing the ethylene concentration by about 70% with no significant differences of irradiation time in the activity against the olefin, while the 40 nm particles were the least active with a reduction of about 45%. On the other hand, the samples treated with VIS light were not able to exceed 40% ethylene removal, with 15 nm samples showing the highest activity, while 100 nm samples were the least active. In contrast to the samples treated with UV light, the samples treated with VIS light showed significant differences with increasing light exposure time, which is in agreement with the low content of UV radiation in VIS light. From this analysis, the photocatalytic effect of titanium dioxide on ethylene was confirmed.

Table 1. Final C_2H_4 degradation of different TiO_2 nanoparticles at different UV irradiation times.

Particle Size (nm)	C_2H_4 Degradation (%)	
	Irradiation Time (min)	
	5	15
15	52 ± 7 [ABa]	51 ± 2 [ABa]
21	73 ± 2 [Ca]	64 ± 1 [Ca]
40	42 ± 4 [Aa]	49 ± 3 [Aa]
100	50 ± 2 [Ba]	58 ± 3 [Ba]

Note: Equal letters mean that there are no statistically significant differences ($p > 0.05$). Upper-case letters: particle-size factor and lower-case letters: irradiation-time factor.

Table 2. Final C_2H_4 degradation of different TiO_2 nanoparticles at different VIS irradiation times.

Particle Size (nm)	C_2H_4 Degradation (%)	
	Irradiation Time (min)	
	5	15
15	23 ± 1 [ABa]	33 ± 1 [ABa]
21	22 ± 2 [ABb]	33 ± 1 [ABa]
40	19 ± 1 [Bb]	40 ± 2 [Ba]
100	18 ± 2 [Ab]	30 ± 1 [Aa]

Note: Equal letters mean that there are no statistically significant differences ($p > 0.05$). Upper-case letters: particle-size factor and lower-case letters: irradiation-time factor.

On the other hand, and unlike the results observed in Section 3.2, VIS light showed an effect on ethylene. Although this activity was lower than that observed with UV light, some effect was evidenced in the ethylene control. This agrees with a previous study carried out in our group where nano-TiO_2 was incorporated into a low-density polyethylene (LDPE) film and its effect on banana ripening was evaluated. In this case, a clear effect of non-irradiated films on physiological aspects of the fruit, such as ripening index and color, was observed [27]. As mentioned above, the major ethylene degradation condition is with UV light, because this kind of light has enough energy to overcome the TiO_2 band energy and cause its photoexcitation, which in turn generates electron–proton pairs (e− - h+) that subsequently generate ROS, responsible for oxidizing ethylene to convert it into CO_2 and H_2O [15]. Based on these results, it was considered that the best condition for the removal of ethylene by nanocomposites was the 21 nm sample with UV irradiation for 5 min.

3.4. Elaboration and Characterization of TiO_2 Nanocomposites through Extrusion Process

To take advantage of the photocatalytic activity and the ethylene-removal ability observed with 21 nm TiO_2 particles, these nanoparticles were used to produce nanocomposites in pellet form. The pellets were produced through a microscale extrusion process to determine the possibility of scaling up the materials developed here to the industrial sector for F&V preservation. The pellets were made with different concentrations of TiO_2 (0, 5 and 10% wt.) in two polymeric matrices (Mater-Bi® (MB) and poly(lactic acid) (PLA)).

3.4.1. Optical Properties

Material color for food packaging is an important factor and determines the first impression of consumers about a product [28]. In addition, color can indicate the quality of the dispersion of the additives incorporated into a polymer matrix. Nano-TiO_2 has a white color and good properties of light refraction, so that the visual color-change of the nanocomposite is usually not significant even with the addition of this oxide. However, if the nanocomposite matrix color is different between samples, the color characteristics of added TiO_2 nanocomposite can vary slightly from one sample to another [15]. To determine if there were differences of color between samples, both for matrix type and nano-TiO_2 concentration factors, CIELAB parameters were determined (Table 3).

Table 3. Colour parameters of PLA and MB nanocomposites with TiO_2.

Matrix	TiO_2 Concentration (%)	L*	a*	b*	ΔE_{00}
PLA	0 (control)	84.8 ± 1.3 [Ba]	−1.0 ± 0.2 [Bb]	10.4 ± 0.1 [Ba]	-
	5	92.8 ± 0.6 [Ba]	−0.2 ± 0.2 [Ba]	6.0 ± 0.1 [Bb]	9.2 ± 0.5 [Ab]
	10	94.8 ± 1.8 [Ba]	0.3 ± 0.2 [Bc]	6.3 ± 0.2 [Bc]	11.0 ± 1.7 [Aa]
MB	0 (control)	94.1 ± 0.2 [Aa]	3.2 ± 0.2 [Ab]	13.5 ± 0.2 [Aa]	-
	5	92.1 ± 0.2 [Aa]	3.0 ± 0.2 [Aa]	11.5 ± 0.2 [Ab]	2.9 ± 0.5 [Bb]
	10	90.5 ± 0.1 [Aa]	2.1 ± 0.2 [Ac]	9.8 ± 0.2 [Ac]	5.3 ± 0.4 [Ba]

Note: Equal letters mean that there are no statistically significant differences ($p > 0.05$). Upper-case letters: matrix factor and lower-case letters: TiO_2 concentration.

It can be observed that nano-TiO_2 had a greater impact in PLA than in MB, especially in the case of parameters a*, b* and ΔE_{00}. High values of L for PLA and MB nanocomposites indicate that all samples tended to white. Likewise, in both materials the variations in parameter a* indicate a shift towards red color, while the decrease in parameter b* reflects a shift towards yellow as the oxide content in the nanocomposites increased. Regarding the ΔE_{00} parameter, it can observed that there were significant differences between polymer matrices and nano-TiO_2 concentrations, however, the change was more important for PLA. As with the other parameters, the white coloration of the MB control and the transparent PLA control would be the cause of the observed changes.

3.4.2. Morphological Properties

The dispersion of TiO_2 into the nanocomposites can play an important role in photocatalytic and ethylene-removal capacity. Figure 5 shows SEM microphotographs of the morphology of different nanocomposites prepared with two polymeric matrixes (PLA and MB) and different nano-TiO_2 content. The microphotographs show that the presence of nano-TiO_2 changed the surface of nanocomposites, and there was a strong interfacial adhesion between the two phases. In addition, the formation of agglomerates can be observed, which would be favored with increasing nano-TiO_2 content.

Moreover, to examine the dispersion of nano-TiO_2 into the polymeric matrix, TEM analysis was carried out. Due to the methodology used to prepare the MB samples, they were very flexible and soft and could not be properly embedded in the epoxy resin and only PLA nanocomposites could be analyzed. The results can be observed in Figure 6. In this figure, it is shown that there was an homogeneous distribution of the groups of nanoparticles through the polymeric matrix and it is also possible to observe that there were differences between TiO_2 concentrations, with PLA 10% having a higher presence of TiO_2. On the other hand, in microphotographs with a greater magnification (5 and 20 K×) it is possible to see that nano-TiO_2 nanoparticles tended to agglomerate, as could be seen in Figure 2 for nanoparticle TEM analysis. However, it should be highlighted that in the nanocomposites there were some individual particles dispersed into the PLA matrix, and this behavior did not occur in the TEM assay of nanoparticles; this could be attributed to extrusion process, which is used to disperse, in a better way, the active compound into the polymeric matrix [29]. In fact, the size of the TiO_2 nanoparticles can be measured from the TEM images at 20 K× and it can be concluded that the TiO_2 nanoparticles exhibited an average particle average size of 16.0 ± 3.2 nm in PLA nanocomposite with 5% wt. TiO_2 (Figure 6C), and the average size particle of TiO_2 in PLA nanocomposite with 10% wt. was 18.2 ± 2.4 nm (Figure 6F), confirming the ability of the extrusion process to disperse such nanoparticles (TiO_2 < 21 nm particle size).

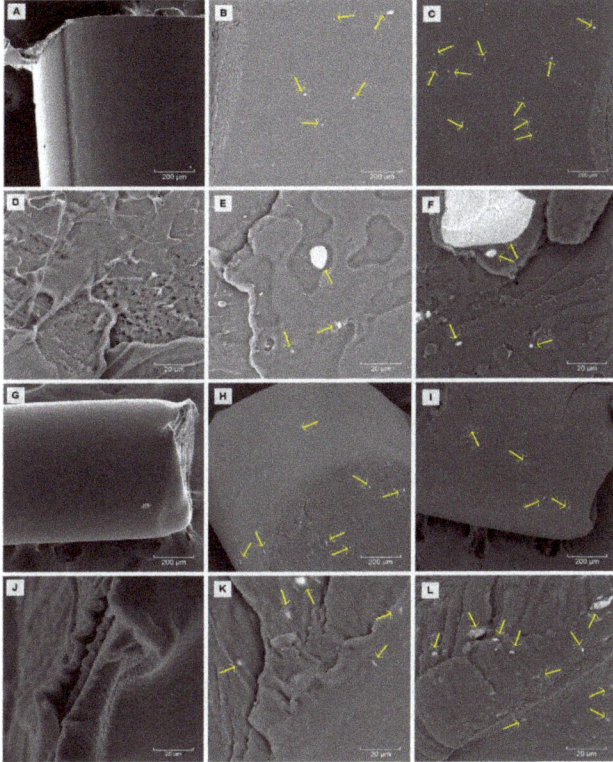

Figure 5. SEM micrographs of nanocomposites: pellet surface of PLA 0%, 5% and 10% (**A–C**) and MB 0%, 5% and 10% (**G–I**) (200×) and pellet tangential cut of PLA 0%, 5% and 10% (**D–F**) and MB 0%, 5% and 10% (**J–L**) (1.5 K×). TiO$_2$ is marked with yellow arrows.

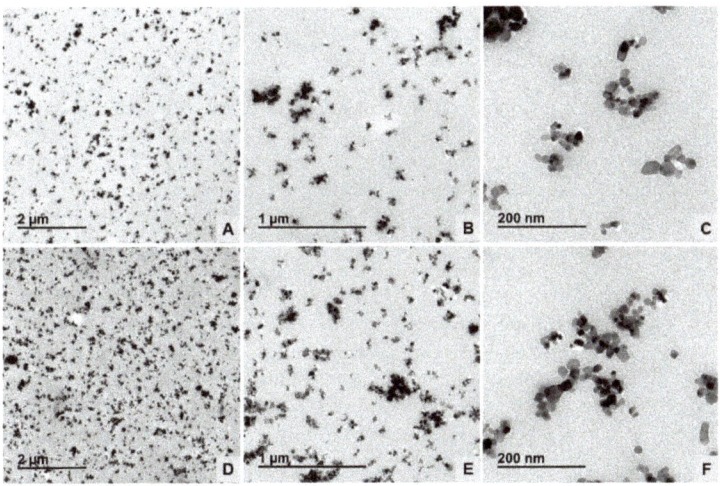

Figure 6. TEM micrographs of PLA nanocomposites: PLA 5% ((**A–C**)—1.5, 5.0 and 20 K×) and PLA 10% ((**D–F**)—1.5, 5.0 and 20 K×).

3.4.3. Evaluation of Ethylene Degradation by PLA and MB Nanocomposites

To determine the ethylene-removal capacity of nanocomposites, the concentration of ethylene was determined over a 6-day period in different samples previously irradiated with UV light (5 min) that were put in contact at 20 °C in hermetically sealed vials. These results can be observed in Figure 7.

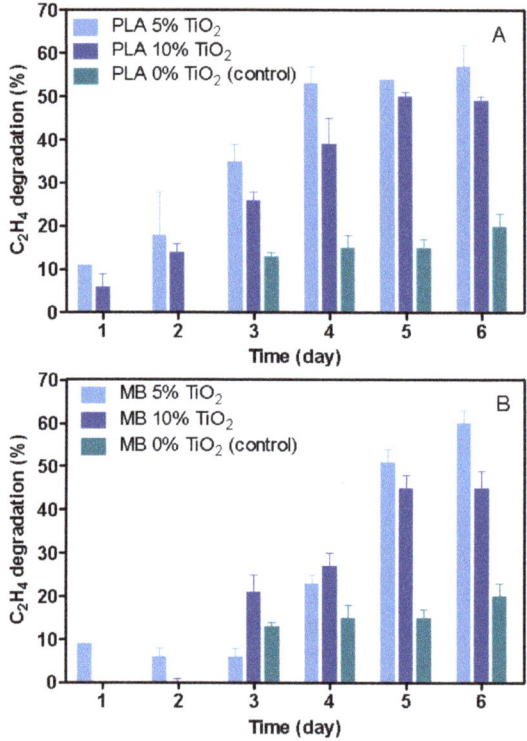

Figure 7. Ethylene degradation of nanocomposites at different TiO_2 concentrations in PLA (**A**) and MB (**B**).

In general, a higher degradation of ethylene was observed as the time increased. Additionally, for both PLA and MB nanocomposites, ethylene degradation was higher for those with 5 wt. % than for those with 10 wt. % of nano-TiO_2. Thus, ethylene degradation for nanocomposites with 5 wt. % of nano-TiO_2 was 51 ± 6% for PLA and 57 ± 2% for MB nanocomposites (Table 4). As discussed in the SEM analysis, the nanocomposites with 10 wt. % oxide were characterized by a higher presence of nano-TiO_2 agglomerates, a situation that generated a less efficient system since it allowed the generation of ROS on the exposed surface and not in the interior of the agglomerate. A similar explanation was given by Shih, Hsin and Lin [30] who found that the photocatalytic degradation of an azo dye from TiO_2 decreased with larger secondary particle sizes due to the higher intracrystalline diffusional resistance across the agglomerated TiO_2 particles. Similarly, [3] found a comparable effect of agglomeration of TiO_2 nanoparticles on their photocatalytic performance in chitosan nanocomposite films, where nanocomposites with a higher concentration of nano-TiO_2 decreased the ethylene-removal capacity. Likewise, the agglomerate formation was used to explain the low activity of ethylene removers based on zeolite and linear LDPE composites. Although in this case, the removal mechanism was linked to an ethylene sorption on the zeolite. The interaction of the ethylene with the mineral only occurred at the surface level of the agglomerate and not on the zeolite located inside the agglomerate [31].

Table 4. Final C_2H_4 degradation by different TiO_2 nanocomposite concentrations.

TiO_2 Concentration (%)	C_2H_4 Degradation (%)	
	Matrix	
	PLA	MB
0 (control)	20 ± 2 [Ac]	18 ± 2 [Ac]
5	51 ± 6 [Aa]	57 ± 2 [Aa]
10	46 ± 2 [Ab]	51 ± 5 [Ab]

Note: Equal letters mean that there are no statistically significant differences ($p > 0.05$). Upper-case letters: matrix factor and lower-case letters: TiO_2 concentration.

Moreover, the geometry of the pellets used in this work could also play an important factor in the ethylene-removal capacity. Disregarding Without considering ethylene diffusion through the material, only the nano-TiO_2 and its agglomerates located on the pellet surface could interact with ethylene to oxidize it. In this respect, changing the extruder configuration to obtain films, and analyzing the distribution of nano-TiO_2 on the film surface and its effect on ethylene removal will be an important step in this research.

Finally, it is important to mention that the results of this study provide a very good approach to developing potential nanocomposites for fresh food preservation. We showed that it is possible to create nanocomposites with eco-friendly material by an extrusion process, which could be scaled to an industrial level, and that different shapes of active packaging, such as films, could be developed.

4. Conclusions

The analyzed TiO_2 nanoparticles showed a crystalline structure typical of anatase; moreover, its photocatalytic activity against methylene blue was dependent on the radiation type used (UV or VIS light) and the particle size. In this way, the nano-TiO_2 21 nm particle size irradiated under UV light showed the highest photocatalytic activity. Regarding the ethylene removal by the different nanoparticles, the results showed that both types of radiation were able to remove ethylene, however, the nanoparticles activated with UV light presented the highest activity against the gas. In addition, the particle size was also a determinant in ethylene degradation; thus, the 21 nm particle showed the best ethylene removal. On the other hand, it was possible to establish that irradiation time had no effect on the ethylene removal.

Finally, it was possible to elaborate TiO_2 nanocomposites with PLA and MB through an extrusion process, scalable to the industrial sector. The nanocomposites showed color differences related to the TiO_2 concentration and the type of polymeric matrix used. In addition, the ethylene-removal study of the nanocomposites showed that all samples were able to reduce the ethylene concentration; however, nanocomposites with 5 wt. % of nano-TiO_2 had the highest activity, reaching a degradation of 51 ± 6 and 57 ± 2% for PLA and MB nanocomposite, respectively. It can be concluded that the development of TiO_2 nanocomposites with biopolymers is a good option for the removal of ethylene. This could be an interesting option for fresh-food preservation; however, future studies should consider changing the structure of the nanocomposite, such as into films, to expose a larger amount of nano-TiO_2 on the surface of the material in order to increase its effectiveness against ethylene.

Author Contributions: Conceptualization, F.R.-M.; methodology, A.M., S.G., N.G., M.D., P.C., X.V. and M.P.A.; software, A.M.; validation, A.M. and F.R.-M.; formal analysis, A.M.; investigation, F.R.-M. and A.M.; resources, F.R.-M., A.G. and M.G.; data curation, A.M.; writing—original draft preparation, A.M. and P.C.; writing—review and editing, F.R.-M., C.H., M.P.A., A.T. and J.B.; visualization, F.R.-M.; supervision, F.R.-M. and C.H.; project administration, F.R.-M.; funding acquisition, F.R.-M. All authors have read and agreed to the published version of the manuscript.

Funding: This research was funded by National Research and Development Agency (ANID) through the National Fund for Scientific and Technological Development (FONDECYT), grant 1211391, Development of International Networking, grant FOVI220086, and the Center of Excellence with Basal Financing (CEDENNA), grant AFB220001.

Institutional Review Board Statement: Not applicable.

Informed Consent Statement: Not applicable.

Data Availability Statement: Not applicable.

Acknowledgments: The authors acknowledge to Agencia Nacional de Investigación and Desarrollo for financial support and the Spanish Ministry of Science and Innovation (MICINN) through PID2021-123753NA-C32 and TED-129920A-C43 funded by MCIN/AEI/10.13039/501100011033 and by ERDF "A way of making Europe" and by the "European Union NextGenerationEU/PRTR". The authors are also grateful to the University of Santiago de Chile for the Foreigner grant (Alba Maldonado).

Conflicts of Interest: The authors declare no conflict of interest.

References

1. Yadav, P.; Ansari, M.W.; Kaula, B.C.; Rao, Y.R.; Al Meselmani, M.; Siddiqui, Z.H.; Brajendra; Kumar, S.B.; Rani, V.; Sarkar, A.; et al. Regulation of Ethylene Metabolism in Tomato under Salinity Stress Involving Linkages with Important Physiological Signaling Pathways. *Plant Sci.* **2023**, *334*, 111736. [CrossRef]
2. Kızıldeniz, T.; Hepsağ, F.; Hayoğlu, İ. Improving Mulberry Shelf-Life with 1-Methylcyclopropene and Modified Atmosphere Packaging. *Biochem. Syst. Ecol.* **2023**, *106*, 104578. [CrossRef]
3. Siripatrawan, U.; Kaewklin, P. Fabrication and Characterization of Chitosan-Titanium Dioxide Nanocomposite Film as Ethylene Scavenging and Antimicrobial Active Food Packaging. *Food Hydrocoll.* **2018**, *84*, 125–134. [CrossRef]
4. English, A. Food and Agriculture Organization of the United Nations. In *The State of Food and Agriculture. 2019, Moving forward on Food Loss and Waste Reduction*; FAO: Rome, Italy, 2019; ISBN 9789251317891.
5. Gaikwad, K.K.; Singh, S.; Lee, Y.S. High Adsorption of Ethylene by Alkali-Treated Halloysite Nanotubes for Food-Packaging Applications. *Environ. Chem. Lett.* **2018**, *16*, 1055–1062. [CrossRef]
6. Wang, H.; Wang, L.; Ye, S.; Song, X. Construction of Bi_2WO_6–TiO_2/Starch Nanocomposite Films for Visible-Light Catalytic Degradation of Ethylene. *Food Hydrocoll.* **2019**, *88*, 92–100. [CrossRef]
7. Böhmer-Maas, B.W.; Fonseca, L.M.; Otero, D.M.; da Rosa Zavareze, E.; Zambiazi, R.C. Photocatalytic Zein-TiO_2 Nanofibers as Ethylene Absorbers for Storage of Cherry Tomatoes. *Food Packag. Shelf Life* **2020**, *24*, 100508. [CrossRef]
8. Keller, N.; Ducamp, M.N.; Robert, D.; Keller, V. Ethylene Removal and Fresh Product Storage: A Challenge at the Frontiers of Chemistry. Toward an Approach by Photocatalytic Oxidation. *Chem. Rev.* **2013**, *113*, 5029–5070. [CrossRef]
9. Nevárez-Martínez, M.C.; Espinoza-Montero, P.J.; Quiroz-Chávez Francisco, J.; Ohtani, B. Fotocatálisis: Inicio, Actualidad y Perspectivas a Través Del TiO_2 Photocatalysis: Beginning, Present and Trends through TiO_2. *Av. Química* **2017**, *12*, 45–59.
10. Zhang, Q.; Ye, S.; Chen, X.; Song, X.; Li, L.; Huang, X. Photocatalytic Degradation of Ethylene Using Titanium Dioxide Nanotube Arrays with Ag and Reduced Graphene Oxide Irradiated by γ-Ray Radiolysis. *Appl. Catal. B* **2017**, *203*, 673–683. [CrossRef]
11. Scaffaro, R.; Sutera, F.; Botta, L. Biopolymeric Bilayer Films Produced by Co-Extrusion Film Blowing. *Polym. Test.* **2018**, *65*, 35–43. [CrossRef]
12. Cruz, R.; Nisar, M.; Palza, H.; Yazdani-Pedram, M.; Aguilar-Bolados, H.; Quijada, R. Development of Bio Degradable Nanocomposites Based on PLA and Functionalized Graphene Oxide. *Polym. Test.* **2023**, *124*, 108066. [CrossRef]
13. Sanyang, M.L.; Sapuan, S.M. Development of Expert System for Biobased Polymer Material Selection: Food Packaging Application. *J. Food Sci. Technol.* **2015**, *52*, 6445–6454. [CrossRef]
14. Scaffaro, R.; Sutera, F.; Mistretta, M.C.; Botta, L.; La Mantia, F.P. Structure-Properties Relationships in Melt Reprocessed PLA/Hydrotalcites Nanocomposites. *Express Polym. Lett.* **2017**, *11*, 555. [CrossRef]
15. Zhang, W.; Rhim, J.W. Titanium Dioxide (TiO_2) for the Manufacture of Multifunctional Active Food Packaging Films. *Food Packag. Shelf Life* **2022**, *31*, 100806. [CrossRef]
16. Kaewklin, P.; Siripatrawan, U.; Suwanagul, A.; Lee, Y.S. Active Packaging from Chitosan-Titanium Dioxide Nanocomposite Film for Prolonging Storage Life of Tomato Fruit. *Int. J. Biol. Macromol.* **2018**, *112*, 523–529. [CrossRef] [PubMed]
17. Luo, M.R.; Cui, G.; Rigg, B. The Development of the CIE 2000 Colour-Difference Formula: CIEDE2000. *Color Res. Appl.* **2001**, *26*, 340–350. [CrossRef]
18. Sunny, N.E.; Mathew, S.S.; Chandel, N.; Saravanan, P.; Rajeshkannan, R.; Rajasimman, M.; Vasseghian, Y.; Rajamohan, N.; Kumar, S.V. Green Synthesis of Titanium Dioxide Nanoparticles Using Plant Biomass and Their Applications—A Review. *Chemosphere* **2022**, *300*, 134612. [CrossRef]
19. Molina Higgins, M.C.; Hall, H.; Rojas, J.V. The Effect of X-Ray Induced Oxygen Defects on the Photocatalytic Properties of Titanium Dioxide Nanoparticles. *J. Photochem. Photobiol. A Chem.* **2021**, *409*, 113138. [CrossRef]

20. Ilyas, M.; Waris, A.; Khan, A.U.; Zamel, D.; Yar, L.; Baset, A.; Muhaymin, A.; Khan, S.; Ali, A.; Ahmad, A. Biological Synthesis of Titanium Dioxide Nanoparticles from Plants and Microorganisms and Their Potential Biomedical Applications. *Inorg. Chem. Commun.* **2021**, *133*, 108968. [CrossRef]
21. Mathan Kumar, P.; Paramasivam, V.; Beemaraj, R.K.; Mathalai Sundaram, C.; Arun Prasath, K. Investigate the Characterization and Synthesis Process of Titanium Dioxide Nanoparticles. *Mater. Today Proc.* **2022**, *52*, 1140–1142.
22. Ortiz-Bustos, J.; Fajardo, M.; del Hierro, I.; Pérez, Y. Versatile Titanium Dioxide Nanoparticles Prepared by Surface-Grown Polymerization of Polyethylenimine for Photodegradation and Catalytic C[Sbnd]C Bond Forming Reactions. *Mol. Catal.* **2019**, *475*, 110501. [CrossRef]
23. Mutuma, B.K.; Shao, G.N.; Kim, W.D.; Kim, H.T. Sol-Gel Synthesis of Mesoporous Anatase-Brookite and Anatase-Brookite-Rutile TiO_2 Nanoparticles and Their Photocatalytic Properties. *J. Colloid. Interface Sci.* **2015**, *442*, 1–7. [CrossRef] [PubMed]
24. He, P.; Zhao, Z.; Tan, Y.; Hengchao, E.; Zuo, M.; Wang, J.; Yang, J.; Cui, S.; Yang, X. Photocatalytic Degradation of Deoxynivalenol Using Cerium Doped Titanium Dioxide under Ultraviolet Light Irradiation. *Toxins* **2021**, *13*, 481. [CrossRef] [PubMed]
25. Alizadeh Sani, M.; Maleki, M.; Eghbaljoo-Gharehgheshlaghi, H.; Khezerlou, A.; Mohammadian, E.; Liu, Q.; Jafari, S.M. Titanium Dioxide Nanoparticles as Multifunctional Surface-Active Materials for Smart/Active Nanocomposite Packaging Films. *Adv. Colloid. Interface Sci.* **2022**, *300*, 102593. [CrossRef]
26. Behnajady, M.A.; Modirshahla, N.; Shokri, M.; Elham, H.; Zeininezhad, A. The Effect of Particle Size and Crystal Structure of Titanium Dioxide Nanoparticles on the Photocatalytic Properties. *J. Environ. Sci. Health A Toxic Hazard. Subst. Environ. Eng.* **2008**, *43*, 460–467. [CrossRef]
27. Pereira, G. Efecto del Uso de una Película Plástica Activa Sobre la Maduración de un Fruto Climatérico tal Como la Banana de Variedad "Cavendish". Bachelor's Thesis, Universidad de Santiago de Chile, Santiago, Chile, 2017.
28. Zhang, W.; Zhang, Y.; Cao, J.; Jiang, W. Improving the Performance of Edible Food Packaging Films by Using Nanocellulose as an Additive. *Int. J. Biol. Macromol.* **2021**, *166*, 288–296. [CrossRef]
29. de Almeida, F.; Costa e Silva, E.; Correia, A.; Silva, F.J.G. Rheological Behaviour of PP Nanocomposites by Extrusion Process. *Procedia Manuf.* **2019**, *38*, 1516–1523. [CrossRef]
30. Shih, Y.H.; Lin, C.H. Effect of Particle Size of Titanium Dioxide Nanoparticle Aggregates on the Degradation of One Azo Dye. *Environ. Sci. Pollut. Res.* **2012**, *19*, 1652–1658. [CrossRef]
31. Rodríguez, F.; Galotto, M.; Guarda, A.; Bruna, J. Active Film That Can Remove Ethylene, Comprising a Modifying Natural Zeolite. WO/2018/094543, 31 May 2018.

Disclaimer/Publisher's Note: The statements, opinions and data contained in all publications are solely those of the individual author(s) and contributor(s) and not of MDPI and/or the editor(s). MDPI and/or the editor(s) disclaim responsibility for any injury to people or property resulting from any ideas, methods, instructions or products referred to in the content.

Article

Pilot-Scale Processing and Functional Properties of Antifungal EVOH-Based Films Containing Methyl Anthranilate Intended for Food Packaging Applications

Alejandro Aragón-Gutiérrez [1,*], Raquel Heras-Mozos [2], Antonio Montesinos [1], Miriam Gallur [1], Daniel López [3], Rafael Gavara [2] and Pilar Hernández-Muñoz [2,*]

1. Grupo de Tecnología de Materiales y Envases, Instituto Tecnológico del Embalaje, Transporte y Logística, ITENE, Unidad Asociada al CSIC, Calle de Albert Einstein 1, 46980 Paterna, Valencia, Spain
2. Instituto de Agroquímica y Tecnología de Alimentos, IATA-CSIC, Calle del Catedrático Agustín Escardino Benlloch 7, 46980 Paterna, Valencia, Spain
3. Instituto de Ciencia y Tecnología de Polímeros, ICTP-CSIC, Calle Juan de la Cierva 3, 28006 Madrid, Spain
* Correspondence: alejandro.aragon@itene.com (A.A.-G.); phernan@iata.csic.es (P.H.-M.); Tel.: +34-961-820-000 (A.A.-G.); +34-963-900-022 (P.H.-M.)

Abstract: Antimicrobial packaging has emerged as an efficient technology to improve the stability of food products. In this study, new formulations based on ethylene vinyl alcohol (EVOH) copolymer were developed by incorporating the volatile methyl anthranilate (MA) at different concentrations as antifungal compound to obtain active films for food packaging. To this end, a twin-screw extruder with a specifically designed screw configuration was employed to produce films at pilot scale. The quantification analyses of MA in the films showed a high retention capacity. Then, the morphological, optical, thermal, mechanical and water vapour barrier performance, as well as the antifungal activity in vitro of the active films, were evaluated. The presence of MA did not affect the transparency or the thermal stability of EVOH-based films, but decreased the glass transition temperature of the copolymer, indicating a plasticizing effect, which was confirmed by an increase in the elongation at break values of the films. Because of the additive-induced plasticization over EVOH, the water vapour permeability slightly increased at 33% and 75% relative humidity values. Finally, the evaluation of the antifungal activity in vitro of the active films containing methyl anthranilate showed a great effectiveness against *P. expansum* and *B. cinerea*, demonstrating the potential applicability of the developed films for active food packaging.

Keywords: EVOH copolymer; methyl anthranilate; melt-extrusion; film properties; antifungal activity; active packaging

1. Introduction

During the last decades, production, processing, and marketing of fresh food have experimented with considerable changes trying to fit consumers' demands. However, wastes due to product spoilage and residues from processing (up to 60% of harvested products) are becoming a serious economic and environmental issue [1]. In this regard, packaging is positioned as one of the main tools to reduce food spoilage and, therefore, many efforts from both academia and the food industry are being directed to improve and optimize food packaging systems [2]. Among others, the development of active antimicrobial packaging has emerged as an innovative technique to extend the shelf life by inhibiting microbial growth in foods while maintaining their quality and safety [3–5]. This system is based on the deliberate incorporation of components into packaging materials for their later progressive release into the packaged product or the head-space of the food, causing an inhibitory effect against pathogens that affect food products [6]. Thanks to their low costs, wide availability and versatile properties, the research and developments on active packaging systems have been mainly focused on the use of plastic materials [7]. In this line, ethylene vinyl alcohol (EVOH) copolymers are one of the best candidates used in the design and development of active materials intended for food packaging applications. EVOH copolymers, which are synthesized through the hydrolysis of ethylene vinyl acetate (EVA), are characterized by a high transparency, great chemical resistance and outstanding oxygen barrier properties [8,9]. In addition, EVOH based films are recyclable in current infrastructure by regrinding processes and several works also report that EVOH copolymers with up to 44 mol% ethylene content can be degraded in specific environmental conditions and biological media [10–12]. Interestingly, its hydrophilic nature, which results in a significant plasticisation of the polymer matrix when it is exposed to high relative humidifies or to food products with high water activity, can be used as the triggering mechanism to allow the release of components from the packaging to the food media [13].

With the aim of providing a more sustainable character to plastic materials and satisfying consumers demands', the food industry is moving towards the use of innovative and natural ingredients such as essential oils (Eos), which have proven antimicrobial and antioxidant properties with great interest in food packaging applications [14,15]. In this context, we considered to extend the use of methyl anthranilate (MA), a natural ingredient already employed in flavouring foods and drugs, to prepare active packaging films based on EVOH copolymer. Methyl anthranilate is a metabolite responsible of the characteristic odour of Concord grapes and is also present in several essential oils such as neroli, ylang-ylang and jasmine [16]. The current commercial MA is prepared by fossil-based chemical processes since the direct extraction and recovery of methyl anthranilate from the aforementioned sources have been proven to be economically unfeasible due to low yields [17]. However, there are already some works in literature dealing with the fermentative production of MA from renewable sources such as glucose, in order to obtain a biobased natural compound via an eco-friendly route [18].

Nowadays, there are two main routes to incorporate natural active compounds into polymeric matrices: the direct incorporation of the natural ingredients into the polymer material and encapsulation technologies. On the one hand, encapsulation technologies of active ingredients have been revealed as a very promising strategy. They present numerous advantages such as protecting active compounds from oxidative processes, photodegradation and thermal conditions or providing a controlled release of the compound from the encapsulating agent [19,20]. Pansuwan et al. reported the encapsulation of methyl anthranilate as an essential oil model in polymer microcapsules by employing two different techniques: microsuspension conventional radical polymerization and microsuspension iodine transfer polymerization using methyl methacrylate (MMA) and ethylene glycol dimethacrylate (EGDMA) copolymer as polymer shells [21]. Buendía et al. encapsulated a mixture of Eos composed of carvacrol, oregano and cinnamon in β-cyclodextrins (β-CD) for the development of an active coated cardboard tray [22]. Similarly, Wen et al. encapsulated cinnamon essential oil in β-CD for the manufacturing of an active electrospun nanofibrous film for food packaging applications [23]. Mohammadi et al. investigated the bioactivity of olive leaf extract encapsulated in soybean oil by nano-emulsion technology [24]. Although these studies showed the potential applicability of different technologies to maintain the beneficial properties of bioactives, there is still a gap of knowledge between the use of certain encapsulation strategies and their effective commercial application. In addition, the food manufacturing industry also needs to face various challenges related to the storability and stability of encapsulated bioactive compounds [25,26]. In this regard, the direct incorporation of active components into polymeric matrices through conventional melt-processing technologies such as extrusion, compounding and injection moulding is still positioned as an explorable and encouraging strategy to prepare active materials scalable in industrial environments. For example, Krzysztof et al. prepared bioactive polypropylene (PP) films incorporated with plasticizers and antimicrobial substances (i.e., oregano oil, rosemary extracted, methylparaben and green tea extract) by cast film extrusion. Authors showed that the presence of plasticizers promoted a gradual release of the active ingredients from the polymer matrix, and thus, the biocidal activity was enhanced [27]. In this study, the effect of plasticizers as release promoters of the active substances can be achieved by tailoring the relative humidity exposure of the bioactive EVOH-based films. More recently, the attention is focused on the development of active materials based on biodegradable polymers. Laorenza et al. produced films of PBAT/PLA blended with carvacrol, citral and α-terpineol essential oils via blown film extrusion [28]. However, the main limitation of these routes is that the high temperatures and the shear forces generated during the process usually result in significant losses of the bioactives due to their degradation or volatilization [29–31]. Moreover, none of the previous research worked on the optimization of process conditions to favour the incorporation of active ingredients into the polymer matrixes in the melt state in order to reduce their potential losses.

Therefore, the objective of this work was to prepare cast-extruded EVOH-based films containing methyl anthranilate as active compound for their use in packaging applications. To this aim, a copolymer composed of 44% ethylene molar content (EVOH$_{44}$) was selected within the EVOH family, since it presents a lower melting temperature due to its crystalline structure and, hence, the extrusion process can be carried out at softer conditions. In addition, a specific screw configuration was designed in order to optimize the melt processing and minimize the potential losses of MA. The influence of different concentrations of methyl anthranilate over the main properties of interest of EVOH films in food packaging applications was intensely studied. Finally, the ability of methyl anthranilate incorporated in the active films to inhibit the growth of *Botrytis cinerea* and *P. expansum* was investigated.

2. Materials and Methods

2.1. Materials

Ethylene vinyl alcohol copolymer composed of 44% ethylene molar content was kindly provided by The Nippon Synthetic Chemical Company (Osaka, Japan). Methyl anthranilate (FCC, Food grade) was purchased from Sigma-Aldrich (Sigma-Aldrich Corp., St. Louis, MO, USA). For the microbiological assays, potato dextrose agar (PDA) was purchased from Scharlau (Scharlab S.L., Barcelona, Spain). The fungal strains *Penicillium expansum* (CECT 2278) and *Botrytis cinerea* (CECT2100) were supplied by the Spanish Type Culture Collection (CECT) and were maintained in potato dextrose broth (PDB) with 20% sterile glycerol at −80 °C. The fungal culture was regenerated and maintained by regular subculture at 26 °C on PDA plates.

2.2. Pilot Scale Production of Active EVOH-Based Films by Extrusion Processing

Film formulations were prepared in a pilot-scale co-rotating twin screw extruder Brabender TSE 20/40 with a 20 mm diameter screw and a length-diameter ratio (L/D) of 40. EVOH pellets, previously dried at 90 °C for 4 h, were fed through the main hopper employing a gravimetric feeder. On the other hand, a loss-in-weight injection pump was used to feed methyl anthranilate into the polymer melt, taking advantage of its low viscosity. The pumping rate (mL/min) was adjusted for each formulation in order to obtain a final content of MA of 3, 5 and 8 wt.% with respect to the polymer. The active ingredient was fed in the fourth barrel of the extruder with the aim of reducing its residence time and avoid potential losses of the compound associated with its volatile nature. The temperature profile along the 6 barrel zones from hopper (zone 1) to die (zone 6) was 175-180-180-180-180-180 °C (see Figure 1), the applied screw rotation speed was set at 40 rpm, and the throughput was 1.5 kg/h. The employed screw configuration (Figure 1) was designed combining different types of elements (i.e., conveying elements, kneading blocks and gear mixing) in order to achieve a proper mixing and dispersion of the components while preventing a potential thermal degradation of the materials during the extrusion process. In particular, MA was added into gear mixing elements, which have the ability to divide and recombine the flow, promoting a better interaction between the polymer matrix in melt state and the active ingredient. Finally, the compounded EVOH—methyl anthranilate formulations were transformed into films of 50 microns in thickness and 10 cm in width using an extrusion roller calender line. Samples were stored in aluminium foils prior to characterization to prevent an undesirable release of the volatile oil.

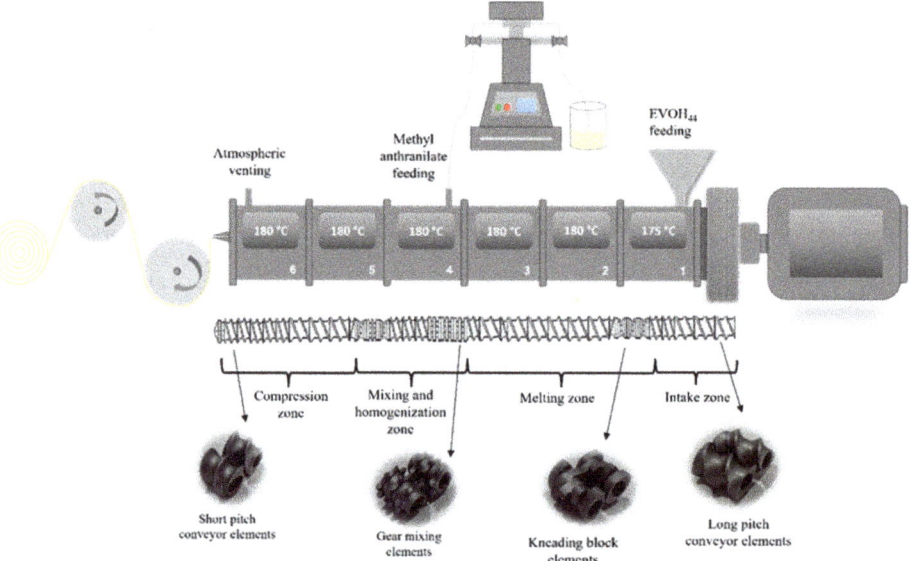

Figure 1. Schematic representation of the film-extrusion process.

2.3. Characterization of Active Films

2.3.1. Identification and Quantification of Methyl Anthranilate in EVOH-Based Films

The presence of methyl anthranilate in EVOH-based films after the extrusion process was qualitatively confirmed by Fourier transform infrared spectroscopy in attenuated total reflectance mode (ATR-FTIR), using a TENSOR 27 Spectrophotometer (Bruker, Massachusetts, MA, USA) equipped with a diamond ATR sampling accessory. A background spectrum was acquired before the analysis to compensate the CO_2 and moisture effect. The spectra of EVOH-based film formulations were then recorded in the 4000–700 cm^{-1} wavenumber range, averaging 64 scans and 4 cm^{-1} resolution.

The final content of MA remaining in the active films was quantitatively determined by solid–liquid extraction followed by gas chromatography mass spectrometry (GC/MS) analysis. In brief, 1 cm^2 of each film formulation containing methyl anthranilate was extracted with 5 mL of ethanol at 40 °C for 24 h. Then, GC/MS analysis was performed employing a triple quadrupole GC/MS instrument from Agilent Technologies (Agilent Technologies, Inc., Santa Clara, CA, USA) equipped with an Agilent DB-624 30 m × 0.32 mm, 0.18 μm column, using hydrogen at 3.5 mL/min as carrier gas with a 10:1 split ratio. For the analysis, 1 μL of the extraction in ethanol was injected. The GC oven was programmed at 85 °C for 1 min and then the temperature was increased by 30 °C min^{-1} to 220 °C, where it was held for 1 min. The quantitative determination of methyl anthranilate was conducted by using a full scan between the m/z values 40–200 and total ion counting. The calibration curve with a $R^2 > 0.99$ was prepared by injecting solutions of methyl anthranilate in ethanol at five known concentrations into the GC/MS instrument. Then, the capacity of EVOH copolymer to retain the active methyl anthranilate compound was determined by employing the following Equation (1) [32].

$$\text{Retention capacity (\%)} = \frac{m_f}{m_i} \times 100 \qquad (1)$$

where m_f is the amount of methyl anthranilate in the packaging material after the extrusion process determined by GC/MS (g methyl anthranilate/g film) and m_i is the nominal content of the active compound initially incorporated to the EVOH copolymer.

2.3.2. Morphological Characterization

The sample surfaces and cross sections, previously coated under vacuum conditions, using a current of 12 mA and for two minutes with a Pd/Au layer, were studied by employing a Hitachi S-4800 (Hitachi Ltd., Tokyo, Japan) microscope operated at 10 kV and a working distance of 10 mm.

2.3.3. Thermal Analysis

The thermal stability and the degradation behaviour of EVOH-based film formulations was evaluated by thermogravimetric analysis (TGA) using a TGA Q-500 thermogravimetric instrument (TA Instruments, New Castle, DE, USA). Typically, 3–8 mg of film sample was placed in alumina pans and then heated from room temperature to 900 °C at a constant heating rate of 10 °C min^{-1} under N$_2$ atmosphere with a flow rate of 50 mL min^{-1}.

The determination of the main thermal parameters and transitions was performed by differential scanning calorimetry (DSC). DSC analyses were carried out by employing a DSC Q-2000 equipment from TA Instruments (New Castle, DE, USA). Approximately 5 mg of each sample was introduced in a sealed 40 µL aluminum pan and was subjected to the following thermal cycles: first heating scan from 0 °C to 200 °C at 10 °C min^{-1}, followed by a cooling to -25 °C and a second heating scan to 200 °C. The temperature was held for 3 min at the end of each cycle. The values of the glass transition temperature, the melting and crystallization temperatures and their corresponding enthalpies were calculated with the TA Universal Analysis software by analyzing the DSC thermograms. The degree of crystallinity was then calculated as described elsewhere [33], taking 217.8 J g^{-1} as the melting enthalpy associated with a pure crystalline EVOH composed of 44% ethylene molar content and following Equation (2) [34].

$$\chi(\%) = \frac{1}{\left(1 - m_f\right)} \left[\frac{\Delta H}{\Delta H_0}\right] \times 100 \qquad (2)$$

where ΔH is the enthalpy for melting or crystallization; ΔH_0 is melting for a 100% crystalline EVOH sample and $\left(1 - m_f\right)$ is the weight fraction of EVOH in the sample.

2.3.4. Mechanical Properties

The tensile characteristics of all films were calculated with an MTS Universal Tensile test machine (MTS Systems Corporation, Minnesota, MN, USA) equipped with a 1 kN load cell. The main mechanical properties (i.e., Young's modulus (E), stress at yield (σ_y), stress at break (σ_b) and elongation at break (ε_b)) were calculated from the stress-strain curves following UNE-EN ISO 527-3:2019 [35]. The dimensions of the film specimens were 170 mm long, 15 mm wide and 50 µm thick and the clamps distance was set at 100 mm. The films were stretched at 100 mm min^{-1} until failure. At least five measurements were conducted of each set of formulations and the results were expressed as the average ± standard deviation. A statistical analysis at a 5% significance level based on the ANOVA test was employed to analyse the results.

2.3.5. Film Thickness, Colour Tests and Transparency

The film thickness of the films was determined employing a micrometre MiniTest 7200FH (ElektroPhysik Dr. Steingroever GmbH & Co, Köln, Germany) with a low range resolution of 0.1 µm. The mean values were calculated from readings taken at 10 different random locations of each film formulation.

The change in the colour properties on EVOH films provoked by the incorporation of the active agent was evaluated by employing a KONICA CM-2500d (Konica Minolta

Sensing Americas, Inc., Ramsey, NJ, USA). The apparatus was calibrated using a white and black standard tile. The tests were run in triplicate in the CIELab space, where L^*, a^* and b^* indicate lightness, redness/greenness, and yellowness/blueness, respectively. The measurements were taken at random positions over the surface of the films and the results were expressed as the average values of at least three measurements. The total color difference (ΔE^*) caused by methyl anthranilate in EVOH-based films was calculated following Equation (3).

$$\Delta E = \left(\Delta L^2 + \Delta a^{*2} + \Delta b^{*2} \right)^{0.5} \quad (3)$$

where ΔL^*, Δa^*, and Δb^* are the distances from the color coordinates of film samples to control EVOH film [36].

The light transmittance of the films was evaluated in the wavelength range of 400 to 800 nm using a Jasco V-630 UV-vis spectrophotometer (Jasco Deutschland GmbH, Pfungstadt, Germany). The transparency of the films was calculated following Equation (4)

$$\text{Transparency} = \frac{A_{600}}{t} \quad (4)$$

in which A_{600}, is the absorbances at 600 nm, and t is the film thickness in millimetres [37].

2.3.6. Barrier Properties

The water vapour permeability tests (WVP) were carried out at room temperature and at two different relative humidity (RH) levels, 33% and 75% RH, following the gravimetrically method based on the standard ASTM E96M [38]. Permeability cells were filled with 4 grams of the desiccant anhydrous calcium chloride, which was dehydrated at 90 °C for 48 h before the tests. Circular shaped film samples with a diameter of 5 cm were fixed to the cell with a flat silicon ring, leaving a permeable surface of 8 cm². Then, the permeation cups were stored at room temperature in chambers containing saturated salt solutions, magnesium chloride, $MgCl_2$, and sodium chloride, NaCl, for 33 and 75% RH, respectively. The weight gain of the cups over time was used to determine the water vapour transferred through the films and absorbed by the anhydrous calcium chloride. Cups were weighted for 21 days with an analytical balance and the water vapour transmission rate (WVTR) was calculated from the plot of the weight increment vs. time following Equation (5).

$$\text{WVTR} \left(g\ h^{-1} m^{-2} \right) = \frac{\Delta m / \Delta t}{A} \quad (5)$$

where $\Delta m / \Delta t$ is the slope of the curve and A is the permeable surface of the film [39]. The obtained WVTR values were then divided by the water pressure gradient and multiplied by the film thickness to determine the water vapour permeability (WVP). Five replicates were tested for each set of formulations.

2.3.7. Influence of Environmental Humidity on the Release of Methyl Anthranilate

In order to study the influence of humidity on the compound release, an extraction technique and GC/MS analyses were performed. Film samples of each formulation were conditioned at three different humidity values, 33%, 75% and >95% at room temperature for 10 days based on the methodology proposed by Kurek and co-workers [40]. After that time, films were recovered from their conditioning chambers, and the remaining amount of methyl anthranilate in the samples was extracted with ethanol. Then, GC/MS was carried out following the procedure described in Section 2.3.1 and the percentage release of methyl anthranilate was calculated as follows:

$$\text{Compound release (\%)} = \frac{c_i - c_f}{c_i} \times 100 \quad (6)$$

where c_i is the real concentration of methyl anthranilate in the films (mg MA/g film) after the extrusion process, and c_f is the final amount of methyl anthranilate in the films after its release promoted by humidity.

2.4. Efficacy of EVOH Films Containing Methyl Anthranilate against Fungal Growth In Vitro

2.4.1. Microbiological Studies

Penicillium expansum (CECT 2278) and *Botrytis cinerea* (CECT 2100) were chosen as fungal models of foodborne pathogens. These strains were obtained from the Spanish type culture collection (CECT). Both fungi were grown and maintained in Potato Dextrose Agar (PDA, Scharlab, Spain) for 7 days at 26 °C. To carry out the antimicrobial assay, a conidial suspension of 10^6 spores/mL was obtained from the fungal surface of PDA plates. For this propose, sterile peptone water with 0.05% (*v/v*) Tween 80 was poured on the fungal plate and scratched with a Digralsky spreader to drag the conidia. The fungal culture solution was transferred to sterile tubes and several dilutions were carried out to obtain the concentration of 10^6 spores/mL, counted with improved Neubauer chamber (Bright-Line Hemacytometer, Hausser Scientific, Horsham, PA, USA).

2.4.2. Antifungal Activity of Methyl Anthranilate

The study of the antifungal capacity of the methyl anthranilate (MA) in vapour phase against *Penicillium expansum* and *Botrytis cinerea* was carried out by determination of values of minimal inhibitory concentration (MIC) and minimal fungicidal concentration (MFC). The MIC value was defined as amount of compound which inhibits growth by at least 50% compared to the control at day 7 of incubation. MFC was defined as the amount of volatile that completely inhibit the fungal growth after 10 days. Both parameters have been expressed as volume (µL) of active compound dosed in each inoculated plate. The PDA plates were inoculated in three equidistant points with 3 µL of conidia suspension previously obtained (10^6 spores/mL). Then, a volume of 1, 2.5, 5, 10, 20, 50 µL of MA was placed in a 50 mm of sterile paper disk and fixed in the lid of the plate. The plates were closed and sealed with parafilm to avoid the volatile loss and incubated during 7 days at 26 °C. The fungal colony diameter was measured after that time to determinate the MIC value (µL/plate). Then, the active compound was removed from the plates and were incubated for 3 days more, and fungal growth was assessed to determinate MFC parameter (µL/plate). All assays were carried out in triplicate.

2.4.3. Antifungal Activity of Cast-Extruded EVOH Films Incorporated with Methyl Anthranilate

Assessment of the antimicrobial activity of cast extruded EVOH films incorporated with methyl anthranilate against both fungal strains was carried out by micro-atmosphere assay, which is based on the effectiveness of the volatile vapour phase generated inside the plate. To conduct this assay, 0.3 g of each film of 8 cm of diameter, with different amount of MA (0, 3, 5, 8 wt.%), was placed on the lid of the petri dish, which contained 15 mL of PDA and was inoculated at 3 equidistant points with conidial suspension. The plates with active film and fungal inoculation were incubated for 10 days at 26 °C. The growth of the fungal colony was monitored by measuring colony diameter at 3, 5, and 7 days, after which the film was removed and incubated for 3 days more to observe the recovery of the fungal colony. A control without film was also carried out. The radial growth of colony was measured in cm and inhibition of fungal growth in respect to the control was calculated and expressed as percentage. All assays were conducted in triplicate.

3. Results and Discussion

3.1. Identification and Quantification of Methyl Anthranilate in EVOH-Based Films

The FTIR spectra in the region of interest of EVOH and EVOH active films are shown in Figure 2. The spectrum of the control EVOH film showed bending C-H vibrations at 1328 cm^{-1} and 1452 cm^{-1}, and the absorbance bands related to stretching vibrations located at 2850 cm^{-1} (C-H), 2930 cm^{-1} (C-H) and 3342 cm^{-1} (O-H) (not shown in the Figure 2). The presence of a significant amount of methyl anthranilate in EVOH copolymer matrix after the extrusion process was confirmed by the observation of several absorbance peaks related to the bioactive compound such as the C=O stretching vibration band at 1690 cm^{-1}, the bending of the N-H at 1620 cm^{-1}, and by the C-O stretching vibration at 1248 cm^{-1} [41]. Moreover, the intensity of these peaks augmented as the concentration of methyl anthranilate increased in the active extruded films.

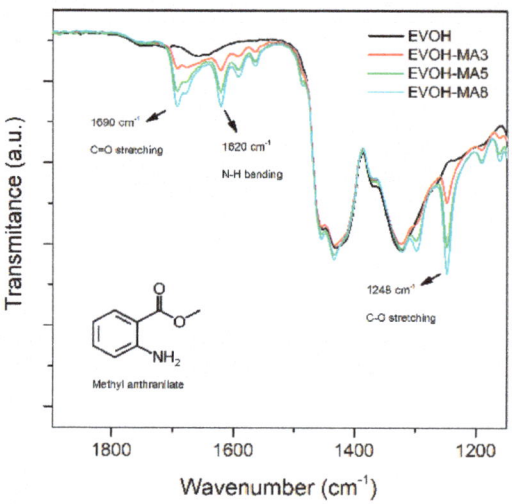

Figure 2. FT-IR spectra of EVOH-MA samples.

To determine the real concentration of MA in the polymeric EVOH films, GC/MS analysis was performed via a previous extraction with ethanol at 40 °C. The final concentrations for each formulation are displayed in Table 1. As shown, the final content of MA in the films was below their initial content, possibly due to losses occurred in the extrusion process caused by the high temperatures and the shear stresses generated during the film manufacturing. Results revealed that less than 25% of the content of MA was lost during processing, indicating a high retention capacity of the volatile oil in the polymer matrix. The used screw design, which favoured a proper mixing of the components, together with the extrusion conditions were the main factors that permitted a considerable amount of the antifungal compound to remain in the material after the processing. Other authors working with volatile organic compounds intended for active food packaging, such as carvacrol, reported a retention capacity lower than 50% when this substance was incorporated into polypropylene by melt extrusion [42]. Considering thymol (THY), another widely employed volatile compound with a boiling point similar to methyl anthranilate (232 °C THY vs. 256 °C MA) that has demonstrated antibacterial and antioxidant activities, a loss of nearly 30% was found by Ramos et al. when it was directly incorporated into poly(lactic acid) by melt processing [43]. Interestingly, they reported that the amount of thymol after the extrusion process was slightly higher in the formulations containing a nanoclay, indicating that the presence of this additive can retard the compound evaporation during the manufacturing process. A similar behaviour was also observed by the incorporation of Ag nanoparticles in PLA/THY formulations produced by melt compounding [44]. A significantly higher loss of thymol was observed by Galotto and co-workers, who reported a loss of 70% when this volatile compound was incorporated with polyethylene, owing to the high temperatures and to the low chemical compatibility between the two components [45]. Therefore, based on the obtained results, it is noticeable that the employed screw configuration and extrusion conditions may have played a crucial role to consider methyl anthranilate a good antifungal compound for the preparation of active packaging films by the melt blending processes.

Table 1. Film formulations and retention capacity of methyl anthranilate quantified by GC-MS analysis.

Formulation	Code	Methyl Anthranilate (wt.%)	Retention Capacity (%)
Neat EVOH	EVOH	n.d.	-
EVOH + 3 wt.% methyl anthranilate	EVOH-MA3	2.52 ± 0.36	84
EVOH + 5 wt.% methyl anthranilate	EVOH-MA5	3.84 ± 0.20	76
EVOH + 8 wt.% methyl anthranilate	EVOH-MA8	6.42 ± 0.17	80

3.2. SEM Analysis

Figure 3 displays the SEM pictures acquired for the surface and cross section of the film samples. The surface of EVOH control film showed, as expected, a clean and homogeneous surface. Regarding the films containing methyl anthranilate, smooth surfaces could be observed for the three active formulations. However, some pores and imperfections could be detected in certain regions along the examined area. This observation may be attributed by an eventual loss of the active ingredient present in the surface of the material after the manufacturing process [46]. On the other hand, the cross sections of the films subjected to cryofracture, revealed a smooth and non-porous morphology, with no appreciable phase separation in the films containing methyl anthranilate, which indicated a good dispersion and compatibility between the polymer matrix and the natural ingredient. In addition, a film thickness around 50 µm could be confirmed for all formulations by SEM observations.

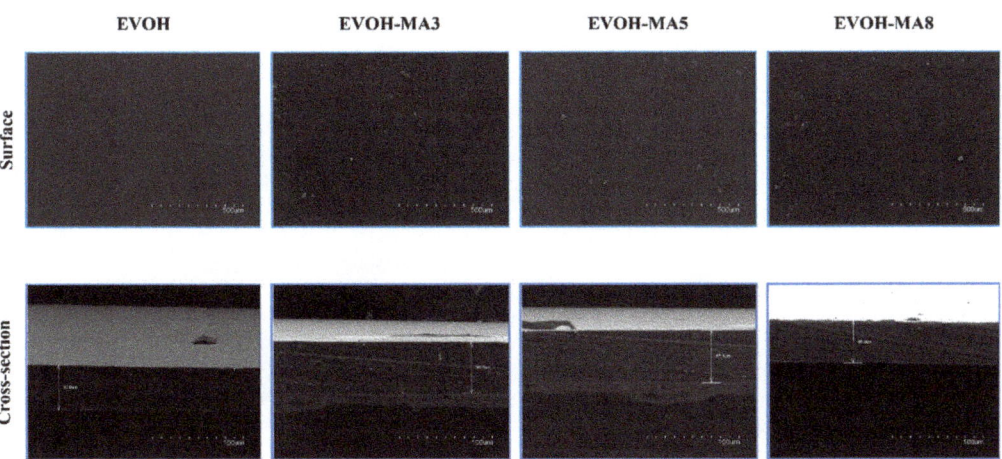

Figure 3. SEM micrographs of sample surfaces and cross sections (×100 and ×500, respectively.).

3.3. Thermal Properties

The DSC thermograms of the different cast-extruded EVOH-based films, corresponding to the heating and cooling stages, are displayed in Figure 4 while the main thermal parameters obtained from the analysis of the DSC curves, are reported in Table 2. In the control EVOH film, it was possible to observe during the first heating scan a second-order transition related to the glass transition temperature of EVOH copolymer, with a value of 52 °C, determined as the midpoint between the onset and the end of the inflectional tangent.

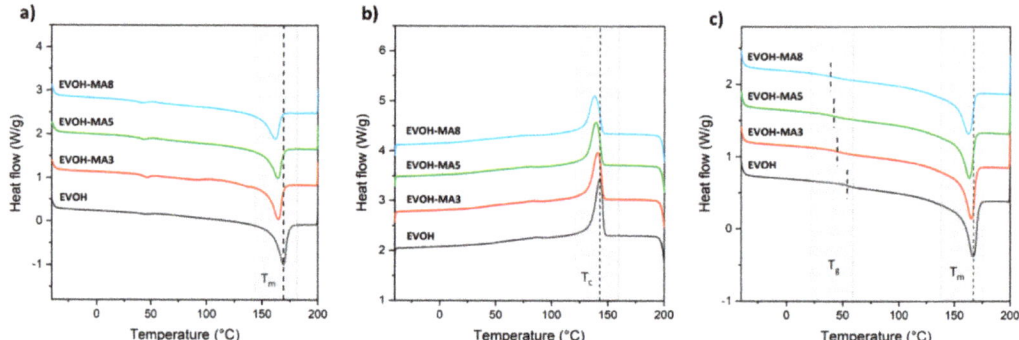

Figure 4. DSC thermograms during (**a**) first heating, (**b**) cooling and (**c**) second heating scan.

Table 2. DSC and TGA parameters.

Formulation	First Heating Scan				Cooling Scan		Second Heating Scan				TGA		
	T_g (°C)	T_m (°C)	ΔH_m (J/g)	χ_c (%)	T_c (°C)	ΔH_c (J/g)	T_g (°C)	T_m (°C)	ΔH_m (J/g)	χ_c (%)	ΔW_{0-200} (%)	T_{maxI} (°C)	T_{maxII} (°C)
EVOH	52	169	68.8	31.5	143	61.8	54	167	66.5	30.5	0.2	399	459
EVOH-MA3	43	165	55.7	26.4	141	56.8	44	165	65.3	30.9	2.25	395	445
EVOH-MA5	40	164	57.4	27.3	139	53.9	42	163	56.9	27.4	4.33	389	452
EVOH-MA8	37	162	50.9	25.4	138	52.9	39	162	52.7	26.3	6.12	392	451

Then, an endothermic process related to the melting of EVOH copolymer was observed at a peak temperature around 162 °C, with a ΔH_m of 68.8 J/g, resulting in a degree of crystallinity of 31.5%. With respect to the bioactive EVOH films in which methyl anthranilate was incorporated, the glass transition temperature values decreased in temperature, ranging between 37 and 43 °C, indicating an additive-induced plasticization over the EVOH copolymer. This phenomenon was more evident with the highest incorporated amount of MA in the films, that is, 8 wt.%, resulting in a decrease in the T_g of 15 °C. The addition of the active ingredient also shifted the melting peak towards lower temperatures with respect to the control EVOH sample. Specifically, the T_m of pristine EVOH was reduced by 4 °C, 5 °C and 7 °C for the 3, 5 and 8 wt.% MA content, respectively. Moreover, the melting enthalpy was seen also to decrease when increasing the MA content, possibly due to alterations in the crystalline structure of EVOH copolymer, and, as a consequence, the degree of crystallinity was reduced from 31.5% to 26.4%, 27.3%, and 25.4% for the film formulations containing 3, 5 and 8 wt.% of MA, respectively. This effect has been observed in previous studies dealing with EVOH copolymer, and was related to interactions occurring between hydroxyl groups present along the copolymer structure and low molecular weight compounds in the amorphous phase [47,48]. Considering the cooling step from the melt (Figure 4b), it was possible to observe, as expected and in good agreement with the heating scan observations, that pristine EVOH presented a crystallization process with a peak temperature centred at 143 °C and a crystallization enthalpy (ΔH_c) of 61.8 J/g. As reported in the literature, depending on both, the content of vinyl alcohol in the copolymer and the cooling rate from the melt state, EVOH copolymers family is able to crystallize in two types of phases: orthorhombic or monoclinic [49,50]. In the case of EVOH$_{44}$, Cerrada et al. concluded that this particular copolymer crystallized, independently of the cooling rate, in the orthorhombic phase [51]. With regard the active films, both, the T_c and ΔH_c, decreased proportionally with the content of MA in the samples. Therefore, it is possible to assume that the presence of the active compound slightly hinders the correct ordering of the copolymer chains, needing higher undercoolings to crystallize and, consequently, shifting the T_c to lower values. During the second heating cycle, the thermal parameters obtained for pristine EVOH were similar to those obtained during the first heating scan. However, the T_g values reported for the active samples were slightly higher when compared to the first heating scan, possibly due to a potential loss of the active ingredient during the different heating and cooling cycles.

The results obtained in the present study are consistent and well supported by those reported in previous works found in literature for compounded EVOH-based materials developed by melt-extrusion and other manufacturing techniques (i.e., solvent casting, electrospinning, etc.) For instance, Luzi et al. reported a decrease in the T_g, T_m and χ_c of EVOH-caffeic acid-based systems prepared by solvent-casting [48]. In another work, Luzi and her co-workers studied the effect of the processing technique, solvent casting and extrusion processing on the properties of EVOH based films incorporated with gallic acid and umbelliferone. They observed a general decrease of the melting temperature, melting enthalpy and degree of crystallinity caused by the presence of the active ingredients [52]. More recently, Meléndez et al. developed electrospun EVOH$_{44}$ fibres with and without cellulose nanocrystals (CNC), and they found lower T_c, T_m and χ_c values in the samples containing CNC compared to the pristine copolymer [9].

With respect to the thermal stability of the samples, EVOH control film presented two major weight loss zones at 300–420 °C and 420–485 °C, corresponding to 90% and 9% mass loss, respectively. The first decomposition step in EVOH film was related to the degradation reactions occurring in the vinyl alcohol component in the copolymer while the second degradation stage, with a maximum peak temperature centred at 459 °C, was associated with the full decomposition of EVOH backbones [53,54]. Films incorporating 3–8 wt.% methyl anthranilate showed a thermal degradation process located in the range of 120–200 °C (see Figure 5), indicating that the thermal volatilization of MA occurs in this temperature interval. As shown in Table 2, the weight loss increment (ΔW_{0-200}) in this range was in good agreement with both the nominal amount of methyl anthranilate incorporated in the films during the cast-extrusion process and the results extracted from the GC-MS analysis. Interestingly, the maximum peak degradation temperatures (T_{maxI} and T_{maxII}) of EVOH copolymer were not significantly affected by the presence of methyl anthranilate, indicating that film formulations maintained a high thermally stable behaviour. Similar results have been observed in previous studies where natural volatile compounds were blended with other thermoplastic polymers [46,55].

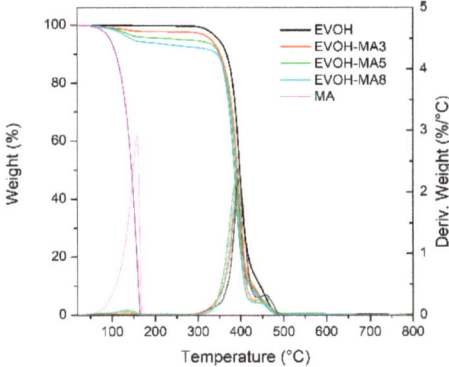

Figure 5. TGA and DTG curves for EVOH-based samples.

3.4. Mechanical Properties

Figure 6 presents the tensile stress-strain curves of the cast-extruded films from which the mechanical properties, i.e., Young Modulus (E), stress at yield (σ_y), stress at break (σ_b) and elongation at break (ε_b) values were calculated. Incorporation of methyl anthranilate slightly decreased the Young Modulus from 2.5 ± 0.2 GPa of EVOH control film to a value of 2.2 ± 0.1 GPa of EVOH film incorporating up to an 8 wt.% of methyl anthranilate, indicating that the bioactive film formulations had less rigid behaviour. After reaching the yield point, it was possible to observe a drop of the stress values, pointing out a necking effect in EVOH-based formulations. The plasticization produced by methyl anthranilate could explain the decreasing tendency in the yield stress from 46 ± 3 MPa in EVOH film to 40 ± 5 MPa in EVOH-MA8 film samples [56]. Beyond the yield point, a strengthening behaviour during the plastic deformation—strain hardening—was detected for all films.

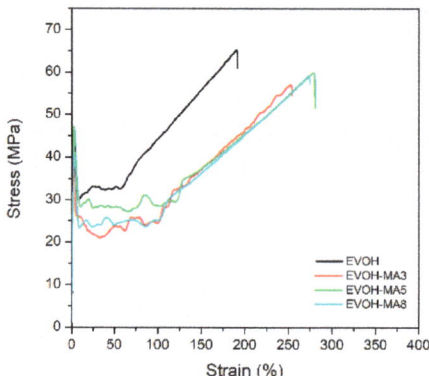

Figure 6. Representative stress-strain curves of EVOH-based films.

This phenomenon, caused by the orientation and alignment of polymer chains, was more remarkable in neat EVOH, owing to its higher degree of crystallinity with respect to bioactive films, as determined in DSC analysis [57]. Consequently, the stress at break value of EVOH film (62 ± 5 MPa) was higher when compared of films incorporating methyl anthranilate, whose values decreased as the content of the active substance increased in the EVOH-based films. Finally, the plasticizing effect of methyl anthranilate over the EVOH matrix resulted in a noticeable increase in the elongation at break values, shifting from 174% in EVOH film to 253% in EVOH-MA8 formulation. A similar behaviour has been reported in previous works in which essential oils or their active constituents were incorporated in thermoplastic polymers by melt processing techniques [58–60]. In this line, the mechanical performances of all film formulations showed their suitability for flexible packaging applications.

3.5. Water Vapour Permeability

The WVP was tested at 33% and 75% relative humidity at room temperature and the values are gathered in Table 3. Regarding the influence of RH on the WVP, all samples showed the same tendency: the permeability was significantly higher for the samples measured at 75% RH, as a consequence of the plasticizing effect of water over the EVOH copolymer chains. The increase in the polymer chain mobility results in a reduction in the cohesive energy in the -OH groups present in EVOH, thus accelerating the molecular diffusion of sorbed water through the film [61]. In particular, the WVP of pure $EVOH_{44}$ shifted from a value of 0.12×10^{-15} kg·m m^{-2} s^{-1} Pa^{-1} at 33% RH to 1.08×10^{-15} kg·m m^{-2} s^{-1} Pa^{-1} at 75%, which are in good concordance with those reported in previous works [9,62]. The incorporation of 3 wt.% of methyl anthranilate slightly reduced the WVP with respect to the control EVOH film at both tested humidity values, possibly due to the low affinity and sensitivity of the active ingredient to water. However, films containing 5 and 8 wt.% of MA seemed to be more water sensitive than the control EVOH film and EVOH-MA3 samples at 33 and 75% RH. This result can be explained as a consequence of the significant plasticizing effect observed in the thermal analysis for the active films and the resulting increase in the amorphous region suitable for water transport in these samples.

Table 3. Mechanical properties and water vapor permeability values of EVOH-based films.

Reference	E (GPa)	Stress @ Yield (MPa)	Stress @ Break (MPa)	Elongation @ Break (%)	WVP·10^{15} (kg·m/m^2 s Pa)	
					33% RH	75% RH
EVOH	2.5 ± 0.2 [a]	46 ± 3 [a]	62 ± 5 [a]	174 ± 16 [a]	0.12 ± 0.01	1.08 ± 0.02
EVOH-MA3	2.4 ± 0.1 [a]	41 ± 4 [a]	54 ± 7 [b]	224 ± 18 [b]	0.11 ± 0.01	0.99 ± 0.04
EVOH-MA5	2.3 ± 0.2 [a]	41 ± 4 [a]	46 ± 10 [b,c]	260 ± 21 [c]	0.15 ± 0.02	1.14 ± 0.02
EVOH-MA8	2.2 ± 0.1 [a]	40 ± 5 [a]	43 ± 8 [c]	253 ± 12 [c]	0.18 ± 0.01	1.38 ± 0.06

[a–c]: different superscript within the same column indicates significant different between film formulations (Turkey test, $p < 0.05$).

3.6. Optical Properties

The visual aspect of packaging films is a key factor since it can influence the consumer perception on a certain end-product. In this line, Table 4 reports the main optical and colour parameters of the cast-extruded active films based on EVOH copolymer incorporated with methyl anthranilate, which all had an average thickness in the range of 50–55 μm. EVOH control film was characterized by a high transparency and with the addition of MA, the film formulations maintained their original appearance. The developed films exhibited very similar L values, 92–93, indicating a high lightness of the colour of the active films. This was attributed to a good dispersion of the active compound in the polymer matrix. Furthermore, the colour parameter a^* of EVOH copolymer film, which indicates the green (negative) or red colour (positive), was not significantly affected by the presence of methyl anthranilate in the active formulations, shifting towards negative values as the content of the active ingredient augmented in the films. On the contrary, a clear tendency to positive values of b^* parameter provoked by the incorporation of methyl anthranilate was observed in the active EVOH-MA films. Therefore, these samples showed a slight yellow tonality that was responsible for the colour change (ΔE) in EVOH-MA formulations, as reported in the Table below. The total colour difference was in the range of 4.1–4.9, which means that the deviation in the optical properties can be noticeable by an unexperienced observer [34]. On the other hand, in terms of transparency, it was observed that the T parameter slightly decreased by increasing the MA content. Thus, EVOH film and EVOH sample with 3 wt.% of MA showed T values of 0.89 and 0.88, respectively, but by increasing the MA content up to 8 wt.%, the transparency value was seen to decrease more evidently (0.77). With all this, it is possible to state that the presence of methyl anthranilate did not significantly affect the great optical properties of EVOH copolymer film, since the colour deviations were moderate and the transparency in the active films remained practically unchanged with respect to the control EVOH film, demonstrating the potential applicability of the developed active films for food packaging.

Table 4. Thickness, colour parameters and transparency of extruded EVOH-based films.

Reference	Thickness (μm)	L	a	b	ΔE	T (A_{600}/t)
EVOH	50 ± 2	93.1	0.04	−0.49	-	0.89 ± 0.01
EVOH-MA3	53 ± 3	92.6 ± 0.1	−0.59 ± 0.02	3.6 ± 0.04	4.1 ± 0.1	0.88 ± 0.01
EVOH-MA5	55 ± 3	92.5 ± 0.1	−0.81 ± 0.01	4.12 ± 0.02	4.7 ± 0.2	0.84 ± 0.01
EVOH-MA8	54 ± 4	92.6 ± 0.1	−0.78 ± 0.05	4.31 ± 0.04	4.9 ± 0.1	0.77 ± 0.02

3.7. Effect of Environmental Humidity on Methyl Anthranilate Release from EVOH-Based Films

One of the main factors that influences the success for developing active packaging materials is related to the trigger for the release of the active from the polymer matrix to achieve a sustained release to the food media. As commented before, EVOH copolymers are characterized by a hydrophilic nature due to their hydroxyl content. Therefore, as the EVOH-based films are exposed to different relative humidity, water molecules begin to penetrate through the polymer structure, leading to a swelling of the matrix, thus enabling the release of methyl anthranilate. As displayed in Figure 7, the % release was greatly influenced by the hydration of the system, with the lowest at 33% RH and the highest at >95% RH. At 33% RH, less than 50% methyl anthranilate was released from the three active formulations. On the other hand, in the highest humidity environment, the release of the active ingredient from the polymer matrix to the head space of the system reached a 65% after 10 days of exposure, associated with the sorption of water in the material and the subsequent hydration of the polymer structure. This result indicated that a significant amount of MA was still present and available to be further released from the active films after ten days of exposure to different RH. This observation was attributed to the fact that EVOH-based films may need longer times to reach equilibrium, as reported by Aucejo and co-workers [63]. In this line, a longer interaction between water and EVOH based films may lead to a greater plasticization and swelling of the polymer matrix and thus to a greater release of methyl anthranilate. In a work conducted by Kurek et al., the effect of relative humidity on carvacrol release from chitosan-based films was intensely studied [40]. They reported that at 0% relative humidity and after 60 days at room temperature, the remaining amount of carvacrol was higher than 45%, while at >96% RH, already after 10 days more than 95% of the carvacrol was released. The results presented herein show that the developed active films may have a great potential to be applied in food products with high water activity and thus where a high relative humidity is generated, permitting the release of methyl anthranilate, and, consequently, inducing the potential antimicrobial effect over the food produce.

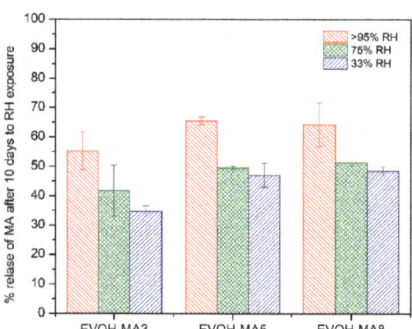

Figure 7. Methyl anthranilate release from EVOH based films after 10 days influenced by relative humidity (33, 75 and >95%).

3.8. Antifungal Activity

3.8.1. Methyl Anthranilate Vapour Inhibition Assay

The antimicrobial activity of methyl anthranilate (MA) against *Penicillium expansum* and *Botrytis cinerea* was evaluated through micro-atmosphere generated into fungal culture plates. The effectiveness of volatile compound expressed as MIC and MFC (μL/plate) are represented in Table 5. Both fungi showed a great inhibition when MA was applied in the plate. *B. cinerea* showed the highest susceptibility to a low amount of MA; thus, 5 μL of active compound exhibited a fungicidal activity. However, a higher amount of compound is required to reach total inhibition of *P. expansum*. Few studies have reported in the literature on the antimicrobial activity of MA. Nidiry and Babu, (2005) evaluated the antifungal effect of some tuberose constituents, including methyl anthranilate, which showed good effectiveness against *Colletotrichum gloeosporioides* [64]. Nevertheless, the antifungal activity was assessed through diluting the compound in the agar culture medium, and not as the vapour phase. Other studies have also reported the antifungal activity of the active ingredient against fungi food pathogens applied as vapour phase, demonstrating the good effectiveness of MA [65,66]. The previous studies and the obtained values for MIC and MFC in the present research (µL/plate) against *P. expansum* and *B. cinerea*, highlight the potential of this compound for its incorporation in active food packaging systems.

Table 5. Antimicrobial effectiveness of methyl anthranilate against *P. expansum* and *B. cinerea* expressed as MIC and MFC (µL/plate) after 10 days of incubation at 26 °C.

	MIC (µL/plate)	MFC (µL/plate)	Control	MIC	MFC
P. expansum	2.5	20			
B. cinerea	1	5			

3.8.2. Antifungal Activity of Cast-Extruded Active EVOH-Based Films

The effectiveness of active films containing methyl anthranilate with 3, 5 and 8 wt.% was evaluated against both fungi by the generation of a micro-atmosphere inside petri dishes. The inhibition of fungal colony was calculated with films containing different amount of MA, and is depicted in Figure 8. As discussed above, EVOH copolymer is a highly hydrophilic material due to its great hydroxyl group content, increasing aroma permeation in presence of high relative humidity. The EVOH-based films in the petri dish were subjected to a high-humidity environment, which allowed MA to be released from the polymer matrix to the plate headspace, allowing it to exert its activity. In good agreement with the obtained data of MIC and MFC, it could be observed that the active films exhibited great effectiveness against *B. cinerea*. Thus, fungicidal activity was observed for films with 5 and 8 wt.% of MA. Moreover, films with 3 wt.% of MA showed an inhibition around 80% during all days of incubation.

On the other hand, EVOH film incorporated with the highest amount of MA (8 wt.%) also resulted in fungicidal activity against *P. expansum*, while this fungi incubated in the presence of EVOH-MA3 and EVOH-MA5 exhibited a great inhibition, and colonies resulted that were white and small, which could be associated with an inhibition of sporulation [67]. The visual aspect of fungal colonies after seven days of incubation is represented in Figure 9, which shows the great effectiveness of the developed active films against both fungi. The antimicrobial activity of thermo-processed films may be compromised by losses occurring during film processing [55,68]. In our study, cast extruded EVOH films incorporated with MA showed great antifungal activity due not only to the effectiveness of MA, but also to the high retention capacity of the active ingredient in the films after the manufacturing process. Thanks to the hydrophilic nature of EVOH copolymer, the compound release from the polymer matrix is mainly due to the exposition of films to high relative humidity since fungal growth medium produces a high humidity atmosphere in plates, which leads to a rapid release of the volatile substance, allowing strong inhibition of microorganisms [5].

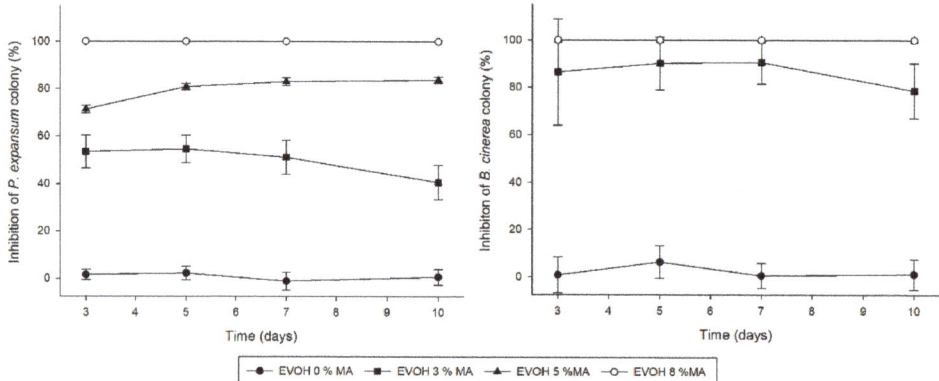

Figure 8. Effectiveness of EVOH incorporated with methyl anthranilate against *P. expansum* and *B. cinerea* expressed as inhibition of fungal colony (%) during 10 days of at 26 °C.

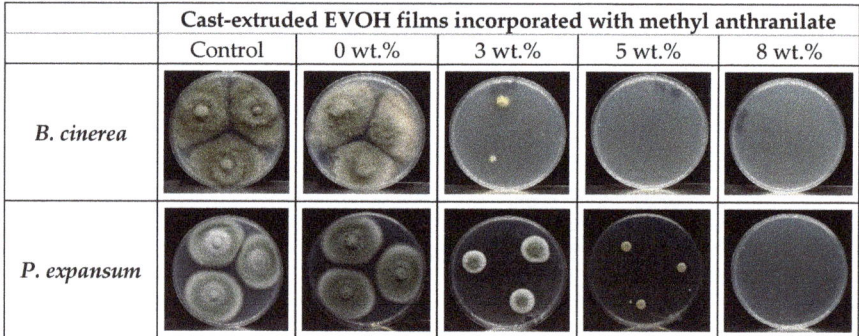

Figure 9. Visual aspect of fungal colony of *P. expansum* and *B. cinerea* co-incubated with cast extruded EVOH films incorporated with methyl anthranilate after 7 days at 26 °C.

Several studies have previously incorporated essential oils in EVOH copolymer matrix aiming to develop antimicrobial polymers with interest in food packaging [5,13,69–71]. Although, to the best of our knowledge, no studies have been reported in which methyl anthranilate is incorporated into a polymer matrix intended for packaging applications, its antifungal activity on fruit pathogens has been previously demonstrated [65,66]. According to these studies and the great fungal inhibition rates of EVOH-based films with MA presented in this work, even with the minor content of the ingredient, the developed films show a great potential to be applied in active food packaging.

4. Conclusions

In the present work, novel active EVOH-based films containing 3 wt.%, 5 wt.%, and 8 wt.% of bioactive methyl anthranilate compound were prepared at pilot scale by melt extrusion processing in a twin-screw extruder with a specifically designed screw configuration, and fully characterized. All active formulations presented a retention capacity of the natural ingredient higher than 75%. Films were optically transparent, but methyl anthranilate provoked a somewhat yellow tonality. DSC analysis indicated a plasticizing effect of MA onto the EVOH matrix, resulting in a decrease in the glass transition temperature of up to 15 °C, while the thermal stability of EVOH-based films evaluated by TGA was not significantly affected by the presence of the antifungal ingredient. The incorporation of different concentrations of methyl anthranilate into EVOH-based films resulted in an increase of the elongation at break values, from 174% in neat EVOH to 253% in EVOH-MA8 formulation, confirming the plasticizing effect of this compound. This behavior explained the slight increase in the water vapor permeability values in the film formulations containing high amounts of methyl anthranilate (5 and 8 wt.%) with respect to the pristine EVOH film. Despite this, all active films showed positive results concerning antifungal activity, exhibiting fungicidal activity against *B. cinerea* and *P. expansum*. The results presented and discussed in this study show the feasibility of the used technique to develop active films with antifungal properties scalable in industrial environments.

Author Contributions: Conceptualization, R.G. and P.H.-M.; methodology, A.A.-G., R.H.-M. and A.M.; formal analysis, A.A.-G., R.H.-M., A.M., D.L., R.G. and P.H.-M.; investigation, A.A.-G., R.H.-M., A.M., M.G., D.L., P.H.-M. and R.G.; resources, M.G., D.L., R.G. and P.H.-M.; data curation, A.A.-G., R.H.-M. and P.H.-M.; writing—original draft preparation, A.A.-G. and R.H.-M.; writing—review and editing, A.A.-G., R.H.-M., M.G., D.L., R.G. and P.H.-M.; supervision, M.G., D.L., R.G. and P.H.-M.; project administration, M.G., R.G. and P.H.-M.; funding acquisition, M.G. and R.G. All authors have read and agreed to the published version of the manuscript.

Funding: The authors wish to thank Generalitat Valenciana (Grant IMAMCM/2021/1) and the Spanish Ministry of Science and Innovation (Grants RTI2018-093452-B-I00 and BES-2016-077380 funded by MCIN/AEI/10.13039/501100011033 and by ERDF A way of making Europe) for the financial support.

Institutional Review Board Statement: Not applicable.

Informed Consent Statement: Not applicable.

Data Availability Statement: The data presented in this study are available on request from the corresponding author.

Conflicts of Interest: The authors declare no conflict of interest.

References

1. FAO. Definitional Framework of Food Loss—Save Food: Global Initiative on Food Loss and Waste Reduction. In *Definitional Framework of Food Loss Working Papers*; FAO: Rome, Italy, 2014; Volume 18.
2. Han, J.-W.; Ruiz-Garcia, L.; Qian, J.-P.; Yang, X.-T. Food Packaging: A Comprehensive Review and Future Trends. *Compr. Rev. Food Sci. Food Saf.* **2018**, *17*, 860–877. [CrossRef] [PubMed]
3. Gavara, R.; López-Carballo, G.; Hernández-Muñoz, P.; Catalá, R.; Muriel-Galet, V.; Cerisuelo, J.P.; Dominguez, I. *Practical Guide to Antimicrobial Active Packaging*; iSmithers Rapra Publishing: Surrey, UK, 2015.
4. Manfredi, M.; Fantin, V.; Vignali, G.; Gavara, R. Environmental assessment of antimicrobial coatings for packaged fresh milk. *J. Clean. Prod.* **2015**, *95*, 291–300. [CrossRef]
5. Muriel-Galet, V.; Cerisuelo, J.P.; Lopez-Carballo, G.; Lara, M.; Gavara, R.; Hernandez-Munoz, P. Development of antimicrobial films for microbiological control of packaged salad. *Int. J. Food Microbiol.* **2012**, *157*, 195–201. [CrossRef] [PubMed]
6. Heras-Mozos, R.; Gavara, R.; Hernández-Muñoz, P. Development of antifungal biopolymers based on dynamic imines as responsive release systems for the postharvest preservation of blackberry fruit. *Food Chem.* **2021**, *357*, 129838. [CrossRef]
7. Ettinger, D.J. Active and Intelligent Packaging: A U.S. and EU Perspective. Available online: https://www.packaginglaw.com/special-focus/active-and-intelligent-packaging-us-and-eu-perspective (accessed on 17 June 2002).
8. Gavara, R.; Catalá, R.; López Carballo, G.; Cerisuelo, J.P.; Dominguez, I.; Muriel-Galet, V.; Hernandez-Muñoz, P. *Use of EVOH for Food Packaging Applications*; Elsevier: Amsterdam, The Netherlands, 2016.
9. Melendez-Rodriguez, B.; Torres-Giner, S.; Zavagna, L.; Sammon, C.; Cabedo, L.; Prieto, C.; Lagaron, J.M. Development and Characterization of Electrospun Fiber-Based Poly(ethylene-co-vinyl Alcohol) Films of Application Interest as High-Gas-Barrier Interlayers in Food Packaging. *Polymers* **2021**, *13*, 2061. [CrossRef] [PubMed]
10. Kaiser, K.M.A.; Ewender, J.; Welle, F. Recyclable multilayer packaging by means of thermoreversibly crosslinking adhesive in the context of food law. *Polymers* **2020**, *12*, 2988. [CrossRef]
11. Arboleda, C.E.; Mejía, A.I.G.; López, B.L.O. Poly (vinylalcohol-co-ethylene) biodegradation on semi solid fermentation by *Phanerochaete chrysosporium*. *Acta Farm. Bonaer.* **2004**, *23*, 123–128.
12. Mejía, G.A.I.; López, O.B.L.; Sierra, L. Biodegradation of poly(vinylcohol-co-ethylene) with the fungus *phanerochaete chrysosporium*. *Mater. Res. Innov.* **2001**, *4*, 148–154. [CrossRef]
13. Muriel-Galet, V.; Cran, M.J.; Bigger, S.W.; Hernández-Muñoz, P.; Gavara, R. Antioxidant and antimicrobial properties of ethylene vinyl alcohol copolymer films based on the release of oregano essential oil and green tea extract components. *J. Food Eng.* **2015**, *149*, 9–16. [CrossRef]

14. Carpena, M.; Nuñez-Estevez, B.; Soria-Lopez, A.; Garcia-Oliveira, P.; Prieto, M.A. Essential Oils and Their Application on Active Packaging Systems: A Review. *Resources* **2021**, *10*, 7. [CrossRef]
15. Chen, H.; Li, L.; Ma, Y.; Mcdonald, T.P.; Wang, Y. Development of active packaging film containing bioactive components encapsulated in β-cyclodextrin and its application. *Food Hydrocoll.* **2019**, *90*, 360–366. [CrossRef]
16. Wang, J.; De Luca, V. The biosynthesis and regulation of biosynthesis of Concord grape fruit esters, including "foxy" methylanthranilate. *Plant J.* **2005**, *44*, 606–619. [CrossRef] [PubMed]
17. Yadav, G.D.; Krishnan, M.S. An Ecofriendly Catalytic Route for the Preparation of Perfumery Grade Methyl Anthranilate from Anthranilic Acid and Methanol. *Org. Process Res. Dev.* **1998**, *2*, 86–95. [CrossRef]
18. Luo, Z.W.; Cho, J.S.; Lee, S.Y. Microbial production of methyl anthranilate, a grape flavor compound. *Proc. Natl. Acad. Sci. USA* **2019**, *116*, 201903875. [CrossRef] [PubMed]
19. Solaberrieta, I.; Jiménez, A.; Cacciotti, I.; Garrigós, M.C. Encapsulation of Bioactive Compounds from Aloe Vera Agrowastes in Electrospun Poly (Ethylene Oxide) Nanofibers. *Polymers* **2020**, *12*, 1323. [CrossRef] [PubMed]
20. Rehman, A.; Tong, Q.; Jafari, S.M.; Korma, S.A.; Khan, I.M.; Mohsin, A.; Manzoor, M.F.; Ashraf, W.; Mushtaq, B.S.; Zainab, S.; et al. Spray dried nanoemulsions loaded with curcumin, resveratrol, and borage seed oil: The role of two different modified starches as encapsulating materials. *Int. J. Biol. Macromol.* **2021**, *186*, 820–828. [CrossRef]
21. Pansuwan, J.; Chaiyasat, A. Innovative and high performance synthesis of microcapsules containing methyl anthranilate by microsuspension iodine transfer polymerization. *Polym. Int.* **2017**, *66*, 1921–1927. [CrossRef]
22. Buendía-Moreno, L.; Soto-Jover, S.; Ros-Chumillas, M.; Antolinos, V.; Navarro-Segura, L.; Sánchez-Martínez, M.J.; Martínez-Hernández, G.B.; López-Gómez, A. Innovative cardboard active packaging with a coating including encapsulated essential oils to extend cherry tomato shelf life. *LWT* **2019**, *116*, 108584. [CrossRef]
23. Wen, P.; Zhu, D.-H.; Wu, H.; Zong, M.-H.; Jing, Y.-R.; Han, S.-Y. Encapsulation of cinnamon essential oil in electrospun nanofibrous film for active food packaging. *Food Control* **2016**, *59*, 366–376. [CrossRef]
24. Mohammadi, A.; Jafari, S.M.; Esfanjani, A.F.; Akhavan, S. Application of nano-encapsulated olive leaf extract in controlling the oxidative stability of soybean oil. *Food Chem.* **2016**, *190*, 513–519. [CrossRef]
25. Trucillo, P.; Campardelli, R.; Reverchon, E. Production of liposomes loaded with antioxidants using a supercritical CO_2 assisted process. *Powder Technol.* **2018**, *323*, 155–162. [CrossRef]
26. Chaudhary, V.; Thakur, N.; Kajla, P.; Thakur, S.; Punia, S. Application of Encapsulation Technology in Edible Films: Carrier of Bioactive Compounds. *Front. Sustain. Food Syst.* **2021**, *5*, 374. [CrossRef]
27. Antosik, A.K.; Kowalska, U.; Stobińska, M.; Dzięcioł, P.; Pieczykolan, M.; Kozłowska, K.; Bartkowiak, A. Development and characterization of bioactive polypropylene films for food packaging applications. *Polymers* **2021**, *13*, 3478. [CrossRef] [PubMed]
28. Laorenza, Y.; Harnkarnsujarit, N. Carvacrol, citral and α-terpineol essential oil incorporated biodegradable films for functional active packaging of Pacific white shrimp. *Food Chem.* **2021**, *363*, 130252. [CrossRef]
29. Nobile, M.A.; Conte, A.; Buonocore, G.G.; Anna Lucia, I.; Massaro, A.; Panza, O. Active packaging by extrusión processing of reciclable and biodegradable polymer. *J. Food Eng.* **2009**, *93*, 1–6. [CrossRef]
30. Nam, S.; Scanlon, M.G.; Han, J.; Izydorczyk, M. Extrusion of Pea Starch Containing Lysozyme and Determination of Antimicrobial Activity. *J. Food Sci.* **2007**, *72*, E477–E484. [CrossRef]
31. Aragón-Gutiérrez, A.; Heras-Mozos, R.; Gallur, M.; López, D.; Hernández-Muñoz, P. Hot-Melt-Extruded Active Films Prepared from Packaging Applications. *Foods* **2021**, *10*, 1591. [CrossRef]
32. Beltrán Sanahuja, A.; Valdés García, A. New Trends in the Use of Volatile Compounds in Food Packaging. *Polymers* **2021**, *13*, 1053. [CrossRef]
33. Faisant, J.B.; Aït-Kadi, A.; Bousmina, M.; Deschênes, L. Morphology, thermomechanical and barrier properties of polypropylene-ethylene vinyl alcohol blends. *Polymer* **1998**, *39*, 533–545. [CrossRef]
34. Aragón-Gutiérrez, A.; Rosa, E.; Gallur, M.; López, D.; Hernández-Muñoz, P.; Gavara, R. Melt-Processed Bioactive EVOH Films Incorporated with Ferulic Acid. *Polymers* **2021**, *13*, 68. [CrossRef]
35. ISO 527:2012; lastics—Determination of Tensile Properties. ISO: Geneva, Switzerland, 2006.
36. Hernández-García, E.; Vargas, M.; Chiralt, A. Starch-polyester bilayer films with phenolic acids for pork meat preservation. *Food Chem.* **2022**, *385*, 132650. [CrossRef] [PubMed]
37. Molinaro, S.; Cruz Romero, M.; Boaro, M.; Sensidoni, A.; Lagazio, C.; Morris, M.; Kerry, J. Effect of nanoclay-type and PLA optical purity on the characteristics of PLA-based nanocomposite films. *J. Food Eng.* **2013**, *117*, 113–123. [CrossRef]
38. Figura, L.O.; Teixeira, A.A. (Eds.) Permeability. In *BT—Food Physics: Physical Properties—Measurement and Applications*; Springer: Berlin/Heidelberg, Germany, 2007; pp. 233–255.
39. Ramos, M.; Fortunati, E.; Peltzer, M.; Jimenez, A.; Kenny, J.; Garrigós, M. Characterization and disintegrability under composting conditions of PLA-based nanocomposite films with thymol and silver nanoparticles. *Polym. Degrad. Stab.* **2016**, *132*, 2–10. [CrossRef]
40. Kurek, M.; Guinault, A.; Voilley, A.; Galić, K.; Debeaufort, F. Effect of relative humidity on carvacrol release and permeation properties of chitosan based films and coatings. *Food Chem.* **2014**, *144*, 9–17. [CrossRef] [PubMed]
41. Khoomsab, R.; Rodkate, N. Establishment of Methyl Anthranilate Quantification by UV-Spectroscopy and Identification Using FTIR. *EAU Herit. J. Sci. Technol.* **2019**, *13*, 94–105.

42. Ramos, M.; Beltrán, A.; Peltzer, M.; Valente, A.J.M.; del Carmen Garrigós, M. Release and antioxidant activity of carvacrol and thymol from polypropylene active packaging films. *LWT Food Sci. Technol.* **2014**, *58*, 470–477. [CrossRef]
43. Ramos, M.; Jiménez, A.; Peltzer, M.; Garrigós, M.C. Development of novel nano-biocomposite antioxidant films based on poly (lactic acid) and thymol for active packaging. *Food Chem.* **2014**, *162*, 149–155. [CrossRef]
44. Ramos, M.; Beltran, A.; Fortunati, E.; Peltzer, M.A.; Cristofaro, F.; Visai, L.; Valente, A.J.M.; Jiménez, A.; Kenny, J.M.; Garrigós, M.C. Controlled release of thymol from poly(Lactic acid)-based silver nanocomposite films with antibacterial and antioxidant activity. *Antioxidants* **2020**, *9*, 395. [CrossRef]
45. Galotto, M.J.; Valenzuela, X.; Rodriguez, F.; Bruna, J.; Guarda, A. Evaluation of the effectiveness of a new antimicrobial active packaging for fresh atlantic salmon (*Salmo Salar* L.) shelf life. *Packag. Technol. Sci.* **2012**, *25*, 363–372. [CrossRef]
46. Ramos, M.; Jiménez, A.; Peltzer, M.; Garrigós, M.C. Characterization and antimicrobial activity studies of polypropylene films with carvacrol and thymol for active packaging. *J. Food Eng.* **2012**, *109*, 513–519. [CrossRef]
47. Vannini, M.; Marchese, P.; Celli, A.; Lorenzetti, C. Strategy To Modify the Crystallization Behavior of EVOH32 through Interactions with Low-Molecular-Weight Molecules. *Ind. Eng. Chem. Res.* **2016**, *55*, 3517–3524. [CrossRef]
48. Luzi, F.; Torre, L.; Puglia, D. Antioxidant Packaging Films Based on Ethylene Vinyl Alcohol Copolymer (EVOH) and Caffeic Acid. *Molecules* **2020**, *25*, 3953. [CrossRef] [PubMed]
49. Duraccio, D.; Mauriello, A.; Cimmino, S.; Silvestre, C.; De Rosa, C.; Pirozzi, B.; Mitchell, G. Structure-property relationships in polyethylene based films obtained by blow molding as model system of industrial relevance. *Eur. Polym. J.* **2014**, *62*, 97–107. [CrossRef]
50. Bunn, C.W. Crystal Structure of Polyvinyl Alcohol. *Nature* **1948**, *161*, 929–930. [CrossRef]
51. Cerrada, M.L.; Pérez, E.; Pereña, J.M.; Benavente, R. Wide-angle X-ray diffraction study of the phase behavior of vinyl alcohol-ethylene copolymers. *Macromolecules* **1998**, *31*, 2559–2564. [CrossRef]
52. Luzi, F.; Puglia, D.; Dominici, F.; Fortunati, E.; Giovanale, G.; Balestra, G.M.; Torre, L. Effect of gallic acid and umbelliferone on thermal, mechanical, antioxidant and antimicrobial properties of poly (vinyl alcohol-co-ethylene) films. *Polym. Degrad. Stab.* **2018**, *152*, 162–176. [CrossRef]
53. Barrera, J.; Rodríguez, J.; Perilla, J.; Algecira, N. A study of poly(vinyl alcohol) thermal degradation by thermogravimetry and differential thermogravimetry. *Ing. Investig.* **2007**, *27*, 100–105.
54. Alvarez, V.A.; Ruseckaite, R.A.; Vázquez, A. Kinetic analysis of thermal degradation in poly(ethylene–vinyl alcohol) copolymers. *J. Appl. Polym. Sci.* **2003**, *90*, 3157–3163. [CrossRef]
55. Krepker, M.; Zhang, C.; Nitzan, N.; Prinz-Setter, O.; Massad-Ivanir, N.; Olah, A.; Baer, E.; Segal, E. Antimicrobial LDPE/EVOH Layered Films Containing Carvacrol Fabricated by Multiplication Extrusion. *Polymers* **2018**, *10*, 864. [CrossRef]
56. Kendall, M.; Siviour, C. Rate dependence of poly(vinyl chloride), the effects of plasticizer and time-temperature superposition. *Proc. R. Soc. A Math. Phys. Eng. Sci.* **2014**, *470*, 20140012. [CrossRef]
57. Gaucher-Miri, V.; Jones, G.K.; Kaas, R.; Hiltner, A.; Baer, E. Plastic deformation of EVA, EVOH and their multilayers. *J. Mater. Sci.* **2002**, *37*, 2635–2644. [CrossRef]
58. Llana-Ruiz-Cabello, M.; Pichardo, S.; Bermudez, J.M.; Baños, A.; Ariza, J.J.; Guillamón, E.; Aucejo, S.; Cameán, A.M. Characterisation and antimicrobial activity of active polypropylene films containing oregano essential oil and *Allium* extract to be used in packaging for meat products. *Food Addit. Contam. Part A* **2018**, *35*, 782–791. [CrossRef] [PubMed]
59. Llana-Ruiz-Cabello, M.; Pichardo, S.; Bermúdez, J.M.; Baños, A.; Núñez, C.; Guillamón, E.; Aucejo, S.; Cameán, A.M. Development of PLA films containing oregano essential oil (*Origanum vulgare* L. virens) intended for use in food packaging. *Food Addit. Contam.* **2016**, *33*, 1374–1386. [CrossRef]
60. Qin, Y.; Li, W.; Liu, D.; Yuan, M.; Li, L. Development of active packaging film made from poly (lactic acid) incorporated essential oil. *Prog. Org. Coat.* **2017**, *103*, 76–82. [CrossRef]
61. López De Dicastillo, C.; Nerín, C.; Alfaro, P.; Catalá, R.; Gavara, R.; Hernández-Muñoz, P. Development of new antioxidant active packaging films based on ethylene vinyl alcohol copolymer (EVOH) and green tea extract. *J. Agric. Food Chem.* **2011**, *59*, 7832–7840. [CrossRef] [PubMed]
62. López-de-Dicastillo, C.; Gómez-Estaca, J.; Catalá, R.; Gavara, R.; Hernández-Muñoz, P. Active antioxidant packaging films: Development and effect on lipid stability of brined sardines. *Food Chem.* **2012**, *131*, 1376–1384. [CrossRef]
63. Aucejo, S.; Marco, C.; Gavara, R. Water effect on the morphology of EVOH copolymers. *J. Appl. Polym. Sci.* **1999**, *74*, 1201–1206. [CrossRef]
64. Nidiry, E.S.J.; Babu, C.S.B. Antifungal activity of tuberose absolute and some of its constituents. *Phytother. Res.* **2005**, *19*, 447–449. [CrossRef]
65. Xing, M.; Zheng, L.; Deng, Y.; Xu, D.; Xi, P.; Li, M.; Kong, G.; Jiang, Z. Antifungal Activity of Natural Volatile Organic Compounds against Litchi Downy Blight Pathogen *Peronophythora litchii*. *Molecules* **2018**, *23*, 358. [CrossRef]
66. Chambers, A.H.; Evans, S.A.; Folta, K.M. Methyl Anthranilate and γ-Decalactone Inhibit Strawberry Pathogen Growth and Achene Germination. *J. Agric. Food Chem.* **2013**, *61*, 12625–12633. [CrossRef]
67. Ma, W.; Zhao, L.; Zhao, W.; Xie, Y. (E)-2-Hexenal, as a Potential Natural Antifungal Compound, Inhibits *Aspergillus flavus* Spore Germination by Disrupting Mitochondrial Energy Metabolism. *J. Agric. Food Chem.* **2019**, *67*, 1138–1145. [CrossRef] [PubMed]

68. González-Estrada, R.R.; Calderón-Santoyo, M.; Ragazzo-Sánchez, J.A.; Peyron, S.; Chalier, P. Antimicrobial soy protein isolate-based films: Physical characterisation, active agent retention and antifungal properties against *Penicillium italicum*. *Int. J. Food Sci. Technol.* **2018**, *53*, 921–929. [CrossRef]
69. Yang, H.; Wang, J.; Yang, F.; Chen, M.; Zhou, D.; Li, L. Active Packaging Films from Ethylene Vinyl Alcohol Copolymer and Clove Essential Oil as Shelf Life Extenders for Grass Carp Slice. *Packag. Technol. Sci.* **2016**, *29*, 383–396. [CrossRef]
70. Mateo, E.M.; Gómez, J.V.; Domínguez, I.; Gimeno-Adelantado, J.V.; Mateo-Castro, R.; Gavara, R.; Jiménez, M. Impact of bioactive packaging systems based on EVOH films and essential oils in the control of aflatoxigenic fungi and aflatoxin production in maize. *Int. J. Food Microbiol.* **2017**, *254*, 36–46. [CrossRef]
71. Heras-Mozos, R.; Muriel-Galet, V.; López-Carballo, G.; Catalá, R.; Hernández-Muñoz, P.; Gavara, R. Active EVOH/PE bag for sliced pan loaf based on garlic as antifungal agent and bread aroma as aroma corrector. *Food Packag. Shelf Life* **2018**, *18*, 125–130. [CrossRef]

Article

Assessment of Thermochromic Packaging Prints' Resistance to UV Radiation and Various Chemical Agents

Sonja Jamnicki Hanzer *, Rahela Kulčar *, Marina Vukoje and Ana Marošević Dolovski

Faculty of Graphic Arts, University of Zagreb, 10000 Zagreb, Croatia
* Correspondence: sonja.jamnicki.hanzer@grf.unizg.hr (S.J.H.); rahela.kulcar@grf.unizg.hr (R.K.)

Abstract: Thermochromic inks, also known as color changing inks, are becoming increasingly important for various applications that range from smart packaging, product labels, security printing, and anti-counterfeit inks to applications such as temperature-sensitive plastics and inks printed onto ceramic mugs, promotional items, and toys. These inks are also gaining more attention as part of textile decorations and can also be found in some artistic works obtained with thermochromic paints, due to their ability to change color when exposed to heat. Thermochromic inks, however, are known to be sensitive materials to the influence of UV radiation, heat fluctuations, and various chemical agents. Given the fact that prints can be found in different environmental conditions during their lifetime, in this work, thermochromic prints were exposed to the action of UV radiation and the influence of different chemical agents in order to simulate different environmental parameters. Hence, two thermochromic inks with different activation temperatures (one being cold and the other being body-heat activated), printed on two food packaging label papers that differ in their surface properties were chosen to be tested. Assessment of their resistance to specific chemical agents was performed according to the procedure described in the ISO 2836:2021 standard. Moreover, the prints were exposed to artificial aging to determine their durability when exposed to UV radiation. All tested thermochromic prints showed low resistance to liquid chemical agents as the color difference values were unacceptable in all cases. It was observed that the stability of thermochromic prints to different chemicals decreases with decreasing solvent polarity. Based on the results obtained after UV radiation, its influence in terms of color degradation is visible on both tested paper substrates, but more significant degradation was observed on the ultra-smooth label paper.

Keywords: intelligent packaging; thermochromic inks; chemical resistance; UV radiation; color degradation

Citation: Jamnicki Hanzer, S.; Kulčar, R.; Vukoje, M.; Marošević Dolovski, A. Assessment of Thermochromic Packaging Prints' Resistance to UV Radiation and Various Chemical Agents. Polymers 2023, 15, 1208. https://doi.org/10.3390/polym15051208

Academic Editor: Swarup Roy

Received: 11 January 2023
Revised: 23 February 2023
Accepted: 24 February 2023
Published: 27 February 2023

Copyright: © 2023 by the authors. Licensee MDPI, Basel, Switzerland. This article is an open access article distributed under the terms and conditions of the Creative Commons Attribution (CC BY) license (https://creativecommons.org/licenses/by/4.0/).

1. Introduction

Today, thermochromic printing inks are used in various fields from the food industry to industrial design. They are printed on the packaging or applied on a label so that the consumer can determine by the color of the ink whether the product is ready for consumption. Active and intelligent packaging has become very popular and in demand in the modern packaging industry. They enable greater added-value, increased security, and additional opportunities for marketing promotion [1–4]. Moreover, thermochromic inks are increasingly interesting to many artists who use these types of inks in their creative expressions [5–13]. Thermochromic inks are part of a group of chromogenic inks that have the characteristic of changing their color due to some external influence. Leuco-dye-based thermochromic (TC) inks change their color due to temperature changes. One of the most important elements of smart packaging is a temperature sensitivity indicator that shows the current temperature of the product. These TC indicators are usually simple in design, and with their dynamic color change, add a functional role to the packaging.

TC color change can go from colorful to colorless, colorless to colorful, or they can change from one color to another. Also, TC colorants can be successfully mixed with conventional inks to obtain a wider range of colors [14–20]. This can expand creative possibilities for designers and artists on varied materials and applications. TC dynamic color change can be reversible or irreversible. The reversible color change can last for 10s of years if the print is protected from direct sunlight, extremely hot temperatures, and aggressive solvents. In recent years, more research has been carried out on TC inks to determine the factors that negatively affect the TC color change effect and to find solutions to preserve it if possible. Each TC ink has its own fixed activation temperature (T_A) at which the color change starts, and today thermochromic inks can be obtained at many different activation temperatures [21]. TC inks based on leuco dyes generally consist of an encapsulated three-part system including a dye that changes color, a color developer, and a solvent [22]. At temperatures below the T_A, the solvent is in a solid state enabling the dye and color developer to form a color which results in a full color effect. When the temperature reaches the T_A, the solvent becomes liquid, keeping color developer and the leuco dye apart making the ink appear transparent or translucent [23–27]. The composition of TC inks includes a mixture of TC microcapsules dispersed in the ink's vehicle [24,28–30]. The main disadvantage of this microencapsulated TC system is its poor stability to some external influences [31,32]. Reversible discoloration can result in ruptures of microcapsules containing dye [33]. How to protect this system, which is responsible for the dynamics of color change, is a broader subject of our research. Some examples include choosing a suitable printing substrate or the use of nanomodified coatings [34–36].

In order to preserve the dynamics of color change and functionality of thermochromic inks as indicators for packaging applications, this paper will investigate the factors that could negatively affect their durability during their life cycle with the aim of maintaining the functional role of TC inks as an additional value on packaging. Thermochromic prints on packaging (labels) could be exposed to the effect of different chemical agents (ethanol in alcoholic products, citric acid in juices, vegetable oils, water, etc. . . .) during the use of the products on which they are applied as indicators, which can affect their functionality. In our previous research, we found that exposure of TC prints to specific liquid agents, such as citric acid and ethanol caused severe damage to the prints which also led to the bleeding of the colorants from the prints [37]. In addition to different chemical agents, the packaging can be exposed to different storage and transportation conditions. Moreover, during their shelf life, packaging can be exposed to UV radiation from light sources in the supermarket as well. All these conditions can negatively affect the packaging properties, in this case the properties of thermochromic indicators [38]. A variety of different complex chemical reactions occur during the natural or artificial degradation of paper/cardboard and prints due to different environmental and storage conditions, especially UV radiation, temperature, and humidity, which can lead to different deteriorations of paper-based packaging materials. In the last 10 years, several studies have been published that dealt with low light fastness issues of TC prints [31,33,39–41], but there is still a lack of research related to the resistance of TC prints to various chemical agents as evidenced by the relatively small number of available scientific papers on the subject.

This research aims to analyze the resistance of UV-cured thermochromic prints to specific liquid chemical agents and UV radiation. For that reason, the prints obtained on two different paper substrates were subjected to artificial aging and were also treated with three liquid agents to prove the prints' resistance to liquids and chemicals. Chemical resistance of prints was observed based on the optical deterioration of prints after treatments with liquid agents compared against untreated prints and by determination of bleeding of the colorants into the receptor surface. Prints' resistance to UV radiation was evaluated by observing the discoloration and any change in the prints.

2. Materials and Methods

2.1. Printing Substrates

As printing substrates, two food packaging label papers of similar basis weight and ISO brightness (R_{457}) values were selected with properties given by the manufacturer, presented in Table 1.

Table 1. Properties of used printing substrates.

Paper Substrate	Abbreviation	Property Unit Method	Basis Weight (g/m^2) ISO 536	Caliper (μm) ISO 534	ISO Brightness (%) ISO 2470, R457	Bekk Smoothness (s) ISO 5627
Niklakett Premium Fashion	NPF		75	74	91	74
Chromolux 700	CHR		80	84	89.3	300

The Niklakett Premium Fashion—NPF (Brigl & Bergmeister GmbH, Niklasdorf, Austria) substrate was a high gloss embossed label paper with a textured surface and about four times lower smoothness compared to a Chromolux 700—CHR (Zanders Paper GmbH, Bergisch Gladbach, Germany) paper characterized by ultra-smooth high gloss surface. Both papers were suitable for all common printing processes. Moreover, Chromolux 700 was approved for direct food contact. Also, both papers had a functional coating on the reverse side as they were designed to be used as labels for beverages.

2.2. Printing Inks

Two thermochromic commercially available UV screen inks were chosen for printing. These inks were colored below a specific temperature and changed to another color as they were heated. One ink was colored in orange below its activation temperature (T_A = 12 °C) and changed to yellow above it (hereinafter OY-12). Due to the low activation temperature, this ink is also called low temperature (or cold activated ink) commonly used for print media to be chilled in the refrigerator or for other cold reactions. Application examples include cold-temperature indicators, cold packaging, and other low temperature applications. Another ink was colored in purple below its activation temperature (T_A = 31 °C) and changed to pink above it (hereinafter PP-31). This ink was the type of body-heat activated ink that is designed to show color at normal room temperature and to change color when warmed up. This can be done by rubbing a finger or breathing warm air on the application. Application examples of this type of ink include media packaging, direct mail pieces, stickers, and labels where a unique interactive consumer experience is desired. Both inks were reversible inks meaning that the original color was restored upon cooling (Figures 1 and 2). These inks were made by mixing two types of colorants: thermochromic leuco dyes encapsulated inside the microcapsule (TC microcapsules) and conventional pigments [30,42]. We can assume that orange (OY-12) ink was made by the addition of red leuco dyes to conventional yellow pigments, and purple (PP-31) ink was made by a combination of conventional pink pigments with blue leuco dyes.

Figure 1. Visual presentation of thermochromic prints obtained on different paper substrates of cold activated OY-12 ink under (8 °C) and above (20 °C) ink's T_A.

Figure 2. Visual presentation of thermochromic prints obtained on different paper substrates of body-heat activated PP-31 ink under (20 °C) and above (45 °C) ink's T_A.

The manufacturer states that these inks are sensitive to adverse environmental conditions. Inks should not be exposed to UV and some fluorescent lights for a long time, nor to direct sunlight for more than a few days, as this can degrade color intensity and color-changing characteristics of the ink. The manufacturer does not recommend exposing these inks to elevated temperatures for extended periods of time, as prolonged exposure to 38 °C (100 °F) or higher, can degrade the color change and intensity of the product. In addition, these inks are sensitive to certain chemicals, that is, wet ink should not come into contact with any solvents.

2.3. Printing Trials

Printing was conducted in laboratory conditions by a semi-automatic screen-printing device Siebdruckgeräte von Holzschuher KG., Wuppertal, using a mesh of 62–64 lin/cm. The prints were printed in full tone. After printing, prints were cured using a Technigraf Aktiprint L 10-1 UV dryer (30 W/cm). The characteristics of inks given by the producer are presented in Table 2.

Table 2. Characteristics of used thermochromic UV screen printing inks.

Property	Value
Viscosity at 25 °C	65–110 poise
Density (Approx.)	8.0 lb./gal
Appearance	Viscous Liquid
Percent Solids (Approx.)	99%
Percent Volatiles (Approx.)	<1.5%

2.4. Assessment of Prints' Resistance to Specific Chemical Agents

Assessment of prints' resistance to specific chemical agents was done following the standard method ISO 2836:2021 [43]. International Standard ISO 2836:2021 in the field of the printing industry defines methods of assessing the resistance of prints to liquid and solid agents, solvents, varnishes, and acids [43]. For this study, thermochromic prints were exposed to distilled water, citric acid, and ethanol, which were chosen as test agents to simulate the exposure of smart beverage labels to water, alcohol, and juice of citrus fruits. For test procedures where water and citric acid were used as liquid agents, the prints were cut to dimensions of approx. 2 cm × 5 cm. For the determination of the resistance to citric acid, the printed test piece was brought into contact with filter papers previously saturated with 5% citric acid and placed under a 1 kg load for 1 h at ambient temperature (23 ± 2 °C). After that, the print was rinsed in deionized water and was left to dry in the oven at 50 °C for 30 min. The strips of filter paper used for the test were left to dry in free air. For assessment of the prints' resistance to water, the printed test piece was brought into contact with filter papers previously saturated with distilled water and was placed under a 1 kg load for 24 h at ambient temperature (23 ± 2 °C). After the treatment, the laboratory prints were dried in an oven for 30 min at a temperature of 40 °C. The strips of filter paper used for the test were left to dry in free air. For ethanol stability assessment, the tests were made with two different concentrations of denatured ethanol (v/v = 96% and v/v = 43%,

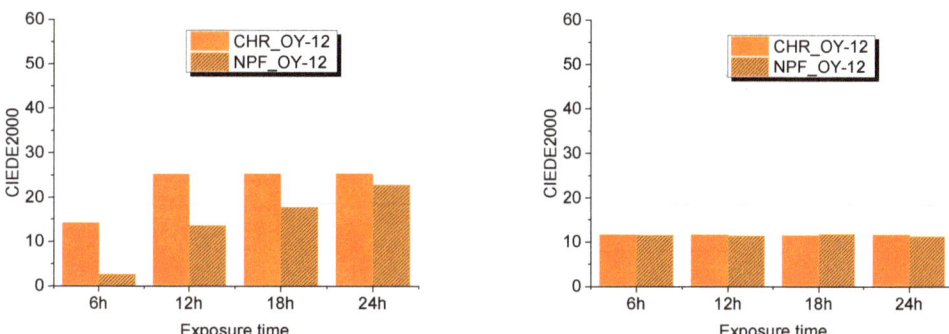

Figure 7. CIEDE2000 difference measured at 8 °C (**left**) and at 20 °C (**right**) on both test substrates printed with OY-12.

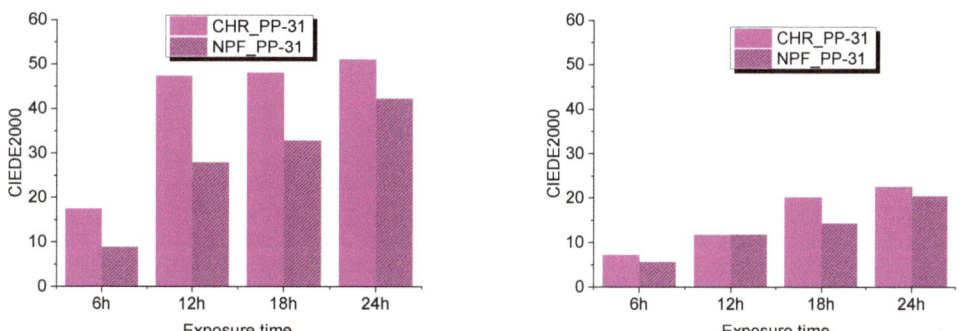

Figure 8. CIEDE2000 difference measured at 20 °C (**left**) and at 45 °C (**right**) on both test substrates printed with PP-31.

Color difference results on both PP-31 and OY-12 prints are significantly smaller when thermochromic colorant is inactive that is, when measured above its activation temperature. Such results could probably mean that UV radiation caused greater degradation of TC colorants, although it is evident from the results that conventional colorants are also strongly affected by UV radiation. When comparing NPF and CHR printing substrates, it can be seen that prints show better stability when they are printed on embossed NPF paper.

Thermochromic printing inks are colored complex mixtures, consisting of colorants (conventional pigments and thermochromic microcapsules), binder (resins), and different additives. Their composition and physical property mostly differ due to the intended printing process. Moreover, thermochromic (TC) microcapsules, which are one of the main components, consist of dye–developer–solvent systems encapsulated inside the polymeric shell. Polymeric shell of TC microcapsule is mostly made of melamine resin, but in this case, we can not make claims about its origin. According to previous research, the binder of used UV-curing TC printing ink is based on polyurethane acrylate (PUA) [52]. In addition, the TC microcapsules in the print, along with conventional pigments, are covered with ink's binder [52]. In the case of TC composite, the intensity of color contrast when mixing TC composit is affected by the intermolecular reactions between color former, color developer, and organic solvent [39,40]. Taking all of that into account, when the UV radiation comes over the surface of TC print, different chemical reactions are involved in this system. UV radiation causes damage to materials through a complex photochemical reactions, where polymeric materials first absorb the energy which is afterwards utilized for the breakage of the molecular bonds. The presence of chromophore groups causes faster materials degradation due to its high capacity of UV absorption [53]. According to Groeneveld et al., that refer to the bleeding of the colorants point out that ink has bled when prints were exposed to water and citric acid ($\Delta E(00) > 1.4$); however, when stains on filter papers were measured above ink's activation temperature, measured CIEDE2000 results were, in most cases, smaller and close to the threshold value ($\Delta E(00) = 1.4$). Also, when compared to the PP-31 ink, OY-12 ink showed a lesser degree of bleeding. Due to the weak contrast of the yellow color on the white filter papers, it was hard to visually notice the bleeding of colorants, which is why the image of bleeding is not presented as in the case of PP-31 ink. Bleeding of the colorants into the ethanol solutions was observed visually (Figure 5) but also to a much lesser extent when compared to the PP-31 ink. Also, a higher concentration of the solvent (96% ethanol) caused the ink to bleed more.

Taking into account the results presented earlier, it can be concluded that CIDE2000 values obtained at temperatures lower than the activation temperature result from both colors of TC microcapsules and conventional pigment, whereas CIDE2000 values obtained at a temperature higher than the activation temperature result mostly from the color degradation of the classic pigment. Also, the stability of TC prints towards different chemicals decreases with decreasing solvent polarity. It is known that water is a polar solvent, while with organic molecules, this influence is reduced due to the hydrocarbon chain. Considering the fact that citric acid is a weak organic acid (carboxylic acid) and that due to the presence of three carboxyl groups and one hydroxyl group, it has a higher polarity than ethanol (one hydroxyl group). The binder present in the used inks and the conventional pigment create an interaction with the used solvents, and in this way are transferred to another medium (filter paper or solvent). Water and citric acid interact mostly with hydrogen bonding, whereas ethanol reacts by polar interactions (Table 3) [32,50]. In 43% ethanol solutions hydrogen interactions predominate, with somewhat lower dispersive interactions. In 96% EtOH solution, the polar interactions are dominant. With that in mind, we can conclude that the used printing ink interacts mostly by high polar interactions as evident from the high color difference. It is also important to emphasize that in the tests in which bleeding was observed, it cannot be asserted with certainty whether it is only a matter of bleeding of the pigment, but it should be emphasized that in this case bleeding of the resin, which is the carrier of the pigment inside the printing ink, is also possible. Therefore, this is one limiting factor of this study that will be considered more in the future. In addition to the influence of chemicals, the influence of the printing substrate on the chemical stability of prints is also visible. A printing substrate with a rougher surface (NPF) shows better stability. This can be attributed to the adhesive properties of the ink, which arise precisely because of the roughness of the surface. Besides physical and chemical adhesion (formation of ionic, covalent, hydrogen bonds) occurring on the paper-printing ink interface, mechanical adhesion occurs by the penetration of printing ink into surface irregularities (pores) of the printing substrate [51]. Thus, the rougher the surface, the better the stability of TC prints.

3.2. Results of Prints' UV Stability

Evaluation of UV degradation of the prints was done by spectroscopic measurements and expressed in color difference values CIEDE2000. This was done by comparing untreated samples with treated samples after 6, 12, 18, and 24 hours of exposure to UV radiation. Exposure to UV radiation diminishes the dynamic of the color of TC prints as is clearly visible from Figures 7 and 8.

photodegradation of organic colorants, which are one component of TC microcapsules, is a dynamic process influenced by a number of different internal (physical state of the dye-absorption spectrum and polarity) and external (spectral distribution and intensity of the light source, presence or absence of oxygen, temperature, humidity, pH, the type of solvent or substrate, and concentration of dye) parameters [54]. When dye adsorbs light, different protective pathways occur and often are followed to lose the excess energy by emitting a photon, through non-radiative relaxation or by molecular reactions (photoreactions). In the case of photoreactions, photodegradations occur due to chemically unstable dye molecules in their photoexcited form, which in the end results in their decomposition through dissociation, intramolecular rearrangement reactions, or redox processes. Due to photoexcited chemically unstable dye molecules with the presence of other reactive substances in the system, both initiated from the singlet or triplet excited state ($^1D^*$, $^3D^*$) of the dye molecules [54]. In indirect photochemical reactions, the compounds with high absorption coefficients and low activation energies (photosensitizers), when excited can react directly with the neighboring dye or pigment or they can react with molecular oxygen ($^3O_2$) to create singlet oxygen ($^1O_2$), which then becomes the reactive species [54]. Friškovec et al. showed that during photooxidation, TC microcapsules shells in the thermochromic printing ink can be destroyed resulting in an irreversible loss of its functional properties because the polymer shell does not protect the TC composite from the environment [33]. Taking into account the ink binder, the photodegradation of PUA probably involves the Norrish type I photocleavage of the excited carbonyl groups in acrylate and carbamate moieties [55–58]. In polyurethane-based materials, the urethane linkages (C-NH) are most susceptible to photodegradation which, in the end, causes the surface erosion of the top layers and the formation of oxidation products consisting of hydroxyl type compounds (alcohols, hydroperoxides) and carbonyl compounds [55]. In addition, to C-NH groups, the methylene (CH) groups are damaged during UV exposure of PUA [55]. The addition of UV absorbers and the piperidine hindered amines (HALS) can be a promising way to increase the UV stability of PUA [56]. Moreover, the photostability of color formers of TC composite can be acieved by the addition of the amphoteric counter-ion (benzophenone or benzotriazole type UV absorbers, naphthalene derivatives of benzotriazole type UV absorbers, and zinc and nickel 2,4-dihydroxybenzophenone-3-carboxylates) according to the literature [39–41].

Keeping that in mind, if the print is exposed to UV radiation, first the binder degrades and results in the formation of oxidative species, which in the end react with the TC microcapsules shell and cause the color degradation. The UV stability of TC printing inks thus is affected by various factors such as the chemical composition of the binder and the TC microcapsules, interactions between the printing ink's binder and microcapsules, and the binder drying mechanism [37]. With that in mind, it can be concluded that TC printing ink is a complex mixture in which during photooxidation, different photochemical reactions occur, causing the formation of oxidation products and fast oxidation of a whole system—the ink binder, TC microcapsules polymer shell, dye-developer-solvent system, and conventional pigment. Due to a complex structure of TC printing inks, the photodegradation should be considered on different levels such as the increased photostability of color formers and color developers in TC composites, TC microcapsules shells, and printing ink binders.

TC inks are considered to work properly if the total color contrast (TCC) between the two states, below and above the activation temperatures, is clearly recognized. TCC was measured as the CIEDE2000 difference between samples at 8 °C and 20 °C for OY-12 ink. For PP-31 ink, TCC was measured as the CIEDE2000 difference between samples at 20° and 45 °C. The higher the TCC value, the more pronounced the thermochromic effect will be.

From Figure 9, it can be seen that the TCC is significantly higher on the embossed NPF paper. Although on untreated samples TCC of PP-31 ink printed on ultra-smooth CHR paper is higher than on NPF paper, already after 6 h of UV radiation its TCC contrast is much lower compared to PP-31 ink printed on NPF paper. In comparison, after 18 hours, TCC on CHR papers is 1.15 CIEDE2000 units for PP-31 and 1.48 units for OY-12. On NPF papers after 18 hours of UV radiation, the TCC is 3.61 CIEDE2000 units for PP-31 and 7.07

units for OY-12. The reason is probably that the NPF paper surface is more structured and the ink penetrates more into the paper structure irregularities compared to the ultra-smooth CHR paper as explained earlier. Photodegradation is strongly influenced by the amount of ink present on a substrate surface which can lead to aggregation of ink concentration. The size of the aggregates is proportional to ink concentration. If the agregates are greater, the lightfastness improves due to the smaller relative surface area accessible to environmental factors responsibe for photofading [54]. Therefore, it can be concluded that if the prints will not be exposed to UV radiation, that is, they will be intended for use in closed spaces, the TCC effect will be more pronounced on smooth substrates, as can be clearly seen in Figure 9. Degradation of prints due to UV radiation on both substrates is quite large; however, due to their structure, embossed papers are a slightly better choice when printing with TC inks.

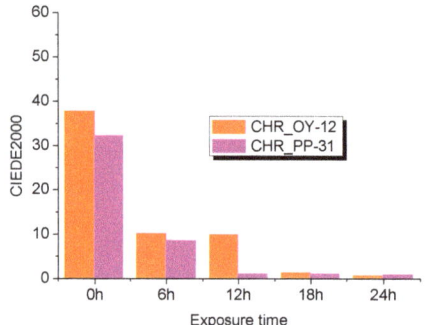 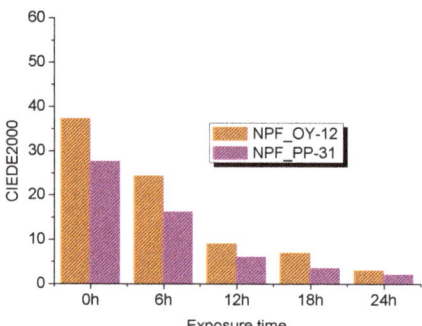

Figure 9. The total color contrast (TCC) between the two states for the OY-12 and PP-31 prints before and after 6, 12, 18, and 24 h exposure to UV radiation.

Visual Evaluation of Prints

The result of the visual evaluation of the samples photographed on the plate on which the samples were cooled and heated for spectrometric measurement can be seen in Figures 10–13. It is evident that the prints, which were exposed to UV radiation, significantly differ from untreated samples in color. Significant changes of color can be seen even after 6 h of exposure to UV radiation, whereas after 18 h of UV radiation, the loss of dynamic color change can be noticed. The degradation of both TC inks is more significantly visible on the smooth CHR substrate On prints obtained with both TC inks (OY-12 and PP-31) the degradation is smaller above the activation temperature when the TC colorants are inactive.

Figure 10. Visual presentation of thermochromic prints of cold activated TC ink OY-12 at 8 °C, before and after exposure of samples to UV radiation unexposed and exposed (6, 12, 18, and 24 h) (above CHR sample, below NMF sample).

Figure 11. Visual presentation of thermochromic prints of cold activated TC ink OY-12 at 20 °C, before and after exposure of samples to UV radiation (6, 12, 18, and 24 h) (above CHR sample, below NMF sample).

Figure 12. Visual presentation of thermochromic print of heat activated TC ink PP-31 at 20 °C, before and after exposure of samples to UV radiation (6, 12, 18, and 24 h) (above CHR sample, below NMF sample).

Figure 13. Visual presentation of thermochromic print of heat activated TC ink PP-31 at 45 °C, before and after exposure of samples to UV radiation (6, 12, 18, and 24 h) (above CHR sample, below NMF sample).

4. Conclusions

The main task of thermochromic inks is to influence the experience of the message or design printed with them through their dynamic color change. Their role can be informative or just creative. By analyzing some factors affecting the degradation of TC prints, that is, the dynamics of color change, an insight into the weaknesses of this complex mixture was provided.

All tested TC prints showed low resistance to liquid chemical agents as color difference values were unacceptable in all cases and bleeding of the colorants was detected as well. It was noticed that the stability of TC prints towards different chemicals decreases with decreasing solvent polarity, and the poorest stability was observed in the case of prints' exposure to a 96% ethanol solution. Prints obtained with body-heat activated TC ink (PP-31) showed higher color difference results when measured below ink's T_A (when both colorants were in their colored state), whereas in the case of prints obtained with cold-activated TC ink (OY-12), higher color differences were detected when prints were measured above ink's T_A (when TC colorant was inactive). Higher bleeding was detected both visually and by spectroscopic measurements from body-heat activated TC PP-31 ink, than from cold-activated TC OY-12 ink.

This research also provided insight into the relationship between TC inks and the substrate on which they are printed. Both inks showed better stability when printed on embossed NPF paper than on ultra-smooth CHR paper. Moreover, in the case of ultra-smooth CHR paper, bleeding of TC PP-31 ink's colorants was visually observed on all filter papers that were put below and above the prints during the experiment, meaning that the migration from the printed side of the test piece to the back side was also noticed.

Due to the sensitivity of TC inks to adverse environmental conditions, the goal of this research was to determine the resistance of TC prints to the influence of liquid chemical agents and UV radiation. Based on the results obtained after UV radiation, its influence in terms of color degradation, that is, reduced dynamics of color change is visible on both tested paper substrates. However, more significant degradation, reflected as a reduced total contrast, was observed on the ultra-smooth paper substrate. The reason is probably that embossed paper, due to its structure, enables better ink penetration into the irregularities of the paper structure compared to ultra-smooth paper. Also, in prints that were not exposed, it was found that the total color contrast is more noticeable on ultra-smooth paper than on embossed paper.

Author Contributions: Conceptualization, S.J.H.; methodology, S.J.H. and R.K.; software, S.J.H. and R.K.; validation, S.J.H. and R.K.; formal analysis, S.J.H., R.K. and M.V.; investigation, S.J.H., R.K., M.V. and A.M.D.; resources, S.J.H., R.K., M.V. and A.M.D.; data curation, S.J.H., R.K. and M.V.; writing—original draft preparation S.J.H., R.K. and M.V.; writing—review and editing, S.J.H., R.K. and M.V.; visualization, S.J.H.; supervision, S.J.H. and R.K.; project administration, S.J.H., R.K. and M.V.; funding acquisition, S.J.H., R.K. and M.V. All authors have read and agreed to the published version of the manuscript.

Funding: The APC was funded by University of Zagreb.

Institutional Review Board Statement: Not applicable.

Data Availability Statement: Not applicable.

Conflicts of Interest: The authors declare no conflict of interest.

References

1. Arman Kandirmaz, E.; Ozcan, A.; Er Ulusoy, D. Production of thermochromic microcapsulated inks for smart packaging and examination of printability properties. *Pigment Resin Technol.* **2020**, *49*, 273–281. [CrossRef]
2. White, M.; LeBlanc, M. Thermochromism in commercial products. *J. Chem. Educ.* **1999**, *76*, 3–7. [CrossRef]
3. Johansson, L. Creation of Printed Dynamic Images. Ph.D. Dissertation, Linköping University, Linköping, Sweden, 2006.
4. Christie, R.; Robertson, S.; Taylor, S. Design Concepts for a Temperature-sensitive Environment Using Thermochromic Colour Change Robert. *Colour Des. Creat.* **2007**, *1*, 1–11.
5. Kerry, J.; Butler, P. *Smart Packaging Technologies for Fast Moving Consumer Goods*; John Wiley & Sons Ltd.: Chichester, UK, 2008.
6. Degrees of Perfect Exibition; Behind the Thermochromic Murals That Morph before Your Eyes. Available online: https://mashable.com/ad/feature/thermochromic-murals-exhibit (accessed on 15 December 2022).
7. Geng, X.; Li, W.; Wang, Y.; Lu, J.; Wang, J.; Wang, N.; Li, J.; Zhang, X. Reversible thermochromic microencapsulated phase change materials for thermal energy storage application in thermal protective clothing. *Appl. Energy* **2018**, *217*, 281–294. [CrossRef]
8. Štaffová, M.; Kučera, F.; Tocháček, J.; Dzik, P.; Ondreáš, F.; Jančář, J. Insight into color change of reversible thermochromic systems and their incorporation into textile coating. *J. Appl. Polym. Sci.* **2021**, *138*, 49724. [CrossRef]
9. Thamrin, E.S.; Warsiki, E.; Bindar, Y.; Kartika, I.A. Thermochromic ink as a smart indicator on cold product packaging—Review. *IOP Conf. Ser. Earth Environ. Sci.* **2022**, *1063*, 012021. [CrossRef]
10. Leite, L.; Nobre, L.; Boticas, I.; Navarro, M.; Bessa, J.; Cunha, F.; Neves, P.; Alves, N.; Moura, C. Temperature-Sensing Inks for Real-Time Monitoring in Food Packaging. *Mater. Proc.* **2022**, *8*, 130.
11. Kavčič, U.; Mraović, M.; Bračko, S.; Muck, D. Printed thermochromic displays. *Color. Technol.* **2019**, *135*, 60–66. [CrossRef]
12. Galagan, Y.; Su, W. Fadable ink for time—Temperature control of food freshness: Novel new time—Temperature indicator. *Food Res. Int.* **2008**, *41*, 653–657. [CrossRef]
13. Liu, G.; Yu, T.; Yao, Z.; Xu, H.; Zhang, Y.; Xu, X.; Xu, X.; Gao, M.; Sun, Q.; Zhang, T.; et al. ViviPaint: Creating Dynamic Painting with a Thermochromic Toolkit. *Multimodal Technol. Interact.* **2022**, *6*, 63. [CrossRef]
14. Zhu, X.; Liu, Y.; Dong, N.; Li, Z. Fabrication and characterization of reversible thermochromic wood veneers. *Sci. Rep.* **2017**, *7*, 16933. [CrossRef] [PubMed]
15. Zhu, C.F.; Wu, A.B. Studies on the synthesis and thermochromic properties of crystal violet lactone and its reversible thermochromic complexes. *Thermochim. Acta* **2005**, *425*, 7–12. [CrossRef]
16. MacLaren, D.C.; White, M.A. Design rules for reversible thermochromic mixtures. *J. Mater. Sci.* **2005**, *40*, 669–676. [CrossRef]

17. Seeboth, A.; Lotzsch, D. *Thermochromic and Thermotropic Materials*; CRC Press by Taylor & Francis Group: Boca Raton, FL, USA, 2013.
18. Tang, H.; MacLaren, D.C.; White, M.A. New insights concerning the mechanism of reversible thermochromic mixtures. *Can. J. Chem.* **2010**, *88*, 1063–1070. [CrossRef]
19. Phillips, G.K. Combining thermochromics and conventional inks to deter dovument fraud. In Proceedings of the SPIE 3973, Optical Security and Counterfeit Deterrence Techniques III, San Jose, CA, USA, 27–28 January 2000; van Renesse, R.L., Vliegenthart, W.A., Eds.; SPIE: Bellingham, WA, USA, 2000; pp. 99–104.
20. Johansson, L.; Kruse, B. The Influence of Paper Properties on Colour Reproduction with Dynamic Inks. In Proceedings of the International Conference on Imaging: Technology and Applications for the 21st Century, Beijing, China, 23–26 May 2005; pp. 214–215.
21. Pu, Y.; Fang, J. Preparation and thermochromic behavior of low-temperature thermochromic microcapsule temperature indicators. *Colloids Surf. A Physicochem. Eng. Asp.* **2022**, *653*, 129889. [CrossRef]
22. Bašnec, K.; Perše, L.S.; Šumiga, B.; Huskić, M.; Meden, A.; Hladnik, A.; Podgornik, B.B.; Gunde, M.K. Relation between colour- and phase changes of a leuco dye-based thermochromic composite. *Sci. Rep.* **2018**, *8*, 5511. [CrossRef]
23. WRAP. *Thermochromic Inks and Reducing Household Food Waste*; WRAP: Banbury, UK, 2013.
24. Panák, O.; Držková, M.; Kaplanová, M. Insight into the evaluation of colour changes of leuco dye based thermochromic systems as a function of temperature. *Dye. Pigment.* **2015**, *120*, 279–287. [CrossRef]
25. Seeboth, A.; Lo, D.; Ruhmann, R.; Muehling, O. Thermochromic Polymers—Function by Design. *Chem. Rev.* **2013**, *114*, 3037–3068. [CrossRef]
26. Maclaren, D.C.; White, M.A. Competition between dye—Developer and solvent—Developer interactions in a reversible thermochromic system. *J. Mater. Chem.* **2003**, *13*, 1701–1704. [CrossRef]
27. MacLaren, D.C.; White, M.A. Dye–developer interactions in the crystal violet lactone–lauryl gallate binary system: Implications for thermochromism. *J. Mater. Chem.* **2003**, *13*, 1695–1700. [CrossRef]
28. Panák, O.; Kashtalyan, K.; Syrový, T. Producing two-component thermochromic pattern by means of offset printing. In Proceedings of the 2015 IEEE Colour and Visual Computing Symposium (CVCS), Gjovik, Norway, 25–26 August 2015; pp. 1–5.
29. Hajzeri, M.; Vatič, N.; Panák, O. *XIth Symposium on Graphic Arts*; University of Pardubice: Pardubice, Czechia, 2013; pp. 109–114.
30. Urbas, R.; Milošević, R.; Kašiković, N.; Pavlović, Ž.; Elesini, U.S. Microcapsules application in graphic arts industry: A review on the state-of-the-art. *Iran. Polym. J. (Engl. Ed.)* **2017**, *26*, 541–561. [CrossRef]
31. Hakami, A.; Srinivasan, S.S.; Biswas, P.K.; Krishnegowda, A.; Wallen, S.L.; Stefanakos, E.K. Review on thermochromic materials: Development, characterization, and applications. *J. Coat. Technol. Res.* **2022**, *19*, 377–402. [CrossRef]
32. Rožić, M.; Vukoje, M.; Kapović, D.; Marošević, L. Solvents interactions with thermochromic print. *J. Graph. Eng. Des.* **2017**, *8*, 19–25. [CrossRef]
33. Friškovec, M.; Kulčar, R.; Klanjšek Gunde, M. Light fastness and high-temperature stability of thermochromic printing inks. *Color. Technol.* **2013**, *129*, 214–222. [CrossRef]
34. Rožić, M.; Kulčar, R.; Jamnicki, S.; Lozo, B.; Gregor-Svetec, D. UV stability of thermochromic ink on paper containing clinoptilolite tuff as a filler. *Cellul. Chem. Technol.* **2015**, *49*, 693–699.
35. Vukoje, M.; Kulčar, R.; Itrić Ivanda, K.; Bota, J.; Cigula, T. Improvement in Thermochromic Offset Print UV Stability by Applying PCL Nanocomposite Coatings. *Polymers* **2022**, *14*, 1484. [CrossRef] [PubMed]
36. Kulčar, R.; Vukoje, M.; Krajnović, I.; Rožić, M. Influence of recycled fibres in paper on the UV stability of thermochromic prints. In Proceedings of the 10th International Symposium GRID 2020, Novi Sad, Serbia, 12–14 November 2020; Dedijer, S., Ed.; University of Novi Sad Faculty of Technical Sciences Department of Graphic Engineering and Design: Novi Sad, Serbia, 2020; pp. 161–168.
37. Jamnicki Hanzer, S.; Kulčar, R.; Vukoje, M.; Širol, P. Mechanical and chemical resistance of thermochromic packaging prints. In Proceedings of the 10th International Symposium GRID, Novi Sad, Serbia, 12–14 November 2020; Dedijer, S., Ed.; University of Novi Sad Faculty of Technical Sciences Department of Graphic Engineering and Design: Novi Sad, Serbia, 2020; pp. 109–118.
38. Vukoje, M.; Kulčar, R.; Itrić, K.; Rožić, M. Spectroscopic evaluation of thermochromic printed cardboard biodegradation. In Proceedings of the Ninth International Symposium GRID 2018, Novi Sad, Serbia, 8–10 November 2018; Pavlović, Ž., Ed.; Grafički Centar GRID: Šabac, Serbia, 2018; pp. 87–96.
39. Oda, H. New developments in the stabilization of leuco dyes: Effect of UV absorbers containing an amphoteric counter-ion moiety on the light fastness of color formers. *Dye. Pigment.* **2005**, *66*, 103–108. [CrossRef]
40. Oda, H. Photostabilization of organic thermochromic pigments: Action of benzotriazole type UV absorbers bearing an amphoteric counter-ion moiety on the light fastness of color formers. *Dye. Pigment.* **2008**, *76*, 270–276. [CrossRef]
41. Oda, H. Photostabilization of organic thermochromic pigments. Part 2: Effect of hydroxyarylbenzotriazoles containing an amphoteric counter-ion moiety on the light fastness of color formers. *Dye. Pigment.* **2008**, *76*, 400–405. [CrossRef]
42. Homola, T.J. *Color-Changing Inks*; AccessScience; McGraw-Hill Education: New York, NY, USA, 2008.
43. *ISO 2836*; Graphic Technology—Prints and Printing Inks—Assessment of Resistance of Prints to Various Agents. International Organization for Standardization: Geneva, Switzerland, 2021.
44. *Astm D3424-09*; Standard Test Methods for Evaluating the Relative Lightfastness and Weatherability of Printed Mater. ASTM International: West Conshohocken, PA, USA, 2010; Volume V, pp. 1–5.

45. European Organisation for Tehnical Approval (EOTA). *Exposure Procedure for Artificial Weathering—TR 010*; EOTA: Brussels, Belgium, 2004.
46. CIE Central Bureau. *Colorimetry*, 3rd ed.; CIE Central Bureau: Vienna, Austria, 2004.
47. Yang, Y.; Ming, J.; Yu, N. Color Image Quality Assessment Based on CIEDE2000. *Adv. Multimed.* **2012**, *2012*, 273723. [CrossRef]
48. Schilling, M. Shift Happens . . . So Measure. Available online: https://inkjetinsight.com/knowledge-base/color-shift-happens-so-measure/ (accessed on 3 February 2023).
49. Kumar, M. *Interdisciplinarnost Barve 2.Del.*; Jeler, S., Kumar, M., Eds.; Društvo koloristov Slovenije: Maribor, Slovenia, 2003; pp. 89–100.
50. Hansen, C.M. *Hansen Solubility Parameters—A User's Handbook*; CRC Press, Taylor & Francis Group: Boca Raton, FL, USA, 2000.
51. Comyn, J. *Adhesion Science*; Royal Society of Chemistry: Cambridge, UK, 2021.
52. Vukoje, M.; Miljanić, S.; Hrenović, J.; Rožić, M. Thermochromic ink–paper interactions and their role in biodegradation of UV curable prints. *Cellulose* **2018**, *25*, 6121–6138. [CrossRef]
53. Nguyen, T.; Gu, X.; Vanlandingham, M.; Byrd, E.; Ryntz, R.; Martin, J.W. Degradation modes of crosslinked coatings exposed to photolytic environment. *J. Coat. Technol. Res.* **2013**, *10*, 1–14. [CrossRef]
54. Groeneveld, I.; Kanelli, M.; Ariese, F.; van Bommel, M.R. Parameters that affect the photodegradation of dyes and pigments in solution and on substrate—An overview. *Dye. Pigment.* **2023**, *210*, 110999. [CrossRef]
55. Decker, C.; Masson, F.; Schwalm, R. Weathering resistance of waterbased UV-cured polyurethane-acrylate coatings. *Polym. Degrad. Stab.* **2004**, *83*, 309–320. [CrossRef]
56. Decker, C.; Moussa, K.; Bendaikha, T. Photodegradation of UV-cured coatings II. Polyurethane–acrylate networks. *J. Polym. Sci. Part A Polym. Chem.* **1991**, *29*, 739–747. [CrossRef]
57. Bénard, F.; Mailhot, B.; Mallégol, J.; Gardette, J.L. Photoageing of an electron beam cured polyurethane acrylate resin. *Polym. Degrad. Stab.* **2008**, *93*, 1122–1130. [CrossRef]
58. Peinado, C.; Allen, N.S.; Salvador, E.F.; Corrales, T.; Catalina, F. Chemiluminescence and fluorescence for monitoring the photooxidation of an UV-cured aliphatic polyurethane-acrylate based adhesive. *Polym. Degrad. Stab.* **2002**, *77*, 523–529. [CrossRef]

Disclaimer/Publisher's Note: The statements, opinions and data contained in all publications are solely those of the individual author(s) and contributor(s) and not of MDPI and/or the editor(s). MDPI and/or the editor(s) disclaim responsibility for any injury to people or property resulting from any ideas, methods, instructions or products referred to in the content.

Review

A Systematic Review of Butterfly Pea Flower (*Clitoria ternatea* L.): Extraction and Application as a Food Freshness pH-Indicator for Polymer-Based Intelligent Packaging

Nur Nabilah Hasanah [1], Ezzat Mohamad Azman [1], Ashari Rozzamri [1], Nur Hanani Zainal Abedin [1] and Mohammad Rashedi Ismail-Fitry [1,2,*]

[1] Department of Food Technology, Faculty of Food Science and Technology, Universiti Putra Malaysia, UPM Serdang 43400, Selangor, Malaysia; nabilah3003@gmail.com (N.N.H.); ezzat@upm.edu.my (E.M.A.); rozzamri@upm.edu.my (A.R.); hanani@upm.edu.my (N.H.Z.A.)

[2] Halal Products Research Institute, Universiti Putra Malaysia (UPM), Putra Infoport, UPM Serdang 43400, Selangor, Malaysia

* Correspondence: ismailfitry@upm.edu.my

Citation: Hasanah, N.N.; Mohamad Azman, E.; Rozzamri, A.; Zainal Abedin, N.H.; Ismail-Fitry, M.R. A Systematic Review of Butterfly Pea Flower (*Clitoria ternatea* L.): Extraction and Application as a Food Freshness pH-Indicator for Polymer-Based Intelligent Packaging. *Polymers* **2023**, *15*, 2541. https://doi.org/10.3390/polym15112541

Academic Editors: Victor G. L. Souza, Lorenzo M. Pastrana and Ana Luisa Fernando

Received: 18 April 2023
Revised: 21 May 2023
Accepted: 25 May 2023
Published: 31 May 2023

Copyright: © 2023 by the authors. Licensee MDPI, Basel, Switzerland. This article is an open access article distributed under the terms and conditions of the Creative Commons Attribution (CC BY) license (https:// creativecommons.org/licenses/by/ 4.0/).

Abstract: The butterfly pea flower (*Clitoria ternatea* L.) (BPF) has a high anthocyanin content, which can be incorporated into polymer-based films to produce intelligent packaging for real-time food freshness indicators. The objective of this work was to systematically review the polymer characteristics used as BPF extract carriers and their application on various food products as intelligent packaging systems. This systematic review was developed based on scientific reports accessible on the databases provided by PSAS, UPM, and Google Scholar between 2010 and 2023. It covers the morphology, anthocyanin extraction, and applications of anthocyanin-rich colourants from butterfly pea flower (BPF) and as pH indicators in intelligent packaging systems. Probe ultrasonication extraction was successfully employed to extract a higher yield, which showed a 246.48% better extraction of anthocyanins from BPFs for food applications. In comparison to anthocyanins from other natural sources, BPFs have a major benefit in food packaging due to their unique colour spectrum throughout a wide range of pH values. Several studies reported that the immobilisation of BPF in different polymeric film matrixes could affect their physicochemical properties, but they could still effectively monitor the quality of perishable food in real-time. In conclusion, the development of intelligent films employing BPF's anthocyanins is a potential strategy for the future of food packaging systems.

Keywords: anthocyanin; *Clitoria ternatea* L.; ultrasonic extraction; pH-responsive indicator; food freshness; intelligent packaging

1. Introduction

Anthocyanins in *Clirotia ternatea* L. are present in the form of polyacrylate anthocyanins also known as Ternatins. They are among the stable forms of anthocyanin [1], and their stability is higher than the non-acylated ones. This plant is primarily found in tropical regions, where it needs intense sunlight and is immune to abiotic stress [2]. In the butterfly pea flower (BPF), the big advantage of polyacrylate anthocyanin is that it is known to be employed as a natural food dye [3]. The most noticeable feature of the BPF is its petal, which has a highly appealing blue colour [4]. In Malaysian dishes, BPFs are used to introduce blue colour to white rice, namely *nasi kerabu*. Other than that, a refreshing lemonade drink is also made using the BPF and is traditionally used in Southeast Asia as a herbal tea. Due to their unique colour properties, BPFs have been utilised for other applications in foods. On top of that, the significant colour-change properties of BPF are also one of the most important criteria for spoilage detection in food products which come from natural sources (anthocyanin) [5]. In this context, the use of intelligent pH-colourimetric packaging is an innovative system for distinguishing, monitoring, securing, and assuring food safety and quality [6].

The increase in consumer awareness of food safety opens a new area of research, with the incorporation of natural colourants from several plant-based sources being recommended as a beneficial alternative to toxic synthetic (chemical) dyes. Generally, consumers frequently use the shelf-life date (expiry date) displayed on the packaging to determine and assess the freshness and quality of perishable foods [7]. However, some perishable food products, such as muscle food, fresh fruits, and vegetables, cannot be assessed for their quality and freshness only based on their shelf-life date [8]. Thus, it has been discovered that polymer-based intelligent packaging can act as an indicator required for monitoring the spoilage of food in real-time and instilling confidence in consumers upon purchase.

According to Poh [8], colour transformations are due to the molecular structure of anthocyanins having an ionic nature. In acidic conditions (pH value < 2), the anthocyanins will appear red, which is the formation of flavylium ions. As the pH increases, the first deprotonation occurs and creates cations, converting flavylium ions into a neutral quinonoid base [9]. Apart from that, anthocyanins in BPFs have a blue colour property in neutral pH; the colour of anthocyanin will change to green-light yellow with an increasing pH condition. The changes occur due to tautomerisation which changes hemiketal structure to cis-chalcone and trans-chalcone as a result of the deprotonation from the C5-OH that has been set aside [10]. According to Rahim et al. [5], microbial and biochemical spoilage causes changes in food pH; thus, intelligent packaging made from polymer infused with active compounds such as anthocyanin can help identify the changes in food pH through a colour response due to reactions between the delphinidin of anthocyanin and volatile amines produced by bacteria and the enzymes [11]. Therefore, this review covers the morphology, anthocyanins extraction, and applications of BPF anthocyanin-rich colourants from various types of polymer matrices as pH indicators in intelligent food packaging systems.

2. Materials and Methods

2.1. Research Strategy

This systematic review paper was written based on the specifications of the Preferred Reporting Items for Systematic Reviews and Meta-Analyses (PRISMA) guidelines [12], as presented in Figure 1, to retrieve articles related to anthocyanins extracted from BPFs (C. ternatea) for application in a pH indicator film to monitor foods freshness. Document searching tools provided by Perpustakaan Sultan Abdul Samad (PSAS), the library of the Universiti Putra Malaysia (UPM), Serdang, Malaysia were used. Data were extracted from full-access articles from EBSCOhost, PROQUEST Dissertations and Theses Global, Scopus, SpringerLink, and ScienceDirect using the electronic database search provided by PSAS. Other than that, information was obtained from the general academic search engine Google Scholar. Moreover, in all databases, PICO strategy was used as terms guided by the specific question created, whereby population (P) referred to the research studies on polymer film matrixes for intelligent packaging applications; intervention (I) referred to polymer-carriers immobilised with natural pigment BPF anthocyanin; comparison (C) can be referred to types of polymer film properties; and outcome (O) referred to the application on food products.

2.2. Keyword Choices

The keywords chosen were butterfly pea flower; anthocyanin; intelligent packaging; polymer matrix; starch; pH film indicator; freshness indicator; application food; monitoring food. The keywords mentioned above were used as the search strategy. Only the first 100 articles that satisfied all the keywords were reviewed when the search engine displayed too many results. When a lower number of search results was obtained, fewer or more general keywords were used to obtain enough search results.

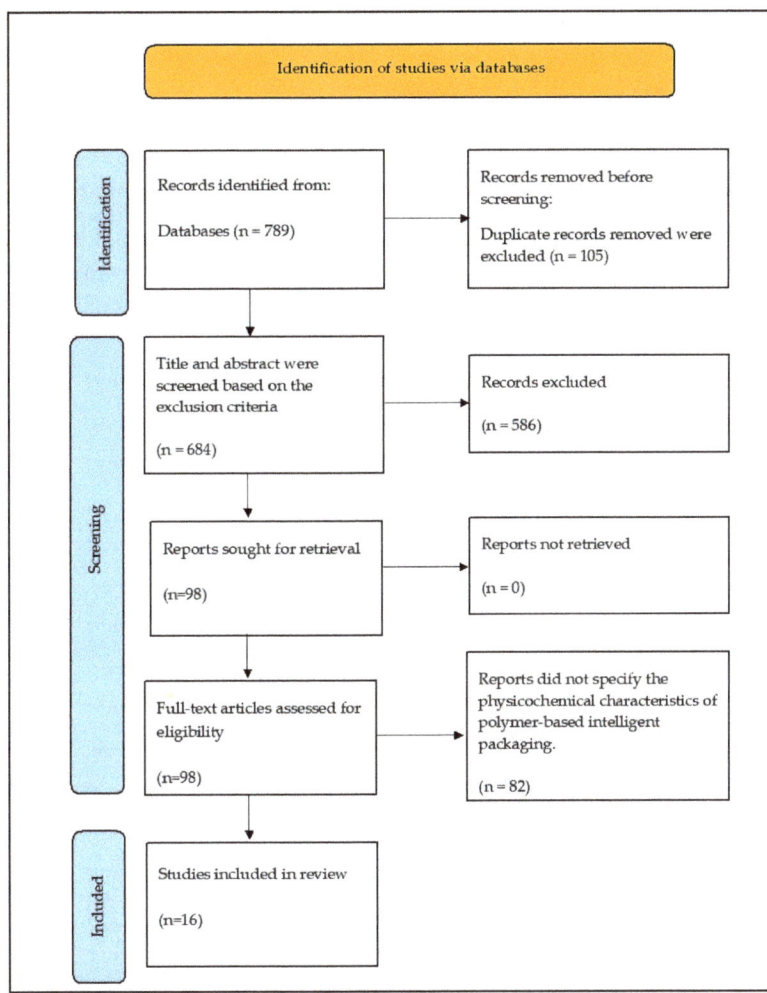

Figure 1. PRISMA flow diagram for article selection process based on the methodology stated [12].

2.3. Inclusion Criteria

The inclusion criteria for this systematic review were determined to screen the eligibility of research articles found in the databases. Studies and articles that fulfilled the following criteria were selected: (a) only articles written in English from research databases were prioritised to save time and to prevent confusion due to mistranslations during the reviewing process; (b) studies that elaborated on the development of intelligent packaging from different types of polymer incorporated with natural pigments as pH indicator films; (c) studies that elaborated on the application of polymer-based strategies for the monitoring of food freshness; (d) articles found in research databases provided by PSAS, UPM, and Google Scholar; (e) articles that were published between the year January 2010 and April 2023, and (f) full-text articles that are accessible.

2.4. Exclusion Criteria

Several exclusion criteria were taken into consideration during the reviewing process; (a) articles that only provided abstracts; (b) studies that did not use butterfly pea flowers for the development of intelligent packaging; and (c) articles that did not focus on the current topic.

2.5. Study Selection and Data Extraction

The articles searched in the databases were chosen based on their title and the keywords in the title. The abstracts of the shortlisted articles obtained were evaluated while also keeping in mind the inclusion criteria that were set (by N.N.H.). The full article was then gone through to identify whether it satisfied the inclusion criteria. Later, M.R.I.-F. and E.MA. cross-checked the shortlisted articles after reviewing the full texts, and disagreements related to conflicting articles were resolved after a discussion between the four authors (M.R.I.-F., E.M.A., A.R., N.H.Z.A.). The data were collected and tabulated and N.N.H. used self-designed tables to tabulate the relevant data. Table 1 is a summary table for all the selected articles, giving information about the author(s), year of publication, polymer material, the composition of polymer film, food application, and conclusion. Moreover, the information extracted from Table 1 was further divided into two categories; Table 2 provides comprehensive information about the physical and mechanical properties tested on the polymer film. Meanwhile, the applications of the butterfly pea flower films as intelligent packaging are listed in Table 3. A PRISMA flow diagram of the article selection process is shown in Figure 1.

2.6. Quality Assessment of the Included Studies

The included research articles were subjected to a quality evaluation step using the Critical Appraisal Skills Programme (CASP) tool [13], which graded each study based on several components, such as the appropriateness of the study design for the research question. Two authors (N.N.H. and M.R.I.-F.) carried this out independently and discussed each item for each study included in this review to reduce the possibility of bias.

Table 1. A summary of articles on BPF immobilised in polymer-based intelligent packaging for monitoring the freshness of food.

Author/Year	Polymer Material	Composition of Polymer Film	Food Application	Conclusion
Hashim et al., 2021 [14]	Sugarcane wax/agar (SW/Agr)	Agr/BF/1% SW, Agr/BF/1.5% SW, Agr/BF/2% SW	Shrimp	Agr/BF/2%SW film was chosen as a sensor for shrimps' freshness due to its overall performance and sensitivity to ammonia vapour.
Wu et al., 2021 [15]	Gellan gum/heat-treated soy protein isolate (G/HSPI)	G/CT, G/HSPI1%/CT, G/HSPI2%/CT	Shrimp	The incorporation of CT extract in G and G/HSPI films successfully reduced the release of CT anthocyanin content.
Ahmad et al., 2020 [16]	Sago (Metroxylon sagu)	SG, SG5/BPF, SG7/BPF, SG10/BPF, SG15/BPF	Chicken	The optimal concentration to formulate the SG film was 5% wt/vol and the surfaces of the films investigated were smooth (complete polymer gelatinisation).
Hidayati et al., 2021 [17]	Chitosan/polyvinyl alcohol (CH/PVA)	CH:PVA (20:80, 40:60, 60:40, and 80:20)	-	CH:PVA (40:60) had the best results for physical and mechanical properties and produced the clearest colour changes with different pH ranges.
Boonsiriwit et al., 2021 [18]	Hydroxypropyl methylcellulose/microcrystalline cellulose biocomposites (HMB)	HMB, 1.0BA-HMB, 1.5BA-HMB, and 2.0BA-HMB	Fish fillet (*Scomber scombrus*)	1.0BA-HMB indicator exhibited the best physical properties; however, 1.5-HMB demonstrated a clear change in the colour response of quality fish (more sensitive).

Table 1. Cont.

Author/Year	Polymer Material	Composition of Polymer Film	Food Application	Conclusion
Sai-Ut et al., 2021 [19]	Gelatin/methylcellulose (G/MC)	G, G/BPE, MC, MC/BPE	-	MC/BPE indicator had improved mechanical and physical properties. Meanwhile, G/BPE showed a clearer response to pH variation.
Koshy et al., 2022 [20]	Soy protein isolate/chitin nanowhisker (SPI/CNW)	SPI, SPI-CNW, SPI-CTE, and SPI-CNW-CTE	-	The addition of CTE to SPI enhanced the mechanical properties. However, the addition of CTE was found to decrease the tensile strength of SPI-CNW film and was found to make the film pH sensitive.
Mary et al., 2020 [21]	Potato starch/nanosized titanium dioxide (S/TiO$_2$)	S, S/BPE, S/TiO$_2$, S/TiO$_2$/BPE	Shrimp	It was observed that the addition of BPE and TiO$_2$ could greatly alter the physical properties of the film. The addition of TiO$_2$ exhibited changes in colour during the spoilage of shrimp.
Yan et al., 2021 [22]	Chitosan (CH)	CH, CH-BP10%, CH-BP15%, CH-BP20%	Tilapia fish	The incorporation of BP extract increased the thickness, WVP, and mechanical properties of CH-BP films, while reducing their moisture content, swelling ratio, and water contact angle.
Kim et al., 2022 [23]	Gelatine/agar/zinc oxide nanoparticles (Gel/Agar/Zno)	Gel/Agar, Gel/Agar/ZnO, Gel/Agar/BA, Gel/Agar/ZnO/BA	Shrimp	The addition of BA and ZnO significantly increased the UV-blocking properties and surface hydrophobicity without significant changes in the film's mechanical, thermal stability, and water vapour barrier properties.
Romruen et al., 2022 [24]	Alginate/agar/cellulose nanosphere (CN)	0% CN, 5% CN, 10% CN, 20% CN, and 30% CN.	Shrimp	CN can improve the mechanical properties of smart bilayer films without affecting their chemical properties and proved it is effectively used to monitor shrimp freshness.
Ahmad et al., 2019 [25]	t-carrageenan	Control, t-carrageenan/BPA	Shrimp and durian	The ability of the developed colourimetric pH sensor film from t-carrageenan shows colour changes on shrimp and durian, which provides a simple way to express the quality of food.
Rawdkuen et al., 2020 [26]	Gelatine	Control, Gelatine/BPA	-	The film with BPA extracts in gelatine films showed the highest antioxidant activity, improved water barrier properties, and showed greater pH sensitivity.
Roy et al., 2021 [27]	Carboxymethyl cellulose/agar (CMC/agar)	CMC/agar, CMCagar/ACN, CMC/agar/SKN	-	The incorporation of anthocyanin in CMC/agar-based films improved physical and functional properties without altering the thermal stability.
Sumiasih et al., 2022 [28]	Chitosan/polyvinyl alcohol (CH/PVA)	CH: PVA (20:80, 40:60, 60:40, and 80:20)	Beef	The best formulation was the composition of 20:80 PVA and chitosan 20:80 with the best thickness and total TVBN analysis

Table 1. *Cont.*

Author/Year	Polymer Material	Composition of Polymer Film	Food Application	Conclusion
Cho et al., 2021 [29]	Corn starch (CS)	CS-BP (9% v/v, 13% v/v, 17% v/v, 20% v/v, and 23% v/v)	Pasteurised milk	The thickness of the films increased with the BP concentration added. Meanwhile, BP solutions incorporating 23% v/v exhibited the greatest ΔE values.

Table 2. pH-indicator composite films based on butterfly pea flower anthocyanin: physical and mechanical changes.

Polymer Film	Main Results after BPFA Incorporation				References
	Physical Properties		Mechanical Properties		
	Thickness	Water Permeability	Tensile Strength	Elongation at Break	
Sugarcane wax and agar matrix	Increase	No significant difference	Low	No significant difference	[13]
Gellan gum and heat-treated soy protein isolate (HSPI)	-	Low	Low	Low	[14]
Chitosan and polyvinyl alcohol (PVA)	No effect	-	High	Low	[16]
Hydroxypropyl methylcellulose biocomposite (HMB)	No effect	-	High	Low	[17]
Gelatine and methylcellulose	No effect	High	Gelatine + BPFA (low) MC + BPFA (high)	Gelatine + BPFA (low) MC + BPFA (high)	[18]
Soy protein isolate (SPI) and chitin nanowhisker (CNW)	No effect	-	Low	Low	[19]
Nanosized TiO_2	Decrease	Low	-	-	[20]
Chitosan	Increase	High	High	Low	[21]
Zinc oxide nanoparticles (ZnO) + gelatine/agar	Increase	No significant difference	Low	High	[22]
Cellulose nanosphere (CN) and alginate/agar	Increase	No significant difference	High	Low	[23]
Gelatine	No effect	Low	Low	High	[25]
Carboxymethyl cellulose (CMC)/agar-based	No effect	Low	No significant difference	High	[26]

Table 3. Application of pH-responsive freshness indicator film of butterfly pea flower in food packaging.

Film Matrix	Foods	Sample Size (g)	Storage (°C)	Visual Colour Change	Final Time (Day)	References
Sugarcane wax and agar matrix	Shrimp	55	25	Deep purple to bluish-green	1.0	[13]
Heat-treated soy protein isolate and gellan gum	Shrimp	-	25	Blue to bluish-green	1.0	[14]
Hydroxypropyl methylcellulose biocomposite (HMB)	Mackerel fish	200	4	Deep purple to violet	6.0	[17]

Table 3. Cont.

Film Matrix	Foods	Sample Size (g)	Storage (°C)	Visual Colour Change	Final Time (Day)	References
Nanosized TiO_2	Prawn	20	4	Pink to green	6.0	[20]
Chitosan	Tilapia fish	-	4	Purple-blue to dark green	6.0	[21]
t-carrageenan	Durian and shrimp	-	28	Shrimp: deep blue to greenish-blue Durian: deep blue to dark purple	Shrimp: 0.5 h Durian: 4.0	[24]
Chitosan and polyvinyl alcohol (PVA)	Beef	60	25	Blue to bluish-green	1.0	[27]
Corn starch	Pasteurised milk	250 (mL)	25	Deep blue to light blue	3.0	[28]

3. Results and Discussion

3.1. Study Characteristics

The PRISMA flow diagram in Figure 1 describes the entire screening process of articles in the review, where we determined the eligibility of research articles found in the database. Firstly, the timeline was set to 13 years, so articles that were published between 2010 and 2023 were included in the review process. In terms of content, the selected research articles were identified and discussed for the studies that focused on the development and investigated the ability of different types of polymer matrix films to be incorporated with butterfly pea flower extract on intelligent packaging specifically as a freshness/pH indicator. Furthermore, the application of butterfly pea flower anthocyanin as a polymer-based pH film indicator was included in the review process where the effectiveness of the pH indicator was in monitoring the freshness of various types of food (e.g., seafood, poultry, milk) stored at room temperature or chiller conditions at 4 °C. The characteristics of all the research involved (summary) are displayed in a data extraction table in Table 1.

3.2. Butterfly Pea Flower (Clitoria ternatea L.)

3.2.1. Plant Morphology

Clitoria ternatea L. belongs to the Fabaceae family. It is also called the Butterfly Pea, Blue Pea, Asian Pigeonwings, and *Telang* (in the Malay language). It is a climbing flowery plant creeping towards other plants in competition for sunlight. The flowers are big and solitary and the most striking feature is the colour of the flowers, which have a vivid deep blue to mauve colour [30] with yellow in the middle and a white spot at the edge. The colours in the flower petals are due to the presence of anthocyanin content such as delphinidins.

According to Suarna and Wijaya [4], BPFs have two different kinds of corollas (the group of petals in the flower) and stamen: (Figure 2a) a normal corolla has four petals, where two petals at the lateral area called wings, and other two petals at the posterior area called carina with diadelphous stamen. On the other hand, a multiple-layer corolla (Figure 2b), which has five petals of corollas, consists of one petal (the biggest) at the anterior area with 10 solitary stamens. Havananda and Luengwilai [31] stated that the reproduction of BPFs is conducted through fruit seeds, and it quickly shows excellent regrowth after cutting or grazing and produces a high yield of blooming flowers. BPFs flourish when lightly grazed throughout the wet season.

 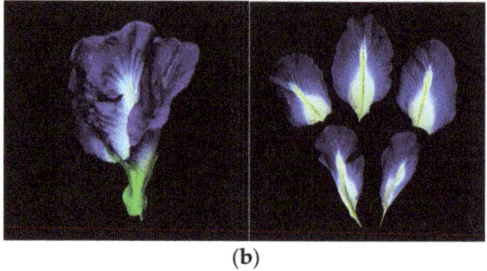

Figure 2. Variation structure of the butterfly pea flower: (**a**) normal corollas; (**b**) multiple layered corollas.

3.2.2. Plant Pigment (Anthocyanin)

Anthocyanins (Anthos is flower and kyanos is blue in Greek) are polar compounds that are water-soluble pigments, belonging to the flavonoids compound, where they are a subclass of the polyphenol family [32]. They provide a colour spectrum in the range of red, blue, and purple, as seen in the higher plants. They are frequently found in flowers and fruits. However, they can also be found in other plants, including vegetables, legumes, and cereals. According to Khoo et al. [33], anthocyanins are glycosides of anthocyanidins, which are often found in a variety of plant species such as cyanidin, delphinidin, pelargonidin, peonidin, petunidin, and malvidin. Moreover, BPFs have medicinal properties. The extracts of anthocyanin have been used over the years as a potential antioxidant, antimicrobial, anti-inflammatory, and antidiabetic activity [34] and, more recently, due to studies indicating anti-cancer properties [35].

In BPFs, a high abundance of polyacylated anthocyanins in the flower is responsible for the plant pigment's stable blue colour [36]. Anthocyanins that are commonly found in BPFs are ternatin and delphinidin [37]. A study by Ahmad et al. [25] found that 12 different compounds of anthocyanin were successfully detected in BPFs where ternatins were the largest anthocyanin groups. Chemically, ternatins are blue and they consist of anthocyanidin or aglycon and sugars with the presence of a third component, which is delphinidin. The primary anthocyanin responsible for this flower's intense blue to purple colour is delphinidin [38]. Jeyaraj et al. [34] reported that the stable polyacylated derivative of delphinidin 3,3′,5′-triglucosides called ternatin is found in the petals of BPFs. It has 3′,5′-side chains with alternate D-glucose and p-coumaric acid units at R and R1. Thus, delphinidin-3,3′,5′-triglucoside is the fundamental structure of all ternatins in BPFs.

Azima et al. [39] reported that when compared with other natural colourants, BPF anthocyanin has a more intense, vibrant, and saturated colour due to significantly higher colour density and chroma values. Furthermore, it was found that cyanidin-3-(p-coumaroyl-glucoside) was the most prevalent anthocyanin present based on its intensity [1] and was identified as the anthocyanin in the aqueous extract of BPFs' blue petals.

3.2.3. pH Sensitivity of Anthocyanin

Anthocyanins, water-soluble compounds that can create a broad range of colours (red, pink, purple, and blue), are commonly extracted from flowers, vegetables, cereals, and fruits. Additionally, Balbinot-Alfaro et al. [40] reported that anthocyanins can be detected in different colours and chemical forms depending on the pH of the solution, and this could be used to track food products over their shelf-life and, eventually, monitor food quality criteria. The source, composition, and structure of anthocyanins are related to the reversible colour characteristics of anthocyanin-rich plants [6]. The ability of anthocyanin to change colour is a unique characteristic. Four different coloured anthocyanin forms can alternately change based on the pH of the solution ranging from 1 to 13, as shown in Figure 3.

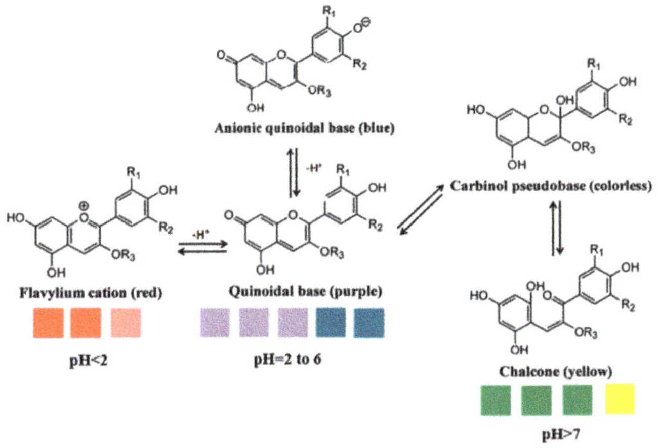

Figure 3. The main forms of anthocyanins at varying pH [6].

Based on Figure 3, the anthocyanin-rich extract's colour changes can be seen when the pH rises gradually from the acidic to the alkaline area. Because of the bathochromic and hyperchromic characteristics [41], the colour of the anthocyanin-rich extract is at pH values pH < 2; where it is at very acidic pH values, the formation of the red flavylium ion (AH+) is favoured. This species is fully protonated and has a delocalised positive charge across the chromophore. With a slight increase in pH, the first deprotonation occurs where kinetic and thermodynamic competition happens between the hydration reaction of the flavylium cation and the proton transfer reactions related to its acidic hydroxyl groups, thus converting into the neutral quinoid base (purple). The colour changes towards a red to purple quinoidal base. At the pH values between six and seven, the quinoid base (purple colour) is deprotonated further, forming the anionic quinonoid base (blue) with a negative delocalised charge, and the colour changes from purple to blue [42]. As the pH is raised from acidic conditions to slightly acidic or nearly neutral circumstances, this blue colour will appear.

The stability of anthocyanins gradually declines as pH increases (pH > 7), where it is dependent on their substituent groups and will generate–a green-light yellow colour as a result of isomerisation into chalcone formation via water catalysed-tautomerisation [43]. It is well-known that anthocyanin properties, including colour expression, are highly influenced by anthocyanin structure and pH. Due to these abilities, anthocyanins extracted from BPFs could potentially act as naturally derived pH dyes for colourimetric indicators in the monitoring of seafood and poultry freshness since they provide significant changes in the colour spectra with the changes in pH, and they are a non-toxic compound and give lower risk to the consumer. Therefore, the pH indicator sensitivity of BPF anthocyanins may be beneficial to be facilitated in intelligent packaging systems.

3.3. Anthocyanin Extraction from BPFs

Extraction is the most important step to isolate polyphenols and anthocyanin as natural colourants from plant sources. Various factors affect the overall extractable yield of anthocyanin in *Clitorria ternatea* that are included in the stability/sensitivity of BPFs such as temperature, pH, light, presence of enzymes, oxygen, metal ions, sulphur dioxide, and phenolic acids [44]. The most important factor to extract anthocyanin from plant sources is the selection of solvent, extraction time and temperature, and the ratio of the extractable substrate (flower) to solvent ratio [39], which will affect the anthocyanin content in the BPF. Traditionally, the direct addition of a powdered form of the dried BPF is a common procedure for the extraction process as shown in Figure 4. Prior to extraction, plant materials are prepared by size reduction using a grinder on either fresh [16] or dried samples [45] to increase the surface area for mixing with solvent. On the other hand, extraction can be divided into two methods, which are either conventional or non-conventional extraction methods to extract anthocyanins from the flower. Therefore, for this review, only extraction techniques that can save time and cost and produce a high extraction yield of anthocyanin from BPF anthocyanin will be discussed.

3.3.1. Extraction Solvent

Plant extraction involves separating the desired plant material components from inactive or undesirable compounds [39], mainly with the help of a liquid solvent. The alcohol-extracted anthocyanins exhibit greater functionality and higher effectiveness in their research on solvent extraction parameters on the quality of BPF extraction [6]. The expansion of the plant matrix by the water increased interactions between the surface area of the plant material and the solvent, and it increases the alcohol's ability to be extracted from the plant material [38]. Examples of alcohol that may be used in anthocyanin extraction are ethanol, methanol, acetone, or mixed solvent with water. The best solvent for natural compounds used for both food and natural medicine is ethanol, which is safe to be consumed by humans. However, using methanol should not be used as a solvent because it is a toxic, water-miscible, and volatile corrosive alcohol.

Tena and Asuero [46] stated that a high amount of ethanol in the solvents promotes the extraction of bioactive chemicals from plant materials, such as anthocyanin in flavanol groups. This is because the presence of aromatic groups and glycosyl residues in anthocyanins causes molecules to dissolve more effectively in a polar solvent which is easily absorbed by cell membranes on the surface of tissue particles in flower petals [41]. In addition, they discovered that using water and ethanol (20–70% v/v) together as a solvent improved the amounts of monomeric anthocyanins compared with using either water or ethanol alone [46]. Jeyaraj et al. [34] found that 50% ethanol (50–50% v/v) was discovered to be the best solvent for BPF extraction with 57.3% extracted yield and a 5.1 mg/g of total anthocyanin content compared with other types of solvent. Therefore, the type of solvent extraction process, specifically, alcohol mixed into water, affects the levels of anthocyanin.

3.3.2. Conventional and Non-Conventional Extraction Methods

The conventional extraction method is a traditional method that has been used since the 1970s for the extraction of BPF. Soxhlet extraction, maceration, and hydro distillation are among the conventional methods that usually involve the use of different solvents with heat and mixing. These methods are effective, but they consume more solvents, time, and thermal energy and are also associated with several disadvantages [47]. Therefore, to overcome the shortcomings of conventional solvent extraction and to increase extraction efficiency, several non-conventional extraction techniques were explored. Non-conventional extraction methods consist of ultrasound-assisted extraction [45], enzyme-assisted extraction, supercritical fluid extraction, microwave-assisted extraction, and pressurised liquid extraction [48]. These techniques are emerging, rapid, eco-friendly, and highly efficient over conventional extraction methods. According to Vidana et al. [36], ultrasound- and microwave-assisted extraction are the best methods to be applied for the extraction of anthocyanin from BPFs. As for now, however, among the non-conventional extraction methods, only ultrasound- and microwave-assisted extraction have been conducted to extract phytochemicals and it has been shown to improve the bioactive compound on *C. ternatea* flowers as shown in the processes in Figure 4.

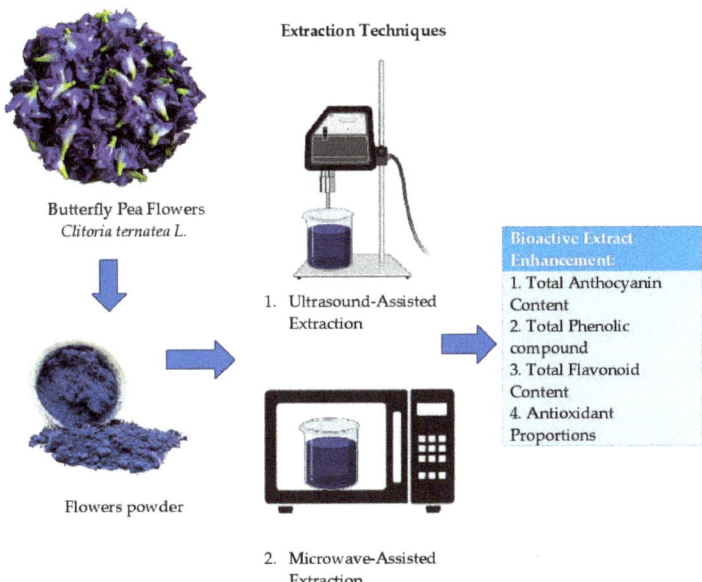

Figure 4. The non-conventional extraction processes of BPF.

3.3.3. Ultrasonic-Assisted Extraction (UAE)

Ultrasonic-assisted extraction (UAE) is a renewable and eco-friendly technology that uses acoustic cavitation to cause molecular movement of the solvent and sample, resulting in the breakdown of plant cell walls and membranes, and thus facilitating their movement to the surrounding solvent [49]. Husin et al. [45] conducted a study using UAE to extract anthocyanin from BPF as a visual indicator for monitoring meat freshness. Apart from that, comparisons were performed between UAE and conventional extraction methods on the extraction of anthocyanins from BPF. It was found that using UAE showed 246.48% better anthocyanin yield with 1.126 mg/g of total anthocyanin content [50].

The UAE procedure can be carried out with a bath ultrasonic (BUE) or probe ultrasonic (PUE). According to Kumar et al. [51], BUE is more practical and cost-effective. However, the energy generated is not evenly spread throughout the bath, which lowers the extraction's effectiveness. Consequently, it is less reproducible than PUE. PUE consists of a probe connected to a transducer, which is applied directly to the sample. The probe is directly immersed in the extraction and disperses the ultrasound in the media with the least amount of energy loss. PUE delivers a more intense ultrasonic sound than the bath device. However, there is not any research showing a comparison between BUE and PUE for BPF extraction. In PUE, the extraction process consists of four steps [52]: (1) when ultrasound waves pass through the solvent, cavitation bubbles are generated near the outermost layer of the plant matrix; (2) the bubbles erupt, releasing a microjet of higher temperature and pressure; (3) the plant matrix surface is ruptured, creating direct contact between the cell membrane and the solvent; (4) the solvent solubilises and transports the intracellular components, making the anthocyanin of the plant during extract more efficient. The ultrasonication process has an impact on the base fluid's agglomerates and nanoparticle sizes [53]. Therefore, probe ultrasonication is frequently chosen to extract bioactive compounds compared with the use of bath ultrasonication.

3.3.4. Microwave-Assisted Extraction (MAE)

Microwave-assisted extraction (MAE) is known as a green technology that uses microwave energy to facilitate the segregation of analytes which are high recovery and solvent blends from the plant matrix into the solvent [54]. The major factor that determines the efficiency of microwave energy is the choice of the solid-to-solvent ratio, plant matrix properties, irradiation time, and irradiated microwave power [55]. Theoretically, the microwave induces dipole rotation of the molecules, which makes electromagnetic energy convert to heat and disrupts the hydrogen bonding, thereby increasing the migration of dissolved ions and promoting solvent penetration into the plant matrix [56]. Various studies have been performed using the microwave-assisted technique to extract natural colourants from BPFs. Gamage and Choo [57] compared the extraction efficiency of natural BPF colourant (anthocyanins) using conventional extraction with water and heating of MAE. MAE was more efficient (9.61 mg/g) than conventional extraction (5.22 mg/g), having a higher total anthocyanin content.

According to the study by Marsin et al. [58], the anthocyanin content of the BPF extract had a short extract time, within 1 min, and obtained 0.541 mg/g of anthocyanin content. This is because Farzaneh and Carvalho [59] mentioned that increasing the time of microwave irradiation might degrade the anthocyanin content and reduce the effectiveness of extracting all of the anthocyanin yields in the plant. In contrast, Romero-díe et al. [60] stated that a longer exposure time to MAE improved extraction yield and simultaneously enhanced processing temperature, which may have contributed to the higher yield of anthocyanin extraction; this study was supported by reference [61], where the anthocyanin content of the 15 min MAE was 0.457 mg/g. Therefore, MAE is considered an efficient approach to extracting valuable active compounds from plant materials, maintaining their natural colourant, and being time efficient and easy to handle.

3.4. Intelligent Packaging

An intelligent packaging system consists of inexpensive components and compact labels or tags that can collect, store, and transmit data about the features and characteristics of packaged food [62]. Intelligent packaging can monitor, capture, and record changes in the product or its external environment and allows it to deliver information to the consumer as an extension of the communication function of conventional food packaging [63,64]. It is crucial to note that intelligent packaging and active packaging are two different things, but some packaging systems may be classified as both. According to Fang et al. [65], active packaging enhances the protection function of conventional packaging, and it is designed to contain a component that enables the release or absorption of substances (e.g., the release of an antimicrobial or antioxidant) into or from the packaged food or the environment surrounding the food to increase shelf-life and food safety and quality. Active packaging and intelligent packaging are compatible. Both packaging systems can work synergistically to achieve the quality of smart packaging. Smart packaging also offers a comprehensive packaging solution that both actively responds to and intelligently analyses changes in the product or environment [66].

3.4.1. Time-Temperature Indicators

Temperature is typically the most important environmental factor determining food spoilage and deterioration [67]. The term "time-temperature indicator or integrator" (TTI) refers to a straightforward, reasonably priced device fastened to shipping containers or individual consumer packages that can display measurably time-dependent changes reflecting the entire or a portion of a food product's temperature history [65]. The main mechanisms of action include enzymatic reactions, polymerisation, and chemical dispersion [66]. Therefore, TTI is useful; it can inform about temperature abuse and could be used as an indirect shelf-life indicator for perishable food products.

3.4.2. Integrity Indicators

Integrity indicators are one of the components of intelligent packaging that function as time indicators and reveal how long a product has been opened. They are also called seal and leak indicators or gas indicators. The application of this intelligent packaging is that foods are packaged with the essential gas composition for storage [66]. This is because the gas composition in the package frequently changes because of the water activity of the food product, leaks, the nature of the package, and external environmental factors [68]. Realini and Marcos [63] stated that when a seal is broken, an integrity indicator activates the label now that it is being consumed, initiates a timer and colour changes to happen throughout time. The functionality of most devices is based on redox dyes, a reducing compound, and an alkaline component [69].

3.4.3. Freshness Indicators

Freshness indicators or pH indicators are such devices that can directly be used to provide an estimate of the remaining shelf-life of perishable products [68]. Most of the pH or freshness indicators cause an alteration in colour in the sensor due to the presence of volatile amine compounds and increasing pH during the spoilage of food products. Horan [70] reported that biogenic amines (putrescine, cadaverine, histamine) are created when proteins in perishable food are broken down into amino acids and then those amino acids are enzymatically decarboxylated. Therefore, the production of biogenic amines from muscle foods can give a direct signal to a pH indicator (colour change), which acts as an indicator of food deterioration.

Bromophenol blue pH dye is the most applied in the muscle food packaging industry to monitor freshness from the production of carbon dioxide caused by microbial growth from protein degradation [69]. Additionally, high carbon dioxide levels cause pH dyes to react and change colour. However, bromophenol blue is a synthetic dye and this synthetic colour is prone to cause problems; it can be hazardous, especially if it is to be used for food purposes. Synthetic dyes such as methyl red and bromocresol green might be toxic (harmful) to humans, and can cause lung disease and skin infection [71]. Therefore, synthetic dyes were replaced with anthocyanin, which is a water-soluble natural pigment (non-toxic) from plant material. There are numerous types of freshness/pH indicators that have been developed recently by immobilising anthocyanin. Apart from that, anthocyanins are responsible for the blue, red, or purple colour based on the anthocyanin's source and composition, where it can visibly change colour when the pH of the surrounding area changes.

3.5. Application of BPF Anthocyanin (BPFA) as a Polymer-Based pH Film Indicator

Studies have developed intelligent packaging with natural pH indicators (BPF) due to their inexpensive cost, safety in contacting food materials, and ability to monitor with the naked eye the pH changes caused by food deterioration. The films can be developed using many types of polymer-carriers with different properties and characteristics. Figure 5 illustrates the fabrication of the BPF–polymer-based film and the colour changes of the film in different pH solutions (pH 1–pH 14).

3.5.1. Effect of BPFA on the Physical Properties of Films
Thickness

The physicochemical, light transmission, and barrier properties (water, gas) of the produced polymer matrix are all influenced by the thickness of packing films, making them a crucial factor [6]. The polymer matrix's composition and the dispersibility of anthocyanin have a significant impact on the pH indicator film. Several studies reported that low concentrations of anthocyanin do not significantly alter the thickness of the films because anthocyanins can distribute evenly throughout the film matrix [26]. Hidayati et al. [17] indicated that the thickness of the film is not significantly influenced by the immobilisation of BPFAs; this is because of the fixed BPFA concentration (ratio), and proper solubility of BPFAs mixed with the chitosan: polyvinyl alcohol (CH: PVA) composite matrix. Similarly, Sai-Ut [19] stated that there is no significant difference in the thickness of the gelatine and methylcellulose films' thickness. This might be caused by the extract's small volume and low moisture content in film compositions.

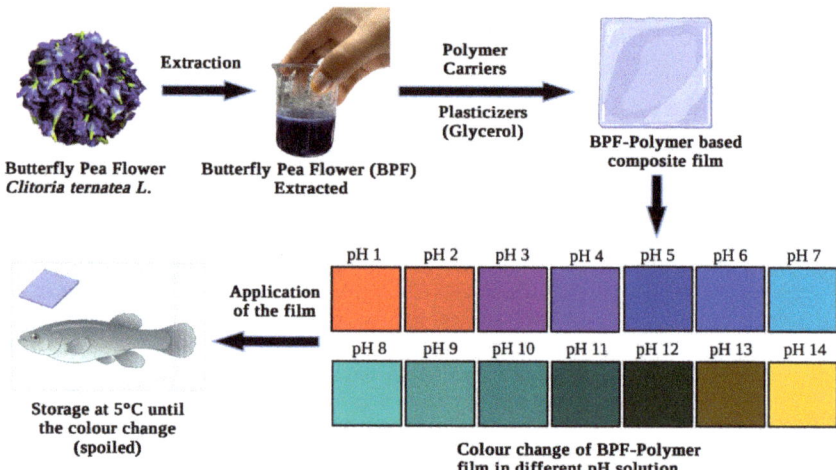

Figure 5. Fabrication of BPF–polymer-based pH indicator film.

On the other hand, the stability of the polymer film and the thickness of the film were unaffected by increasing anthocyanin content. On this matter, the thickness of pH-colourimetric hydroxypropyl methylcellulose (HM) films recorded no significant difference with the addition of BPF anthocyanin content [18]. In contrast, Yan et al. [22] stated that increases in the amount of BPFA-loaded chitosan films have significantly increased compared with the control films without BPFA. In line with these findings, the incorporation of BPFA resulted in a higher thickness value of colourimetric corn-starch-based films, and sago-starch-based films recorded the lowest value [17].

Following these studies, based on the composition of the polymer matrix and the amount of BPFA properties, it can be concluded that the addition of BPFAs had a variety of effects on the thickness (thicker or thinner) of pH indicator intelligent films. Table 2 shows the type of polymer used and their physical and mechanical properties when incorporated with BPFA.

Water Vapour Permeability

Water vapour permeability (WVP) is an important aspect of evaluating the effectiveness of polymeric materials used in packaging films for preventing moisture transfer (permeability) attributes into food packaging structures to act as barrier properties and is considered an important criterion in monitoring food quality and safety [72]. The prevention and minimising of oxygen and moisture transfer between the packaged food and the external environment are crucial factors of food packaging systems to prevent food spoilage from happening quickly [26]. Yan et al. [21] stated that the WVP of the film matrix was significantly higher due to the hydrophilicity of chitosan and that with the increasing amount of BPFA, the wettability of the film surface was enhanced. Polyphenolic molecules of BPFA possibly caused low intermolecular interactions in the film network. Thus, the hydroxyl groups of BPFA molecules might cooperate with water and intervene with the network formation of hydrogen bonds.

Moreover, Sai-Ut et al. [18] stated that there was an increased WVP in gelatine and methylcellulose composite films when they were infused with BPFA. The value of WVP can be influenced by the incorporation of bioactive compounds. On the other hand, Wu et al. [15] reported that the incorporation of BPFAs and gellan gum/heat-treated soy protein isolate (G/HSPI) matrix films significantly decreased the WVP value. The research reported that these changes may be linked to the addition of anthocyanin compounds (polyphenols) into polysaccharide- (gellan gum) or protein-based (soy protein isolated) films, which resulted in the interactions between the polymer-based matrix film and BPFA due to the hydrogen bonds and noncovalent hydrophobic interactions were formed when polyphenol (anthocyanin) were added to polysaccharide- or protein-based films [73], Moreover, the BPFA structure's aromatic rings have an impact on the development of a stronger microstructure link, which lowers the polymer chain's binding affinity to water molecules.

Mary et al. [20] also reported that the WVP value of starch with titanium dioxide (TiO_2) was decreased with the addition of BPFA. The probable cause for this result was the restriction of water vapour pathways in the polymer film matrix by TiO_2 nanoparticles. Moreover, titanium oxide particles bind with the starch hydroxyl group to form a complex network that inhibits the hydrophilic groups' capability for the absorption of water vapour. However, Hashim et al. [14] stated that the incorporation of BPFA into sugarcane wax/agar matrix films does not change the WVP value of the films. Many variables affect the permeability of the film, such as how the concentration of plasticiser should be compatible with the film-forming polymer [74], the different proportions of biopolymer during the development of the film, as well as the drying and storage time effect [72].

3.5.2. Effect of BPFA on the Mechanical Properties of Films

Food products' sustainability and integrity could be ensured by matrix films with sufficient mechanical strength [6]. There are two major mechanical criteria for judging the durability and flexibility of packaging films: tensile strength (TS) and elongation at break (EAB). According to Yong et al. [5], TS is defined as the maximum tolerance of composite films against the applied stress while being pulled or stretched before breaking occurs. Moreover, EAB is defined as the maximum capability of composite films to maintain alterations in the length and shape of the films deprived of any crack formation. Generally, the mechanical properties (TS and EAB) of BPFA-rich films can be affected by various factors. In this regard, Koshy et al. [19] reported that the TS of the BPFA mixed with soy protein isolates/chitin nanowhisker (SPI/CNW) composite films decreased with the addition of BPFA. This might be due to the occurrence of several interactions between BPFA and the composite matrix due to the plasticisation impact of BPFA, which breaks the compact, stiff structure between SPI and CNW by promoting molecular mobility of the polymer chains and destroying the hydrogen connection between SPI and CNW, despite the EAB value showing a significant increase. This could be explained by the extract's phenols and flavonoids operating where they are repulsive forces to each other, decreasing the attraction between the chains of proteins.

Nevertheless, in research by Wu et al. [14], the addition of BPFA in the gellan gum/heat-treated soy protein isolate G/HSPIs composite significantly decreased the values in both mechanical TS and EAB. The presence of the anthocyanin-rich extract may weaken the intermolecular interaction and discontinuous microstructure of gellan gum and heat-treated soy protein isolate film. In contrast, Yan et al. [21] stated that up to 20% concentration of BPF anthocyanin pigment was added to cationic chitosan matrix films, both TS and EAB values were significantly increased because of the many hydroxyl groups in anthocyanin that create hydrogen bonds and result in significant interfacial adhesion between the polymer and BPFA extract. The additional anthocyanin compound may increase the volume of molecules that can move freely, enhancing the flexibility of the composite film.

Moreover, the difference in anthocyanin source could have a significant impact on the mechanical properties of the polymer-based matrix films loaded with BPFAs. For instance, in Boonsiriwit et al.'s [17] work, the hydroxypropyl methylcellulose, HMB-based intelligent packaging films contained an increasing amount of anthocyanins demonstrated an increasing TS; however, the tensile strength of the highest BPFA-HMB was not significantly different from that of the control (without BPFA). The changes in TS can be clarified based on the strength of the electrostatic contacts between anthocyanin molecules and the HMB composite, as well as the intermolecular interactions [74], while a lower EAB was found with a significant increase in the BPFA concentration. This was due to the high concentration of phenolic chemicals in BPFA, which restricted polymer chain motion and prevented chain–chain connections of polymeric backbones; this was attributed to the interactions of BPFA with the HMB matrix, which decreased the flexibility of the films. Overall, the type of polymer used, co-film forming elements (plasticiser) and BPFA concentration could influence the mechanical properties of BPFA-loaded matrix films [69].

3.6. Evaluation of BPFA pH Indicator Potential Tested on Food

Natural biopolymers, particularly those based on polysaccharides and proteins, have been utilised extensively to create colourimetric intelligent packaging films because of their biodegradability, nontoxicity, safety, stability, and good film-forming capabilities [6]. Figure 6 illustrates the typical freshness indicator system and demonstrates the colour changes when the pH increases from blue (fresh) to green (spoiled) inside the food packaging, while Table 3 shows the types of film matrices used, foods and their storage conditions, and colour changes observed in the films.

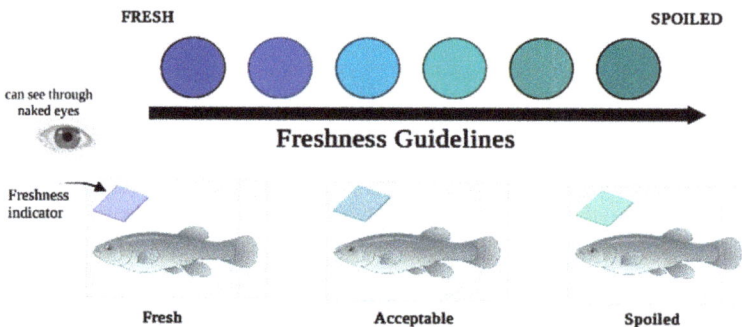

Figure 6. Typical freshness indicator system applied to food.

Ahmad et al. [25] reported good performance of an edible BPFA pH indicator immobilized in t-carrageenan, as it was employed to track the pH alterations of durian at room temperature. In addition, the pH-sensitive indicator can be used to observe pasteurised milk, as it can easily distinguish between fresh and deteriorated milk using colour detection and will show a deep blue to the light blue colour range [29]. Boonsiriwit [17] developed a pH-colourimetric indicator film using a hydroxypropyl methylcellulose biocomposite and BPFA. It was observed that the colour of the anthocyanin compound on the indicator film changed from deep purple to violet on mackerel fish storage. Approximately 1.5 g of BPFA indicator exhibited the most efficient indicator in terms of NH_3 reactivity. The 1.5 BPFA-HMB indicator clearly changed colour in response to variations in mackerel fish quality.

Chitosan is the most common natural biopolymer that has been used in the fabrication of smart films with BPFA. Yan et al. [20] monitored tilapia fish deterioration using chitosan film immobilised with BPFA. According to the research, colour alterations in the film were seen as a result of the fish sample's various storage conditions at room temperature, and it was observed that no colour change was seen on the freshness indicator film after day two. However, after day four of storage, the colour began to shift to green, which showed initial spoilage and pH increase. After day six of storage, the colour completely turned to dark green, indicating that the tilapia had undergone complete spoilage. Sumiasih [27] developed an intelligent pH indicator film using chitosan and PVA and evaluated a BPFA to monitor beef spoilage. As a result of the film's high sensitivity to pH changes, the results showed that the indicator film's initial blue colour drastically altered to a greenish colour within 24 h at room temperature storage, which may undergo a decomposition process. Generally, as the seafood and poultry deteriorated, it produced an unpleasant aroma from the formation of volatile alkaline compounds such as ammonia, dimethylamine (DMA), and trimethylamine (TMA) from enzymatic reactions [75]. Volatile alkaline was produced as bacteria degraded proteins into amino acids [70] and oxidation of unsaturated fatty acids in the muscle food body occurred [76]. The BPF's anthocyanin characteristics are sensitive to variations in acidity. Therefore, the pH change from acidic to alkaline can be detected by the anthocyanin in the BPF extract changing colour.

4. Conclusions

Recently, the immobilisation of butterfly pea flowers' anthocyanin (BPFAs) into polymer-based films has provided advancements for intelligent food packaging application systems. Incorporating BPFA into the starch composite of packaging films often causes different changes in their physicochemical properties, mainly due to interactions between natural polymer-carriers and hydroxyl groups in anthocyanins. Furthermore, the BPF is a plant that is easy to obtain, is not seasonal, and has potential features such as antioxidant capabilities in addition to pH sensors in food packaging systems to prolong shelf-life. To conclude, the BPF has a promising future for intelligent polymeric films using BPFAs due to their multifunctional utilisation such as monitoring and improving food products and consumer safety.

Author Contributions: Conceptualisation, M.R.I.-F. and N.N.H.; methodology, N.N.H.; software, N.N.H.; validation, M.R.I.-F., E.M.A. and A.R.; formal analysis, N.N.H.; investigation, N.N.H.; resources, M.R.I.-F.; data curation, N.N.H.; writing—original draft preparation, N.N.H.; writing—review and editing, M.R.I.-F., E.M.A., A.R. and N.H.Z.A.; supervision, M.R.I.-F., E.M.A., A.R. and N.H.Z.A.; project administration, M.R.I.-F.; funding acquisition, M.R.I.-F. All authors have read and agreed to the published version of the manuscript.

Funding: This research was funded by the Malaysian Ministry of Higher Education, grant number FRGS/1/2020/WAB04/UPM/01/3, and the APC was funded by the Research Management Centre, Universiti Putra Malaysia.

Institutional Review Board Statement: Not applicable.

Informed Consent Statement: Not applicable.

Data Availability Statement: Not applicable.

Conflicts of Interest: The authors declare no conflict of interest.

References

1. Thuy, N.M.; Minh, V.Q.; Ben, T.C.; Thi Nguyen, M.T.; Ha, H.T.N.; Tai, N.V. Identification of anthocyanin compounds in butterfly pea flowers (*Clitoria ternatea* L.) by ultra-performance liquid chromatography/ultraviolet coupled to mass spectrometry. *Molecules* **2021**, *26*, 4539. [CrossRef] [PubMed]
2. Jamil, N.; Pa'ee, F. Antimicrobial activity from leaf, flower, stem, and root of *Clitoria ternatea*–A review. In Proceedings of the AIP Conference, Yogyakarta, Indonesia, 15 August 2018; Volume 2002, p. 020044.
3. Marpaung, A.M.; Lee, M.; Kartawiria, I.S. The development of butterfly pea (*Clitoria ternatea*) flower powder drink by co-crystallization. *Indones. Food Sci. Technol. J.* **2020**, *3*, 34–37. [CrossRef]
4. Suarna, I.W.; Wijaya, I.M.S. Butterfly pea (*Clitoria ternatea* L.: Fabaceae) and its morphological variations in Bali. *J. Trop. Biodivers. Biotechnol.* **2021**, *6*, 63013. [CrossRef]
5. Rahim, M.Z.A.; Husin, N.; Mohd Noor, M.A.; Yet, Z.R.; Ismail-Fitry, M.R. Screening of natural colours from various natural resources as potential reusable visual indicators for monitoring food freshness. *Malaysian J. Anal. Sci.* **2020**, *24*, 288–299.
6. Abedi-Firoozjah, R.; Yousefi, S.; Heydari, M.; Seyedfatehi, F.; Jafarzadeh, S.; Mohammadi, R.; Rouhi, M.; Garavand, F. Application of red cabbage anthocyanins as pH-sensitive pigments in smart food packaging and sensors. *Polymers* **2022**, *14*, 1629. [CrossRef]
7. Ghaani, M.; Cozzolino, C.A.; Castelli, G.; Farris, S. An overview of the intelligent packaging technologies in the food sector. *Trends Food Sci. Technol.* **2016**, *51*, 1–11. [CrossRef]
8. Poh, N.A.B.A. *Investigation of the Colour Changes of Clitoria ternatea in Different pH Conditions*; School of Science and Technology: Singapore, 2019.
9. Roy, S.; Rhim, J.W. Anthocyanin food colorant and its application in pH-responsive color change indicator films. *Crit. Rev. Food Sci. Nutr.* **2021**, *61*, 2297–2325. [CrossRef]
10. Omar, S.R.; Noh, A.A.; Rohaizan, A.N. The Potential Use of Anthocyanin in Butterfly Pea (*Clinoteria ternatea*) Petals as Colorimetric Indicator in Intelligent Food Packaging Article. *Malaysian J. Sci. Health Technol.* **2022**, *8*, 71–76. [CrossRef]
11. Oladzadabbasabadi, N.; Nafchi, A.M.; Ghasemlou, M.; Ariffin, F.; Singh, Z.; Al-Hassan, A.A. Natural anthocyanins: Sources, extraction, characterization, and suitability for smart packaging. *Food Packag. Shelf Life* **2022**, *33*, 100872. [CrossRef]

12. Page, M.J.; McKenzie, J.E.; Bossuyt, P.M.; Boutron, I.; Hoffmann, T.C.; Mulrow, C.D.; Shamseer, L.; Tetzlaff, J.M.; Akl, E.A.; Brennan, S.E.; et al. The PRISMA 2020 Statement: An Updated Guideline for Reporting Systematic Reviews. *BMJ* **2021**, *372*, n71. [CrossRef]
13. Critical Appraisal Skills Programme. CASP (Qualitative Studies) Checklist. 2018. Available online: https://casp-uk.net (accessed on 6 May 2023).
14. Hashim, S.B.; Tahir, H.E.; Liu, L.; Zhang, J.; Zhai, X.; Mahdi, A.A.; Awad, F.N.; Haassan, M.M.; Xiaobo, Z.; Jiyong, S. Intelligent colorimetric pH sensoring packaging films based on sugarcane wax/agar integrated with butterfly pea flower extract for optical tracking of shrimp freshness. *Food Chem.* **2022**, *373*, 131514. [CrossRef] [PubMed]
15. Wu, L.T.; Tsai, I.L.; Ho, Y.C.; Hang, Y.H.; Lin, C.; Tsai, M.L.; Mi, F.L. Active and intelligent gellan gum-based packaging films for controlling anthocyanins release and monitoring food freshness. *Carbohydr. Polym.* **2021**, *254*, 117410. [CrossRef] [PubMed]
16. Ahmad, A.N.; Abdullah Lim, S.; Navaranjan, N. Development of sago (metroxylon sagu)-based colorimetric indicator incorporated with butterfly pea (*Clitoria ternatea*) anthocyanin for intelligent food packaging. *J. Food Saf.* **2020**, *40*, e12807. [CrossRef]
17. Hidayati, N.A.; Wijaya, M.W.; Bintoro, V.P.; Mulyani, S.; Pratama, Y. Development of biodegradable smart packaging from chitosan, polyvinyl alcohol (PVA) and butterfly pea flower's (*Clitoria ternatea* L.) anthocyanin extract. *Food Res.* **2021**, *5*, 307–314. [CrossRef]
18. Boonsiriwit, A.; Lee, M.; Kim, M.; Inthamat, P.; Siripatrawan, U.; Lee, Y.S. Hydroxypropyl methylcellulose/microcrystalline cellulose biocomposite film incorporated with butterfly pea anthocyanin as a sustainable pH-responsive indicator for intelligent food-packaging applications. *Food Biosci.* **2021**, *44*, 101392. [CrossRef]
19. Sai-Ut, S.; Suthiluk, P.; Tongdeesoontorn, W.; Rawdkuen, S.; Kaewprachu, P.; Karbowiak, T.; Degraeve, P.; Degraeve, P. Using anthocyanin extracts from butterfly pea as pH indicator for intelligent gelatin film and methylcellulose film. *Curr. Appl. Sci. Technol.* **2021**, *21*, 652–661.
20. Koshy, R.R.; Reghunadhan, A.; Mary, S.K.; Pillai, P.S.; Joseph, S.; Pothen, L.A. pH indicator films fabricated from soy protein isolate modified with chitin nanowhisker and *Clitoria ternatea* flower extract. *Curr. Res. Food Sci.* **2022**, *5*, 743–751. [CrossRef]
21. Mary, S.K.; Koshy, R.R.; Daniel, J.; Koshy, J.T.; Pothen, L.A.; Thomas, S. Development of starch based intelligent films by incorporating anthocyanins of butterfly pea flower and TiO_2 and their applicability as freshness sensors for prawns during storage. *RSC Adv.* **2020**, *10*, 39822–39830. [CrossRef]
22. Yan, J.; Cui, R.; Qin, Y.; Li, L.; Yuan, M. A pH indicator film based on chitosan and butterfly pudding extract for monitoring fish freshness. *Int. J. Biol. Macromol.* **2021**, *177*, 328–336. [CrossRef]
23. Kim, H.J.; Roy, S.; Rhim, J.W. Gelatin/agar-based color-indicator film integrated with *Clitoria ternatea* flower anthocyanin and zinc oxide nanoparticles for monitoring freshness of shrimp. *Food Hydrocoll.* **2022**, *124*, 107294. [CrossRef]
24. Romruen, O.; Kaewprachu, P.; Karbowiak, T.; Rawdkuen, S. Development of smart bilayer alginate/agar film containing anthocyanin and catechin-lysozyme. *Polymers* **2022**, *14*, 5042. [CrossRef]
25. Ahmad, N.A.; Yook Heng, L.; Salam, F.; Mat Zaid, M.H.; Abu Hanifah, S. A colorimetric pH sensor based on *Clitoria* sp and *Brassica* sp for monitoring of food spoilage using chromametry. *Sensors* **2019**, *19*, 4813. [CrossRef]
26. Rawdkuen, S.; Faseha, A.; Benjakul, S.; Kaewprachu, P. Application of anthocyanin as a color indicator in gelatin films. *Food Biosci.* **2020**, *36*, 100603. [CrossRef]
27. Roy, S.; Kim, H.J.; Rhim, J.W. Effect of blended colorants of anthocyanin and shikonin on carboxymethyl cellulose/agar-based smart packaging film. *Int. J. Biol. Macromol.* **2021**, *183*, 305–315. [CrossRef]
28. Sumiasih, I.H. Indicator film of natural coloring of butterfly pea (*Clitoria ternatea* L.) as detection of beef damage: Indicator film of natural coloring of butterfly pea (*Clitoria ternatea* L.) as detection of beef damage. *Int. J. Appl. Biol.* **2022**, *6*, 79–92.
29. Cho, T.F.; Yassoralipour, A.; Lee, Y.Y.; Tang, T.K.; Lai, O.M.; Chong, L.C.; Kuan, C.-H.; Phuah, E.T. Evaluation of milk deterioration using simple biosensor. *J. Food Meas. Charact.* **2021**, *16*, 258–268. [CrossRef]
30. Chauhan, N.; Rajvaidhya, S.; Dubey, B.K. Pharmacognostical, phytochemical and pharmacological review on *Clitoria ternatea* for antiasthmatic activity. *Int. J. Pharm. Sci.* **2012**, *3*, 398.
31. Havananda, T.; Luengwilai, K. Variation in floral antioxidant activities and phytochemical properties among butterfly pea (*Clitoria ternatea* L.) germplasm. *Genet. Resour. Crop. Evol.* **2019**, *66*, 645–658. [CrossRef]
32. Jamil, N.; Zairi, M.N.M.; Nasim, N.A.I.M.; Pa'ee, F. Influences of environmental conditions to phytoconstituents in *Clitoria ternatea* (butterfly pea flower)—A review. *J. Sci. Technol.* **2018**, *10*, 208–228. [CrossRef]
33. Khoo, H.E.; Azlan, A.; Tang, S.T.; Lim, S.M. Anthocyanidins and anthocyanins: Colored pigments as food, pharmaceutical ingredients, and the potential health benefits. *Food Nutr. Res.* **2017**, *61*, 1361779. [CrossRef]
34. Jeyaraj, E.J.; Lim, Y.Y.; Choo, W.S. Extraction methods of butterfly pea (*Clitoria ternatea*) flower and biological activities of its phytochemicals. *J. Sci. Technol.* **2021**, *58*, 2054–2067. [CrossRef] [PubMed]
35. Madhu, K. Phytochemical screening and antioxidant activity of in vitro grown plants *Clitoria ternatea* L., using DPPH assay. *Asian J. Pharm. Clin. Res.* **2013**, *6*, 38–42.
36. Vidana-Gamage, G.C.; Lim, Y.Y.; Choo, W.S. Anthocyanins from *Clitoria ternatea* flower: Biosynthesis, extraction, stability, antioxidant activity, and applications. *Front. Plant. Sci.* **2021**, *12*, 792303. [CrossRef] [PubMed]
37. Adisakwattana, S.; Pasukamonset, P.; Chusak, C. Clitoria ternatea beverages and antioxidant usage. In *Pathology*; Academic Press: Cambridge, MA, USA, 2020; pp. 189–196.

38. Campbell, S.M.; Pearson, B.; Marble, S.C. *Butterfly Pea* (Clitoria ternatea) *Flower. Extract (BPFE) and Its Use as a pH-Dependent Natural Colorant*; University of Florida: Gainesville, FL, USA, 2019.
39. Azima, A.S.; Noriham, A.; Manshoor, N. Phenolics, antioxidants and color properties of aqueous pigmented plant extracts: *Ardisia colorata var. elliptica, Clitoria ternatea, Garcinia mangostana* and *Syzygium cumini*. *J. Funct. Foods* **2017**, *38*, 232–241. [CrossRef]
40. Balbinot-Alfaro, E.; Craveiro, D.V.; Lima, K.O.; Costa, H.L.G.; Lopes, D.R.; Prentice, C. Intelligent packaging with pH indicator potential. *Food Eng. Rev.* **2019**, *11*, 235–244. [CrossRef]
41. Wiyantoko, B. Butterfly pea (*Clitoria ternatea* L.) extract as indicator of acid-base titration. *Indones. J. Chem.* **2020**, *3*, 22–32. [CrossRef]
42. Basílio, N.; Pina, F. Chemistry and photochemistry of anthocyanins and related compounds: A thermodynamic and kinetic approach. *Molecules* **2016**, *21*, 1502. [CrossRef]
43. Saptarini, N.M.; Suryasaputra, D.; Nurmalia, H. Application of Butterfly Pea (*Clitoria ternatea Linn*) extract as an indicator of acid-base titration. *J. Chem. Pharm. Res.* **2015**, *7*, 275–280.
44. Sarwar, S.; Rahman, M.R.; Nahar, K.; Rahman, M.A. Analgesic and neuropharmacological activities of methanolic leaf extract of *Clitoria ternatea Linn*. *J. Pharmacogn. Phytochem.* **2014**, *2*, 110–114.
45. Husin, N.; Rahim, M.Z.A.; Zulkhairi, M.; Azizan, M.; Mohd Noor, M.A.; Rashedi, M.; Ismail-Fitry, M.R.; Hassan, N. Real-time monitoring of food freshness using delphinidin-based visual. *Malaysian J. Anal. Sci.* **2020**, *24*, 558–569.
46. Tena, N.; Asuero, A.G. Up-to-date analysis of the extraction methods for anthocyanins: Principles of the techniques, optimization, technical progress, and industrial application. *Antioxidants* **2022**, *11*, 286. [CrossRef]
47. Ijod, G.; Musa, F.N.; Anwar, F.; Suleiman, N.; Adzahan, N.M.; Azman, E.M. Thermal and nonthermal pretreatment methods for the extraction of anthocyanins: A review. *J. Food Process. Preserv.* **2022**, *46*, e17255. [CrossRef]
48. Wen, C.; Zhang, J.; Zhang, H.; Dzah, C.S.; Zandile, M.; Duan, Y.; Ma, H.; Luo, X. Advances in ultrasound assisted extraction of bioactive compounds from cash crops—A review. *Ultrason. Sonochem.* **2018**, *48*, 538–549. [CrossRef]
49. Chemat, F.; Khan, M.K. Applications of ultrasound in food technology: Processing, preservation, and extraction. *Ultrason. Sonochem.* **2011**, *18*, 813–835. [CrossRef]
50. Chong, F.C.; Gwee, X.F. Ultrasonic extraction of anthocyanin from *Clitoria ternatea* flowers using response surface methodology. *Nat. Prod. Res.* **2015**, *29*, 1485–1487. [CrossRef]
51. Kumar, K.; Srivastav, S.; Sharanagat, V.S. Ultrasound assisted extraction (UAE) of bioactive compounds from fruit and vegetable processing by-products: A review. *Ultrason. Sonochem.* **2021**, *70*, 105325. [CrossRef]
52. Panja, P. Green extraction methods of food polyphenols from vegetable materials. *Curr. Opin. Food Sci.* **2018**, *23*, 173–182. [CrossRef]
53. Sandhya, M.; Ramasamy, D.; Sudhakar, K.; Kadirgama, K.; Harun, W.S.W. Ultrasonication an intensifying tool for preparation of stable nanofluids and study the time influence on distinct properties of graphene nanofluids—A systematic overview. *Ultrason. Sonochem.* **2021**, *73*, 105479. [CrossRef]
54. Nour, A.H.; Oluwaseun, A.R.; Nour, A.H.; Omer, M.S.; Ahmad, N. Microwave-assisted extraction of bioactive compounds. In *Theory and Practice*; Springer: Berlin/Heidelberg, Germany, 2021; pp. 1–31.
55. Uzel, R.A. Microwave-assisted green extraction technology for sustainable food processing. In *Emerging Microwave Technologies in Industrial, Agricultural, Medical and Food Processing*; IntechOpen: London, UK, 2018; pp. 159–178.
56. Alupului, A.; Calinescu, I.; Lavric, V. Microwave extraction of active principles from medicinal plants. *UPB Sci. Bull. B Chem. Mater. Sci.* **2012**, *74*, 129–142.
57. Gamage, G.C.V.; Choo, W.S. Hot water extraction, ultrasound, microwave, and pectinase-assisted extraction of anthocyanins from blue pea flower. *Food Chem. Adv.* **2023**, *2*, 100209. [CrossRef]
58. Marsin, A.M.; Jusoh, Y.M.M.; Abang, D.N.; Zaidel, Z.H.; Yusof, A.H.M.; Muhamad, I.I. Microwave-assisted encapsulation of blue pea flower (*Clitoria ternatea*) colourant: Maltodextrin concentration, power, and time. *Chem. Eng.* **2020**, *78*, 199–204. [CrossRef]
59. Farzaneh, V.; Carvalho, I.S. Modelling of microwave assisted extraction (MAE) of anthocyanins (TMA). *J. Appl. Res. Med. Aromat. Plants* **2017**, *6*, 92–100. [CrossRef]
60. Romero-díez, R.; Matos, M.; Rodrigues, L.; Bronze, M.R.; Rodríguez-rojo, S. Microwave and ultrasound pre-treatments to enhance anthocyanins extraction from different wine lees. *Food Chem.* **2019**, *272*, 258–266. [CrossRef] [PubMed]
61. Izirwan, I.; Munusamy, T.D.; Hamidi, N.H.; Sulaiman, S.Z. Optimization of microwave-assisted extraction of anthocyanin from *Clitoria ternatea* flowers. *Int. J. Mech. Eng. Robot. Res.* **2020**, *9*, 1246–1252. [CrossRef]
62. Pereira de Abreu, D.A.; Cruz, J.M.; Paseiro Losada, P. Active and intelligent packaging for the food industry. *Food Rev. Int.* **2012**, *28*, 146–187. [CrossRef]
63. Realini, C.E.; Marcos, B. Active and intelligent packaging systems for a modern society. *Meat Sci.* **2014**, *98*, 404–419. [CrossRef]
64. Heising, J.K.; Dekker, M.; Bartels, P.V.; Van Boekel, M.A.J.S. Monitoring the quality of perishable foods: Opportunities for intelligent packaging. *Crit. Rev. Food Sci. Nutr.* **2014**, *54*, 645–654. [CrossRef]
65. Fang, Z.; Zhao, Y.; Warner, R.D.; Johnson, S.K. Active and intelligent packaging in meat industry. *Trends Food Sci. Technol.* **2017**, *61*, 60–71. [CrossRef]
66. Vanderroost, M.; Ragaert, P.; Devlieghere, F.; De Meulenaer, B. Intelligent food packaging: The next generation. *Trends Food Sci. Technol.* **2014**, *39*, 47–62. [CrossRef]
67. Müller, P.; Schmid, M. Intelligent packaging in the food sector: A brief overview. *Foods* **2019**, *8*, 16. [CrossRef]

68. Kuswandi, B. Freshness sensors for food packaging. In *Reference Module in Food Science*; Elsevier: Amsterdam, The Netherlands, 2017.
69. Jarray, A.; Gerbaud, V.; Hemati, M. Polymer-plasticizer compatibility during coating formulation: A multi-scale investigation. *Prog. Org. Coat.* **2016**, *101*, 195–206. [CrossRef]
70. Horan, T.J. Method for Determining Deleterious Bacterial Growth in Packaged Food Utilizing Hydrophilic Polymers. U.S. Patent 6,149,952, 21 November 2000.
71. Ledakowicz, S.; Paździor, K. Recent achievements in dyes removal focused on advanced oxidation processes integrated with biological methods. *Molecules* **2021**, *26*, 870. [CrossRef]
72. Grzebieniarz, W.; Tkaczewska, J.; Juszczak, L.; Kawecka, A.; Krzyściak, P.; Nowak, N.; Guzik, P.; Kasprzak, M.; Janik, M.; Jamróz, E. The influence of aqueous butterfly pea (*Clitoria ternatea*) flower extract on active and intelligent properties of furcellaran Double-Layered films-in vitro and in vivo research. *Food Chem.* **2023**, *413*, 135612. [CrossRef]
73. Du, Y.; Sun, J.; Wang, L.; Wu, C.; Gong, J.; Lin, L.; Pang, J. Development of antimicrobial packaging materials by incorporation of gallic acid into Ca^{2+} crosslinking konjac glucomannan/gellan gum films. *Int. J. Biol. Macromol.* **2019**, *137*, 1076–1085. [CrossRef]
74. Jamróz, E.; Kulawik, P.; Guzik, P.; Duda, I. The verification of intelligent properties of furcellaran films with plant extracts on the stored fresh Atlantic mackerel during storage at 2 °C. *Food Hydrocoll.* **2019**, *97*, 105211. [CrossRef]
75. Boziaris, I.S.; Stamatiou, A.P.; Nychas, G.J.E. Microbiological aspects and shelf life of processed seafood products. *J. Sci. Food Agric.* **2013**, *93*, 1184–1190. [CrossRef]
76. Liu, Y.; Huang, Y.; Wang, Z.; Cai, S.; Zhu, B.; Dong, X. Recent advances in fishy odour in aquatic fish products, from formation to control. *Int. J. Food Sci. Technol.* **2021**, *56*, 4959–4969. [CrossRef]

Disclaimer/Publisher's Note: The statements, opinions and data contained in all publications are solely those of the individual author(s) and contributor(s) and not of MDPI and/or the editor(s). MDPI and/or the editor(s) disclaim responsibility for any injury to people or property resulting from any ideas, methods, instructions or products referred to in the content.

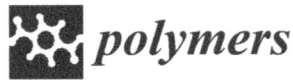

Article

Colorimetric Freshness Indicator Based on Cellulose Nanocrystal–Silver Nanoparticle Composite for Intelligent Food Packaging

Seongyoung Kwon and Seonghyuk Ko *

Department of Packaging, Yonsei University, Wonju 26493, Korea
* Correspondence: s.ko@yonsei.ac.kr

Abstract: In this study, a colorimetric freshness indicator based on cellulose nanocrystal-silver nanoparticles (CNC-AgNPs) was successfully fabricated to offer a convenient approach for monitoring the quality of packaged food. AgNPs were directly synthesized and embedded in CNC via a one-pot hydrothermal green synthesis, and CNC-AgNP composited indicator films were prepared using a simple casting method. The AgNPs obtained were confirmed by UV-Vis diffuse reflectance spectroscopy and X-ray diffraction. The ability of the as-prepared CNC-AgNP film to indicate food quality was assessed in terms of the intensity of its color change when in contact with spoilage gases from chicken breast. The CNC-AgNP films initially exhibited a yellowish to dark wine-red color depending on the amount of AgNPs involved. They gradually turned colorless and subsequently to metallic grey. This transition is attributed to the reaction of AgNPs and hydrogen sulfide (H_2S), which alters the surface plasmon resonance of AgNPs. Consequently, the color change was suitably discernible to the human eye, implying that the CNC-AgNP composite is a highly effective colorimetric freshness indicator. It can potentially serve as an accurate and irreversible food quality indicator in intelligent packaging during distribution or storage of products that emit hydrogen sulfide when deteriorating, such as poultry products or broccoli.

Keywords: cellulose nanocrystal; silver nanoparticle; hydrothermal green synthesis; colorimetric freshness indicator; intelligent packaging

Citation: Kwon, S.; Ko, S. Colorimetric Freshness Indicator Based on Cellulose Nanocrystal–Silver Nanoparticle Composite for Intelligent Food Packaging. *Polymers* 2022, 14, 3695. https://doi.org/10.3390/polym14173695

Academic Editors: Victor G. L. Souza, Lorenzo M. Pastrana and Ana Luisa Fernando

Received: 22 July 2022
Accepted: 28 August 2022
Published: 5 September 2022

Publisher's Note: MDPI stays neutral with regard to jurisdictional claims in published maps and institutional affiliations.

Copyright: © 2022 by the authors. Licensee MDPI, Basel, Switzerland. This article is an open access article distributed under the terms and conditions of the Creative Commons Attribution (CC BY) license (https://creativecommons.org/licenses/by/4.0/).

1. Introduction

Intelligent packaging is an advanced technique that can be used to monitor the external or internal environment of packaging and provide appropriate information to customers [1–3]. It is classified according to the type of smart device used, such as data carriers, sensors, and indicators. Barcodes and radio frequency identification tags (RFID) are typical data carriers that can provide information pertinent to supply chain management, facilitating inventory control and product identification [4]. Sensors typically include a receptor, transducer, signal-processing electronics, and signal display [5] and can be categorized as gas sensors or biosensors based on the type of receptor [6,7]. Time-temperature and shock-vibration indicators are representative indicators that are placed on the exterior of the packaging to record the surroundings. Other sensors include gas or freshness indicators placed within the packaging for the direct detection of gas, pathogens, pH, or their derivatives induced by food decay within the packaging. They primarily comprise redox dyes, pH dyes, or dyes that sense metabolites and convey information on food quality through color changes [1].

Freshness indicators based on gas or biosensors have been extensively studied because they can directly monitor the quality of packaged food. Recently, the replacement of typical dyes with metal nanoparticles (NPs) has attracted significant attention. Metal NPs change their color when a surface chemical reaction occurs due to specific chemicals that are released during food decay [8,9]. The highly sensitive and selective sensing characteristics of metal

NPs, such as gold (Au) and silver (Ag), enable their application in gas detection [10–13] as well as biosensing and bioimaging [14,15].

Silver nanoparticles (AgNPs) are well-known nanomaterials used in sensing systems owing to their selective detection of hydrogen sulfide (H_2S) and unique optical properties. When the diameter of the AgNP is considerably smaller than the wavelength of the incident light ($R/\lambda \ll 0.1$), the electron cloud on the AgNP oscillates owing to the attraction of protons and repulsion of the electromagnetic field [16]. If the oscillation is enhanced by incident light of a specific wavelength, the light scattering and absorption of AgNPs are maximized, resulting in a unique color expression [16,17]. This phenomenon, known as localized surface plasmon resonance (LSPR), is dependent on the density of the electric field located on the AgNPs, which can be varied by manipulating the size and shape of the AgNPs [18]. The surface chemical reaction between AgNPs and hydrogen sulfide (H_2S), called sulfidation, induces a change in electron cloud density and intrinsic color of AgNPs. When AgNPs react with H_2S under moist aerobic conditions at room temperature, the AgNP surface is transformed to silver sulfide (Ag_2S) [19], resulting in a silver-silver sulfide ($Ag@Ag_2S$) core-shell structure. Furthermore, the Ag_2S layer changes the LSPR and refractive index of AgNPs in the visible range, thereby inducing a drastic color change in AgNPs [20].

H_2S is the first derivative of the enzymatic degradation of sulfur-containing amino acids. Moreover, its odor can be sensed at low concentrations. Thus, it is considered an indicator of the freshness of meat [21]. Zhai et al. [12] studied AgNPs as an indicator to detect H_2S and recorded food quality using the principle of the optical change exhibited by AgNPs. AgNPs were synthesized using gellan gum (GG) to form a composite, which was subsequently transformed into a hydrogel using agar. The GG-AgNP hydrogel selectively reacted with the H_2S present in the spoilage gas from meat deterioration and transformed from yellow to colorless.

Cellulose nanocrystals (CNCs) obtained from natural resources such as wood, cotton, algae, tunicates, and bacteria have been widely applied in the food packaging industry. Especially, CNCs based on wood have remarkable surface chemical properties and mechanical strength compared to other cellulose-based materials, such as cellulose nanofibers (CNFs) and microcrystalline cellulose (MCC) [22–25]. In addition, the well-ordered cellulose molecules enable CNCs to form a rod-like structure with a high specific surface area. Furthermore, the numerous sulfate groups (OSO_3^-) on the CNC rods act as promising electron donors for the reduction of silver ions (Ag^+) to silver (Ag^0) [26].

The primary purpose of this study was to characterize and evaluate a freshness indicator that can detect H_2S in chicken breast spoilage gas using a CNC-AgNP composite film prepared in a facile hydrothermal synthesis. The H_2S detection properties and color changes of the CNC-AgNP films were investigated using a mechanism for the surface chemical reaction between AgNPs and H_2S.

2. Materials and Methods

2.1. Materials

CNCs hydrolyzed by sulfuric acid were purchased as a freeze-dried powder from Celluforce™ (Montreal, QC, Canada). Sodium hydroxide (NaOH) was obtained from Sigma-Aldrich (St. Louis, MO, USA). Silver nitrate ($AgNO_3$) was obtained from Alfa Aesar (Haverhill, MA, USA). H_2S standard gas at a concentration of 2 ppm was purchased from RIGAS Co., Ltd. (Daejeon, Korea). The raw chicken breast used in this study was purchased from a local grocery store as slaughtered the day before. All materials were used without further purification.

2.2. Experiments

2.2.1. Preparation of CNC-AgNP Solutions and Composite Films

CNC-AgNP solutions with various concentrations of AgNPs were synthesized using the following steps: the CNC powder was mixed with deionized water and vigorously stirred overnight. The CNC suspension was further dispersed using a probe sonicator (BCX 750, Sonics & Materials, Inc., Newtown, CT, USA) at 8000 J. NaOH solution (5 N) was added to the CNC suspension until a pH of 11 was attained. Various concentrations of $AgNO_3$ solution were mixed with the CNC suspension. Subsequently, the hydrothermal synthesis of AgNPs was performed in an oven at 95 °C for 1 h. Finally, the CNC-AgNP solution was cooled overnight at room temperature.

CNC-AgNP films were prepared using a solution casting method [27]. Approximately 10 g of the as-prepared CNC-AgNP solution was poured into a disposable round polystyrene plate (60 mm in diameter and 15 mm in height) and dried in a desiccator for 7 days. Subsequently, the dried CNC-AgNP films were carefully retrieved from the plate and stored in a PET pouch. The nominal Ag content in the composite films was calculated using the weight ratio of Ag^+ to CNC [28] and the samples were named accordingly (Table 1).

Table 1. Nominal weight ratio of AgNPs in CNC-AgNP solution.

2 wt.% CNC Suspension (g) 48			Water (g) 47.04	CNC (g) 0.96	$AgNO_3$ (g) 2	CNC-AgNP Solution (g) 50	
Conc. (mM)	$AgNO_3$ (g)	Mol	Mass (g)	Ag^+ mass (g)	CNC-AgNP ratio in CNC-AgNP solution (wt.%)		Sample code
		CNC only					CNC
10		2.0×10^{-5}	3.40×10^{-3}	2.16×10^{-3}	0.22		C-0.22Ag
50	2	1.0×10^{-4}	1.70×10^{-2}	1.08×10^{-2}	1.12		C-1.12Ag
100		2.0×10^{-4}	3.40×10^{-2}	2.16×10^{-2}	2.25		C-2.25Ag
200		4.0×10^{-4}	6.79×10^{-2}	4.31×10^{-2}	4.49		C-4.49Ag

2.2.2. Analysis of Sulfuric Compounds in the Spoilage Gas

Qualitative analysis of sulfur compounds was performed using a gas-chromatography-pulsed flame photometric detector (PFPD, Varian 450-GC, Bruker, Billerica, MA, USA). The spoilage gas in a 5 L Tedlar aluminum bag was concentrated at −15 °C using a Series 2 Unity equipped with an air server (Markes International Ltd., Llantrisant, UK), and inserted into the GC inlet at a flow rate of 15 mL/min of helium gas at 270 °C in split mode. CP-Sil 5CB (60 m × 0.32 mm inner diameter, 5 μm, Agilent J&W GC columns, Santa Clara, CA, USA) was used to separate sulfur compounds from the spoilage gas. The column was operated at a programming temperature of 80 °C for 3.5 min, increased at a rate of 6 °C/min to 150 °C, and maintained for 18 min.

To analyze H_2S quantitatively, the spoilage gas was collected in a 250 mL glass jar with 5 g of raw chicken breast at 25 °C for 48 h. A gas sampling pump with an H_2S gas detector tube extracted 100 mL of headspace gas through a septum on the glass jar cap. Gases were sampled six times for a predetermined storage time, and three samples were considered each time to obtain the final average value. The detector tubes used (GASTEC, Seoul, Korea) were 4LT (0.05–4.0 ppm), 4LB (0.5–12 ppm), and 4LK (1–40 ppm).

2.2.3. Characterization of CNC-AgNP Solutions and Composite Films

The light absorption properties of the samples were measured using a UV-Vis spectrophotometer (V-650, Jasco, Tokyo, Japan) in the wavelength range 300–800 nm. The surfaces of the neat CNC and CNC-AgNP films were observed with field emission scanning electron microscopy (FE-SEM, Quanta FEG 250, FEI Co., Ltd., Hillsboro, OR, USA) at magnifications up to 8000×. X-ray diffraction (XRD, D2 Phaser model system, Bruker, Billerica, MA, USA) with Cu Kα radiation (λ = 1.5418°) was used to identify the crystallinity of the films. Finally, XRD patterns were obtained in the 2θ range of 5–80°.

2.2.4. Evaluation of CNC-AgNP Composite Films as a Colorimetric Freshness Indicator

The H_2S sensing test was conducted with the C-0.22Ag film in contact with H_2S standard gas. The film was placed under the cap of a 250 mL glass jar with 30 mL of water. The headspace was evacuated of gases over a period of 1 min and then filled with H_2S standard gas over a period of 25 min at a flow rate of 10 mL/min. The tested film was analyzed using a UV-Vis spectrophotometer, and its color change was observed as a function of contact time.

The ability of the film to detect spoilage gas from raw chicken breast decomposition was evaluated. The film sample (3 cm × 4 cm) was installed in the same glass jar with 5 g of raw chicken breast and was maintained at 25 °C for 48 h. In addition, photographs of the change in film color with storage time were taken. The light absorption properties of the films were measured using a UV-Vis spectrophotometer. Finally, the changes in the morphology and crystallinity of the films were investigated using FE-SEM and XRD.

3. Results and Discussion

3.1. Confirmation of AgNPs Formation

The appearances of the as-prepared CNC and CNC-AgNP composite samples are presented in Figure 1. A distinct color was observed when $AgNO_3$ was added to the CNC suspension, resulting in an orange or dark brown color. From the UV-Vis spectra of the samples illustrated in Figure 2, both the CNC-AgNP solutions and composite films displayed strong absorption peaks at approximately 418 nm, which were not detected in the CNC suspension and film. Previous studies have reported that silver nanospheres with a diameter of approximately 40 nm absorb light of wavelength 410 nm and exhibit a unique color [17,29]. The CNC-AgNP samples exhibited similar light absorption properties and colors, indicating the presence of spherical AgNPs.

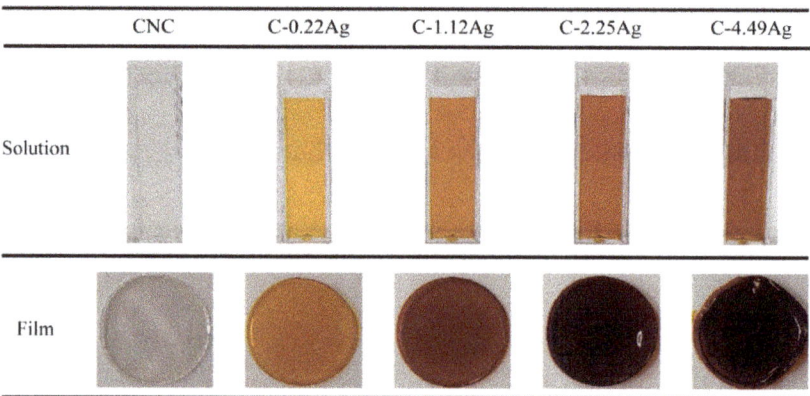

Figure 1. Photographs of CNC and CNC-AgNP composite samples.

Figure 3 depicts the XRD patterns of the neat CNC and CNC-AgNP composite films. Two remarkable peaks at 2θ = 15.4° and 22.7°, corresponding to CNC(101) and CNC(002) [22], were observed in all films. New peaks at approximately 2θ = 38.1° and 44.3° were detected in the CNC-AgNP composite films, corresponding to Ag(111) and Ag(200), respectively [30]. Thus, the conclusion can be drawn that AgNPs were successfully synthesized through green hydrothermal synthesis using CNCs.

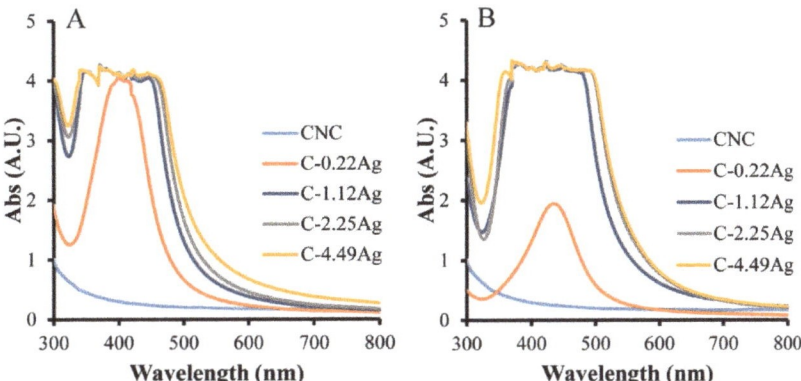

Figure 2. UV-Vis spectra of neat CNC and CNC-AgNP composites; liquids (**A**) and composite films (**B**).

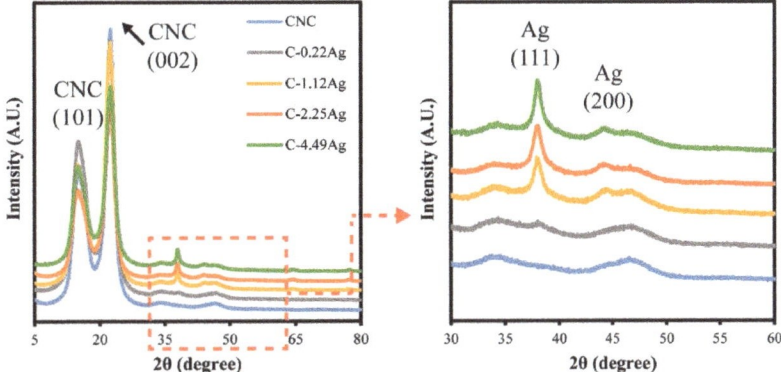

Figure 3. XRD patterns of neat CNC and CNC-AgNP composite films.

3.2. Sulfur Compounds in Spoilage Gas

The qualitative analysis of sulfur compounds from the deterioration of raw chicken breast revealed that six different sulfuric chemicals were produced: hydrogen sulfide (H_2S), methanethiol (CH_3SH), carbon disulfide (CS_2), dimethyl sulfide (($CH_3)_2S$), dimethyl disulfide (($CH_3)_2S_2$), and dimethyl trisulfide (($CH_3)_2S_3$). The presence of these chemicals can be attributed to the high protein content of raw chicken breast [31]. Previous studies have reported that the enzymatic degradation of cysteine, which is a sulfur-containing amino acid in proteins, generates diverse sulfides, from low molecular sulfides to polysulfides [21].

Figure 4 shows the H_2S generation tendency against various storage times at room temperature using a gas detector tube. H_2S was not detected in the initial 8 h, after which its concentration progressively increased up to 32.55 ppm. Typically, H_2S is generated from the degradation of meat products by microbes under anaerobic conditions and the depletion of glucose [32]. Sukhavattanakul and Manuspiya reported that minced pork stored in a closed vial at 4 °C produced H_2S, which was first detected after eight days and eventually peaked at a concentration of 250 ppm [33]. In this study, the transition at 8 h can be attributed to the fact that suitable conditions for microbial decomposition were formed at this point, resulting in an accelerated increase in H_2S concentration.

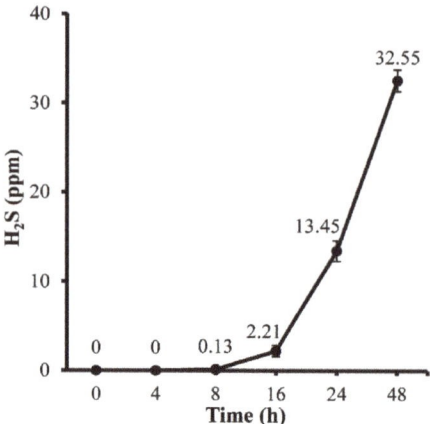

Figure 4. H_2S concentration along with storage time at room temperature.

3.3. Reaction Property of CNC-AgNP Films to H_2S Standard Gas

The UV-Vis spectra and photographs of the C-0.22Ag film after contact with H_2S standard gas are presented in Figures 5 and 6, respectively. The intensity of color change peaked at approximately 430 nm; it started to decline at 5 h and completely disappeared after 30 h of reaction. The change in color was first observed at 5 h; the film gradually turned dark brown and eventually disintegrated. This phenomenon can be attributed to the formation of Ag_2S, which causes the LSPR of AgNPs to change. Park et al. reported that the surface chemical reaction between AgNPs and H_2S results in the formation of the Ag@Ag_2S core-shell structure. Consequently, the diameter, LSPR, and refractive index of AgNPs in the visible region change [20].

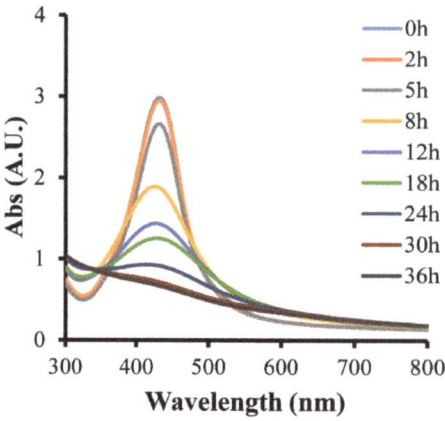

Figure 5. UV-Vis spectra of C-0.22Ag film after reaction with 2 ppm H_2S.

Figure 6. Photographs of C-0.22Ag film after reaction with 2 ppm H_2S.

As illustrated in Figure 7, Lilienfeld and White suggested that the sulfidation of AgNPs was induced by H_2S [19]. When H_2S molecules contact the AgNP surface under moisturized conditions, they are oxidized to water and sulfur by the oxygen molecules on the AgNPs. The oxidation of H_2S immediately causes the Ag and sulfur atoms to combine and form Ag_2S. The sulfidation process can be summarized as follows:

$$H_2S + \frac{1}{2}O_2 \rightarrow H_2O + S \tag{1}$$

$$S + 2Ag \rightarrow Ag_2S \tag{2}$$

$$2Ag + H_2S + \frac{1}{2}O_2 \rightarrow Ag_2S + H_2O \text{ (under moisture condition)} \tag{3}$$

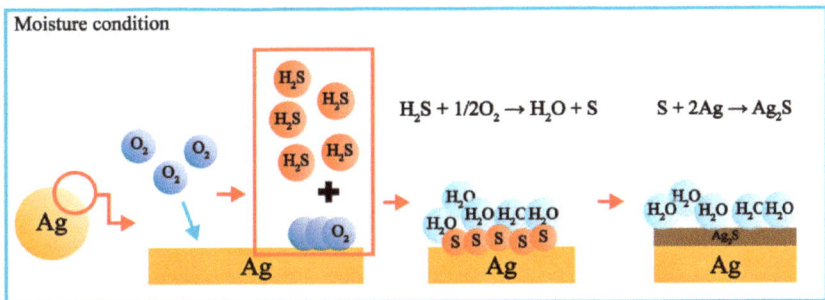

Figure 7. Schematic diagram of sulfidation of AgNP.

3.4. Quality Indicating Performance of CNC-AgNP Film

Figure 8 shows the color change of the neat CNC and CNC-AgNP films after contact with the chicken breast spoilage gas. The initial color remained unchanged up to 8 h, following which a weak purple color appeared after 16 h in the case of the neat CNC film. In addition, studies have reported a color change in the visible-light range if moisture is absorbed in the chiral nematic structure of CNC rods [34]. Thus, the color change of the neat CNC film can be attributed to the presence of moisture, and not H_2S. However, no distinct color change was observed for any of the CNC-AgNP films up to this stage because no H_2S had been generated, as shown in Figure 4. Color change in the C-0.22Ag film was first detected at 8 h into the experiment. The color change from reddish-brown to grayish yellow was completed at 24 h. However, the films containing over 1.12 wt.% AgNP behaved differently compared to the C-0.22Ag film. The C-1.12Ag film brightened progressively by a marginal amount until 16 h, and a grayish stain was observed after 24 h. In the case of the C-2.25Ag and C-4.49Ag films, the color intensity decreased progressively until 8 h. Metal-like surfaces were observed in both films at 16 h, and a glossy surface was observed in the C-4.49Ag film after 24 h. As shown in Figure 8, all the CNC-AgNP composite films reacted sensitively with H_2S and changed color. However, the C-0.22Ag film is the most promising for practical application owing to its higher Ag content, which results in a larger metallic surface.

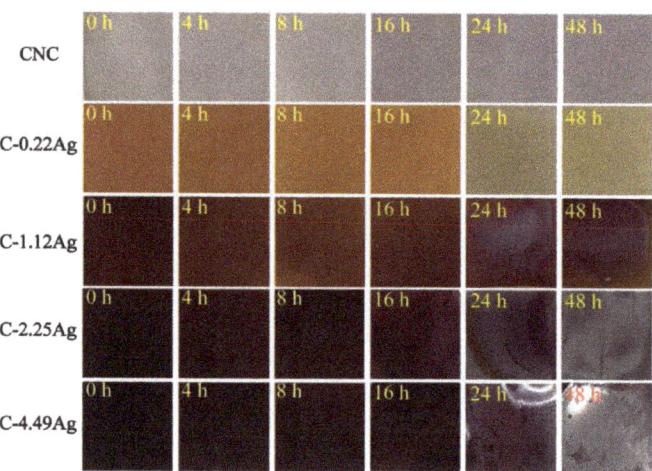

Figure 8. Photographs of neat CNC and CNC-AgNP films after contact with the chicken breast spoilage gas.

Figure 9 displays the surface morphologies of the neat CNC and CNC-AgNP films after reacting with H_2S for 48 h. The surface of the C-0.22Ag film transformed into a dotted pattern, which was observed at a magnification of 8000×. When the weight ratio of AgNP was increased, crack formation was initiated and a white spot-like pattern began to appear on the C-1.12Ag film surface after reaction with the spoilage gas. This tendency was intensified in the C-2.25Ag film, which exhibited larger crack patterns and white markings. In the case of the highest concentration of AgNPs, white particles were observed on the severely damaged surface. In the XRD patterns displayed in Figure 10, two peaks at $2\theta = 15.4°$ and $22.7°$, corresponding to CNCs [22], were intact after contact with the spoilage gas. When the CNC-AgNP films were exposed to the spoilage gas, several peaks at $2\theta = 28.9°, 31.8°, 33.6°, 34.4°, 36.8°, 38.1°, 40.6°$, and $52.7°$ matched to Ag_2S [35] were observed. These peaks were intensified and readily discerned when the Ag content in the CNC-AgNP films was increased.

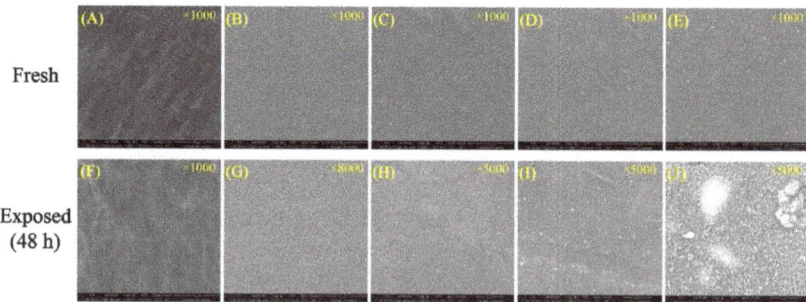

Figure 9. FE-SEM images of fresh (**A–E**) and exposed films to spoilage gas (**F–J**); neat CNC (**A,F**), C-0.22Ag (**B,G**), C-1.12Ag (**C,H**), C-2.25Ag (**D,I**), and C-4.49Ag films (**E,J**).

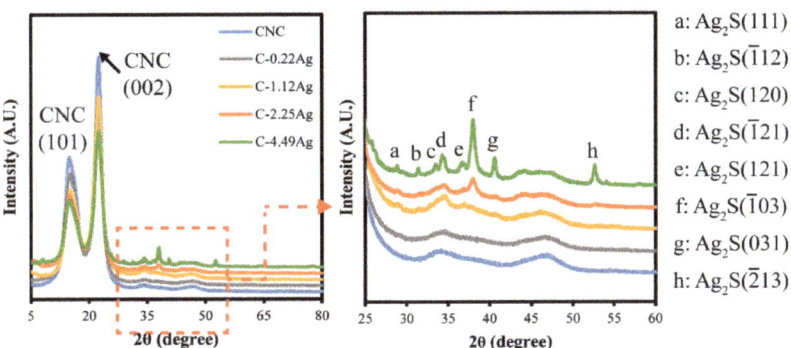

Figure 10. XRD patterns of tested CNC and CNC-AgNP films.

The FE-SEM and XRD analyses revealed that H_2S did not affect the crystallinity of CNCs, but only reacted with AgNPs, resulting in the formation of Ag_2S. Ma et al. [36] presented a similar result: Ag_2S appeared as a small white particle on the film. In this study, Ag_2S particles appeared in a dotted pattern on the C-0.22Ag film after the reaction, as shown in Figure 9. It evolved into crack-patterned and damaged surfaces in the case of films with Ag content over 1.12 wt.%. This phenomenon was observed simultaneously with an increase in the intensity of the XRD patterns assigned to Ag_2S, which implies that a larger amount of Ag_2S particles was initially formed. The particles, subsequently, aggregated owing to the high concentration of AgNPs [37] and eventually covered the film surface. The severely damaged surface observed in the tested C-4.49Ag film surface was a result of chemical corrosion caused by the high Ag_2S content [38]. Therefore, the conclusion was drawn that sulfidation of AgNPs was induced by the H_2S in the spoilage gas from the decaying raw chicken breast, which was dependent on the Ag content of the CNC-AgNP films.

4. Conclusions

We successfully prepared CNC-AgNP nanocomposites that can be used as colorimetric freshness indicators of packaged food, owing to the LSPR of AgNPs. The presence of AgNPs synthesized within the CNCs was confirmed using UV-Vis spectroscopy and XRD patterns. The ability of the CNC-AgNP films to indicate food quality was evaluated in terms of how effectively they detected spoilage gas emitted from decomposing raw chicken breast. In addition, UV-Vis, FE-SEM, and XRD were used to characterize this ability. Prior to testing, spoilage gases were analyzed specifically for sulfuric compounds, and the generation of H_2S during raw chicken breast decay was confirmed using GC-PFPD. The CNC-AgNP films exhibited an exceptional ability in detecting the spoilage gases, with a noticeable color change from yellowish-dark wine red to colorless metallic silver depending on the AgNP content. Furthermore, the results revealed that the sulfidation of AgNPs by H_2S induced LSPR changes in the AgNPs under moisturized and low-oxygen conditions, resulting in a color change in the CNC-AgNP nanocomposite film. Finally, the XRD results confirmed that Ag_2S was formed, and SEM revealed that the surface of films with a higher Ag content was damaged by chemical corrosion caused by Ag_2S.

Author Contributions: Conceptualization, S.K. (Seongyoung Kwon) and S.K. (Seonghyuk Ko); methodology, S.K. (Seongyoung Kwon) and S.K. (Seonghyuk Ko); investigation, S.K. (Seongyoung Kwon); writing—original draft preparation, S.K. (Seongyoung Kwon); writing—review and editing, S.K. (Seonghyuk Ko); supervision, S.K. (Seonghyuk Ko); funding acquisition, S.K. (Seonghyuk Ko). All authors have read and agreed to the published version of the manuscript.

Funding: This research was funded by the National Research Foundation of Korea (NRF) supported by the Korea Government (MSIT), grant number 2020R1F1A1068893.

Institutional Review Board Statement: Not applicable.

Informed Consent Statement: Not applicable.

Data Availability Statement: The original contributions presented in the study are included in the article, further inquiries can be directed to the corresponding author.

Conflicts of Interest: The authors declare no conflict of interest.

References

1. Fang, Z.; Zhao, Y.; Warner, R.D.; Johnson, S.K. Active and intelligent packaging in meat industry. *Trends Food Sci. Technol.* **2017**, *61*, 60–71. [CrossRef]
2. European Commission. Commission Regulation (EC) No 450/2009 of 29 May 2009 on active and intelligent materials and articles intended to come into contact with food. *Off. J. Eur. Union* **2009**, *135*, 3–11.
3. Yam, K.L.; Takhistov, P.T.; Miltz, J. Intelligent packaging: Concepts and applications. *J. Food Sci.* **2005**, *70*, R1–R10. [CrossRef]
4. Hurme, E.; Thea, S.-M.; Ahvenainen, R.; Nielsen, T. 5-Active and intelligent packaging. In *Minimal Processing Technologies in the Food Industries*, 1st ed.; Ohlsson, T., Bengtsson, N., Eds.; Woodhead Publishing: Cambridge, UK, 2002; Volume 1, pp. 87–123.
5. Hanrahan, G.; Patil, D.G.; Wang, J. Electrochemical sensors for environmental monitoring: Design, development and applications. *J. Environ. Monit.* **2004**, *6*, 657–664. [CrossRef]
6. Kerry, J.P.; O'Grady, M.N.; Hogan, S.A. Past, current and potential utilisation of active and intelligent packaging systems for meat and muscle-based products: A review. *Meat Sci.* **2006**, *74*, 113–130. [CrossRef]
7. Ghaani, M.; Cozzolino, C.A.; Castelli, G.; Farris, S. An overview of the intelligent packaging technologies in the food sector. *Trends Food Sci. Technol.* **2016**, *51*, 1–11. [CrossRef]
8. Sukhavattanakul, P.; Manuspiya, H. Fabrication of hybrid thin film based on bacterial cellulose nanocrystals and metal nanoparticles with hydrogen sulfide gas sensor ability. *Carbohydr. Polym.* **2020**, *230*, 115566. [CrossRef]
9. Yuan, Z.; Lu, F.; Peng, M.; Wang, C.-W.; Tseng, Y.-T.; Du, Y.; Cai, N.; Lien, C.-W.; Chang, H.-T.; He, Y.; et al. Selective colorimetric detection of hydrogen sulfide based on primary amine-active ester cross-linking of gold nanoparticles. *Anal. Chem.* **2015**, *87*, 7267–7273. [CrossRef]
10. Chen, R.; Nuhfer, N.T.; Moussa, L.; Morris, H.R.; Whitmore, P.M. Silver sulfide nanoparticle assembly obtained by reacting an assembled silver nanoparticle template with hydrogen sulfide gas. *Nanotechnology* **2008**, *19*, 455604. [CrossRef]
11. Heli, B.; Morales-Narváez, E.; Golmohammadi, H.; Ajji, A.; Merkoçi, A. Modulation of population density and size of silver nanoparticles embedded in bacterial cellulose via ammonia exposure: Visual detection of volatile compounds in a piece of plasmonic nanopaper. *Nanoscale* **2016**, *8*, 7984–7991. [CrossRef]
12. Zhai, X.; Li, Z.; Shi, J.; Huang, X.; Sun, Z.; Zhang, D.; Zou, X.; Sun, Y.; Zhang, J.; Holmes, M. A colorimetric hydrogen sulfide sensor based on gellan gum-silver nanoparticles bionanocomposite for monitoring of meat spoilage in intelligent packaging. *Food Chem.* **2019**, *290*, 135–143. [CrossRef]
13. Li, H.; Gan, J.; Yang, Q.; Fu, L.; Wang, Y. Colorimetric detection of food freshness based on amine-responsive dopamine polymerization on gold nanoparticles. *Talanta* **2021**, *234*, 122706. [CrossRef]
14. Tan, P.; Li, H.; Wang, J.; Gopinath, S.C.B. Silver nanoparticle in biosensor and bioimaging: Clinical perspectives. *Biotechnol. Appl. Biochem.* **2020**, *68*, 1236–1242. [CrossRef]
15. Benson, J.; Fung, C.M.; Lloyd, J.S.; Deganello, D.; Smith, N.A.; Teng, K.S. Direct patterning of gold nanoparticles using flexographic printing for biosensing applications. *Nanoscale Res. Lett.* **2015**, *10*, 127. [CrossRef]
16. Hutter, E.; Fendler, J.H. Exploitation of localized surface plasmon resonance. *Adv. Mater.* **2004**, *16*, 1685–1706. [CrossRef]
17. Hong, Y.C.; Huh, Y.M.; Yoon, D.S.; Yang, J.M. Nanobiosensors based on localized surface plasmon resonance for biomarker detection. *J. Nanomater.* **2012**, *2012*, 111. [CrossRef]
18. Eustis, S.; El-Sayed, M.A. Why gold nanoparticles are more precious than pretty gold: Noble metal surface plasmon resonance and its enhancement of the radiative and nonradiative properties of nanocrystals of different shapes. *Chem. Soc. Rev.* **2006**, *35*, 209–217. [CrossRef]
19. Lilienfeld, S.; White, C.E. A study of the reaction between hydrogen sulfide and silver. *J. Am. Chem. Soc.* **1930**, *52*, 885–892. [CrossRef]
20. Park, G.R.; Lee, C.H.; Seo, D.H.; Song, H.J. Full-color tuning of surface plasmon resonance by compositional variation of Au@Ag core–shell nanocubes with sulfides. *Langmuir* **2012**, *28*, 9003–9009. [CrossRef]
21. Varlet, V.; Fernandez, X. Review. Sulfur-containing volatile compounds in seafood: Occurrence, odorant properties and mechanisms of formation. *Food Sci. Technol. Int.* **2010**, *16*, 463–503. [CrossRef]
22. Hamad, W.Y. *Cellulose Nanocrystals: Properties, Production and Applications*, 1st ed.; Wiley: Chichester, UK, 2017; pp. 1–312.
23. Moon, R.J.; Martini, A.; Nairn, J.; Simonsen, J.; Youngblood, J. Cellulose nanomaterials review: Structure, properties and nanocomposites. *Chem. Soc. Rev.* **2011**, *40*, 3941–3994. [CrossRef] [PubMed]
24. Henriksson, M.; Berglund, L.A.; Isaksson, P.; Lindström, T.; Nishino, T. Cellulose nanopaper structures of high toughness. *Biomacromolecules* **2008**, *9*, 1579–1585. [CrossRef] [PubMed]

25. Shak, K.P.Y.; Pang, Y.L.; Mah, S.K. Nanocellulose: Recent advances and its prospects in environmental remediation. *Beilstein J. Nanotechnol.* **2018**, *9*, 2479–2498. [CrossRef] [PubMed]
26. Brown, T.L.; LeMay, H.E.; Bursten, B.E.; Murphy, C.J.; Woodward, P.M. *Chemistry: The Central Science*, 12th ed.; Pearson Prentice Hall: Hoboken, NJ, USA, 2012; pp. 1–1248.
27. Oksman, K.; Mathew, A.P. Processing of bionanocomposites: Solution casting. In *Handbook of Green Materials*, 1st ed.; Oksman, K., Mathew, A.P., Bismarck, A., Rojas, O., Sain, M., Eds.; World Scientific: Singapore, 2013; Volume 5, pp. 35–52.
28. Bumbudsanpharoke, N.; Ko, S.H. In-situ green synthesis of gold nanoparticles using unbleached kraft pulp. *BioResources* **2015**, *10*, 6428–6441. [CrossRef]
29. Anker, J.N.; Hall, W.P.; Lyandres, O.; Shah, N.C.; Zhao, J.; Van Duyne, R.P. Biosensing with plasmonic nanosensors. *Nanosci. Technol.* **2009**, *7*, 308–319.
30. Chowdhury, S.; Basu, A.; Kundu, S. Green synthesis of protein capped silver nanoparticles from phytopathogenic fungus Macrophomina phaseolina (Tassi) Goid with antimicrobial properties against multidrug-resistant bacteria. *Nanoscale Res. Lett.* **2014**, *9*, 365. [CrossRef]
31. Probst, Y. Nutrient values for Australian and overseas chicken meat. *Nutr. Food Sci.* **2009**, *39*, 685–693. [CrossRef]
32. Borch, E.; Kant-Muermans, M.-L.; Blixt, Y. Bacterial spoilage of meat and cured meat products. *Int. J. Food Microbiol.* **1996**, *33*, 103–120. [CrossRef]
33. Sukhavattanakul, P.; Manuspiya, H. Influence of hydrogen sulfide gas concentrations on LOD and LOQ of thermal spray coated hybrid-bacterial cellulose film for intelligent meat label. *Carbohydr. Polym.* **2021**, *254*, 117442. [CrossRef]
34. Bumbudsanpharoke, N.; Kwon, S.Y.; Lee, W.S.; Ko, S.H. Optical response of photonic cellulose nanocrystal film for a novel humidity indicator. *Int. J. Biol. Macromol.* **2019**, *140*, 91–97. [CrossRef]
35. Tian, C.; Kang, Z.; Wang, E.; Mao, B.; Li, S.; Su, Z.; Xu, L. 'One-step' controllable synthesis of Ag and Ag_2S nanocrystals on a large scale. *Nanotechnology* **2006**, *17*, 5681–5685. [CrossRef] [PubMed]
36. Ma, Y.; Wan, H.; Ye, Y.; Chen, L.; Li, H.; Zhou, H.; Chen, J. In-situ synthesis of size-tunable silver sulfide nanoparticles to improve tribological properties of the polytetrafluoroethylene-based nanocomposite lubricating coatings. *Tribol. Int.* **2020**, *148*, 106324. [CrossRef]
37. Shrestha, S.; Wang, B.; Dutta, P. Nanoparticle processing: Understanding and controlling aggregation. *Adv. Colloid Interface Sci.* **2020**, *279*, 102162. [CrossRef]
38. Elechiguerra, J.L.; Larios-Lopez, L.; Liu, C.; Garcia-Gutierrez, D.; Camacho-Bragado, A.; Yacaman, M.J. Corrosion at the nanoscale: The case of silver nanowires and nanoparticles. *Chem. Mater.* **2005**, *17*, 6042–6052. [CrossRef]

Article

Active and pH-Sensitive Nanopackaging Based on Polymeric Anthocyanin/Natural or Organo-Modified Montmorillonite Blends: Characterization and Assessment of Cytotoxicity

Tomy J. Gutiérrez [1,*], Ignacio E. León [2], Alejandra G. Ponce [3] and Vera A. Alvarez [1]

[1] Grupo de Materiales Compuestos Termoplásticos (CoMP), Instituto de Investigaciones en Ciencia y Tecnología de Materiales (INTEMA), Facultad de Ingeniería, Universidad Nacional de Mar del Plata (UNMdP) y Consejo Nacional de Investigaciones Científicas y Técnicas (CONICET), Colón 10850, Mar del Plata B7608FLC, Argentina

[2] Centro de Química Inorgánica "Dr. Pedro J. Aymonino" (CEQUINOR), Facultad de Ciencias Exactas, Universidad Nacional de La Plata (UNLP) y Consejo Nacional de Investigaciones Científicas y Técnicas (CONICET), Blvd. 120 No. 1465, La Plata 1900, Argentina

[3] Grupo de Investigación en Ingeniería en Alimentos (GIIA), Instituto de Ciencia y Tecnología de Alimentos y Ambiente (INCITAA, CIC-UNMDP), Facultad de Ingeniería, Universidad Nacional de Mar del Plata, Juan B. Justo 4302, Mar del Plata B7602AYL, Argentina

* Correspondence: tomy.gutierrez@fi.mdp.edu.ar; Tel.: +54-223-6260627; Fax: +54-223-481-0046

Abstract: Polymeric anthocyanins are biologically active, pH-sensitive natural compounds and pigments with beneficial functional, pharmacological and therapeutic properties for consumer health. More recently, they have been used for the manufacture of active and pH-sensitive ("intelligent") food nanopackaging, due to their bathochromic effect. Nevertheless, in order for polymeric anthocyanins to be included either as a functional food or as a pharmacological additive (medicinal food), they inevitably need to be stabilized, as they are highly susceptible to environmental conditions. In this regard, nanopackaging has become a tool to overcome the limitations of polymeric anthocyanins. The objective of this study was to evaluate their structural, thermal, morphological, physicochemical, antioxidant and antimicrobial properties, as well as their responses to pH changes, and the cytotoxicity of blends made from polymeric anthocyanins extracted from Jamaica flowers (*Hibiscus sabdariffa*) and natural or organo-modified montmorillonite (Mt), as active and pH-sensitive nanopackaging. This study allowed us to conclude that organo-modified Mts are efficient pH-sensitive and antioxidant nanopackaging systems that contain and stabilize polymeric anthocyanins compared to natural Mt nanopackaging and stabilizing polymeric anthocyanins. However, the use of these polymeric anthocyanin-stabilizing organo-modified Mt-based nanopackaging systems are limited for food applications by their toxicity.

Keywords: active compounds; food additives; food toxicity; functional foods; nano-encapsulation

Citation: Gutiérrez, T.J.; León, I.E.; Ponce, A.G.; Alvarez, V.A. Active and pH-Sensitive Nanopackaging Based on Polymeric Anthocyanin/Natural or Organo-Modified Montmorillonite Blends: Characterization and Assessment of Cytotoxicity. *Polymers* 2022, 14, 4881. https://doi.org/10.3390/polym14224881

Academic Editors: Victor G. L. Souza, Lorenzo M. Pastrana and Ana Luisa Fernando

Received: 27 October 2022
Accepted: 10 November 2022
Published: 12 November 2022

Publisher's Note: MDPI stays neutral with regard to jurisdictional claims in published maps and institutional affiliations.

Copyright: © 2022 by the authors. Licensee MDPI, Basel, Switzerland. This article is an open access article distributed under the terms and conditions of the Creative Commons Attribution (CC BY) license (https://creativecommons.org/licenses/by/4.0/).

1. Introduction

Polymeric anthocyanins are biologically active and pH-sensitive natural pigments that have attracted the attention of food scientists and technologists globally, as they have important beneficial effects on consumer health [1,2]. In particular, polymeric anthocyanins extracted from Jamaica flowers (*Hibiscus sabdariffa*) have been shown to reduce cholesterol levels, increase lipid peroxidation inhibition capacity [3], prevent kidney diseases [4] and reduce hypertension in patients with type 2 diabetes [5]; they have antimicrobial [6] and antioxidant [3] properties, and crucially, they are non-toxic [3]. These characteristics make Jamaica flower extract (JFE) a promising food additive for the development of functional and/or medicinal foods. In addition, polymeric anthocyanins, in general, have contributed to the manufacture of pH-sensitive composite food packaging [7–10]. These pH-sensitive packaging materials that contain polymeric anthocyanins have been normally classified as "intelligent packaging", since they can act as sensors that are capable of providing information on the status of food quality and safety [11]. For example, Merz et al. [12] utilized anthocyanins that were extracted from jambolan (*Syzygium cumini*) fruit to develop pH-sensitive packaging materials (colorimetric indicator/chromatic sensor) for the purpose of monitoring shrimp quality and safety. This fact was justified, since biogenic amines (e.g., histamine–allergenic compounds) that are generated as secondary metabolites from histidine (essential amino acid) during the decomposition of fishery products can modify the structure and coloration of polymeric anthocyanins, thus providing early warnings for food quality and safety [13–18]. Notably, Jamaica flowers are traditionally marketed to make polymeric anthocyanin-rich infusions or teas, and the main polymeric anthocyanin contained in JFE is delphinidin-3-O-sambubioside [3].

However, in order for polymeric anthocyanins to be useful to the food industry, they must be stabilized, as they are susceptible to light, oxygen, temperature and the presence of enzymes [19,20]. One way of doing this is by nanopackaging the active compound using ionic gelation, with pectin as a wall biopolymer [21]. In fact, the nanopackaging of various active compounds with different biopolymers has frequently been employed, in order to develop safe functional foods [22]. Notwithstanding, the stabilization and nanopackaging of polymeric anthocyanins extracted from blueberries [23] and acerola [20] has been carried out more recently through the use of nano-clays, mainly montmorillonite (Mt). Nonetheless, these bio-nanocomposite systems have not yet been evaluated in terms of their toxicity.

According to the literature, up until now there has been only one study that has evaluated the toxicity of Mt-based nanopackaging composite materials as food additives [24], and we feel that more research is required to evaluate their true potential as polymeric anthocyanin nanopackaging materials for the food industry.

Polymeric anthocyanins can be stabilized through the use of Mt by intercalating the active compound between the aluminosilicate layers. In this way, bio-nanocomposite systems are generated that protect the polymeric anthocyanins from different processing conditions [20,23].

The two main aims of this study were as follows: (1) to evaluate and characterize different systems of chemically modified Mts, regarding their ability to stabilize and nanopackage polymeric anthocyanins extracted from Jamaica flowers; and (2) to assess possible toxicological effects in order to determine whether these bio-nanocomposite systems can be applied as nanopackaging materials in the food industry.

2. Experimental

2.1. Materials

Polymeric anthocyanins (antioxidant and pH-sensitive pigment) obtained from Jamaica flower (*Hibiscus sabdariffa*) extract (JFE) were added to three different montmorillonites (Mts), with the aim of developing potentially active and pH-sensitive nanopackaging. All of the Mts tested were supplied by Laviosa Chimica Mineraria S.p.A. (Livorno, Italy): natural Mt (NMt, or HPS according to the manufacturer), Mt modified with dimethyl benzylhydrogenated tallow ammonium (MtMB, denominated 43B by the manufacturer) and Mt modified with dimethyl dihydrogenated tallow ammonium (MtMD, or 72T according to the manufacturer). Following the manufacturer's descriptions, these modified Mts were nano-clays that were prepared from a naturally occurring Mt, purified and modified with quaternary ammonium salts (dimethyl benzylhydrogenated tallow ammonium and dimethyl dihydrogenated tallow ammonium). All Mts were used as received. The cation exchange capacity (CEC) of the Mts, as measured by the methylene blue method [25], yielded a CEC of 128 meq/100 g clay. JFE (100% polymeric anthocyanin—average molecular weight ranging from 600 to 10,000 Da) was obtained according to the methodology proposed by Dai et al. [26] using ethanol as a solvent, since it maintains the properties of polymeric anthocyanin-rich extracts. Briefly, dehydrated Jamaica flowers were purchased from a local market in the Ciudad Autónoma de Buenos Aires, Provincia de Buenos Aires, Argentina. The flowers were marketed by an Argentinian company, and were labelled as harvested in the Jardín America (Street Amado Nervo 478, Provincial de Misiones, Argentina; geographical coordinates—latitude S 27°2'25.597" and longitude W 55°14'22.906"). A total of 30 g of Jamaica flowers were weighed and immediately immersed in 200 mL of ethanol (Aldrich, St. Louis, MO, USA—product code: 34923), applying a slight pressure on the immersed flowers in order to extract the polymeric anthocyanins. The extract was then decanted to remove the solid fragments of the flower petals. The JFE and the JFE-containing Mts were developed on the same day, and kept refrigerated at 5 °C in a dark container until further processing, in order to avoid oxidative damage.

2.2. Manufacture of Potential Active and pH-Sensitive Nanopackaging

JFE was added to the Mts by manually blending 20 g of each Mt with 50 mL of JFE at room temperature (21 °C). The blend was then frozen at −20 °C for 48 h, after which it was lyophilized at 13.33 Pa (100 mTorr) and then frozen again at −50 °C for 72 h, utilizing a Gland type Vacuum Freeze Dryer, Columbia International, Model FD-1B-50 (Shaan Xi, China), in order to obtain a solvent-free product. Lyophilization also preserves the polymeric anthocyanins (active compound) in the JFE, and ensures JFE-containing nano-sized Mt particles are obtained. The resulting JFE-containing Mts were conditioned in containers with a saturated solution of NaBr (a_w~0.575 at 25 °C) for seven days before testing, in order to maintain controlled and known conditions. During this period, the containers were protected from light in a dark room, in order to avoid the photodegradation of the polymeric anthocyanins. Six Mt nanosystems were manufactured and labeled as follows: natural Mt (NMt), natural Mt-containing JFE (NMt + JFE), Mt modified with dimethyl benzylhydrogenated tallow ammonium (MtMB), Mt modified with dimethyl benzylhydrogenated tallow ammonium containing JFE (MtMB + JFE), Mt modified with dimethyl dihydrogenated tallow ammonium (MtMD) and Mt modified with dimethyl dihydrogenated tallow ammonium containing JFE (MtMD + JFE).

2.3. Characterizations of Potential Active and pH-Sensitive Nanopackaging

2.3.1. X-Ray Diffraction (XRD)

An X-Pert Pro diffractometer (Almelo, Netherlands) operating at 40 kV and 40 mA, with Cu K_α radiation (λ = 1.5406 Å), was used to obtain the XRD patterns of the Mts. Samples of finely ground Mt powder were placed in horizontal glass holders. XRD patterns were recorded at a scanning speed of 0.5°/min in an angular range of 2θ = 2° to 10°. The distances between the planes of the crystals d (Å) were then calculated from the diffraction angles (°) measured from the XRD patterns, following Bragg's law [13]:

$$d = n * \lambda * (2 * \sin\theta)^{-1} \qquad (1)$$

where n is the order of reflection, and λ the wavelength of Cu K_α radiation. For the calculations, n = 1 was used. The differences between the interlayer distances (Δ_{id}) of the samples tested were determined, considering as a reference the interlayer distance of the NMt (d_{NMt}) as follows:

$$\Delta_{id} = d_c - d_{NMt} \qquad (2)$$

where d_c is the interlayer distance of each Mt sample:

2.3.2. Thermogravimetric Analysis (TGA)

A thermal analyzer (TA Instruments) Model TGA Q500 (Hüllhorst, Germany), at a heating rate of 10 °C/min from room temperature (approx. 30 °C) to 900 °C under nitrogen atmosphere, was used to carry out the TGA essays. The Mt mass varied between 10 and 28 mg. The mass fraction of JFE (X_{JFE}) contained in the Mts was determined as follows:

$$X_{JFE} = \frac{Rw_{n+JFE} - Rw_n}{Rw_{JFE} - Rw_n} \qquad (3)$$

where Rw_{n+JFE} is the residual mass of the JFE-containing Mts, Rw_n is the residual mass of the Mts without JFE, and R_{wJFE} is the residual mass of the JFE. Residual mass values were recorded at 900 °C where the decomposition of the JFE was stable (~11.77%). Analyses were conducted *per* triplicate to guarantee repeatability, and the data were reported as mean values ± standard deviation (SD).

2.3.3. Field Emission Scanning Electron Microscopy (FESEM)

An FESEM Supra55, Zeiss (Oberkochen, Germany) operating at an acceleration voltage of 3 kV was used to acquire the FESEM micrographs of the Mts. ImageJ software was utilized to determine the average size of the Mt particles by randomly choosing at least 5 FESEM images for each nanosystem. All of the Mt samples were sputter coated with a thin layer of gold for 35 s, using an Ar^+ ion beam at an energy level of 3 kV and sputter rate of 0.67 nm/min, to guarantee electrical conduction and to diminish surface charging during the essays. The sputter rate was determined by employing a Ni/Cr multilayer standard.

2.3.4. Moisture Content (MC)

A Moisture Analyzer, Model MA150 (Goettingen, Germany) was used to determine the MC of each of the Mts. Samples (~0.5 g) were dried at 105 °C to a constant mass. Three measurements *per* nanosystem of Mt were obtained, and the values were reported as % average moisture ± SD.

2.3.5. Attenuated Total Reflectance Fourier Transform Infrared (ATR/FTIR) Spectroscopy

The infrared spectra of the Mt samples were recorded on a Nicolet 8700 (Thermo Scientific Instrument Co., Madison, WI, USA) equipped with a diamond ATR probe at an incident angle of 45°, over the range of 4000–600 cm^{-1}, from 32 co-added scans at 4 cm^{-1} resolution. Approximately 10 mg of each of the finely ground Mt samples were placed on the sample holder. Each sample was scanned three times, and showed good reproducibility.

2.3.6. Raman Spectroscopy

An Invia Reflex confocal Raman microscope (Renishaw, Wotton-under-Edge, UK), with an argon laser operating at a power level of 3 mW, was set to obtain the Raman spectra of the Mts. The Raman spectra were obtained at 785 nm. The microscope that was used worked with a 50× objective lens to focus the beam onto the sample. The integration time was 0.5 s, and the number of accumulations were 200. Spectral resolution and repeatability were better than 1 cm^{-1} and 0.2 cm^{-1}, respectively. The Raman spectrum for each sample was acquired as a number of evenly distributed points. No thermal effects were observed in the samples during the measurements.

2.3.7. Confocal Laser Scanning Microscopy (CLSM)

An inverted microscope (Nikon, Eclipse T*i* series, Tokyo, Japan) operating at 515 nm was used to further examine the developed nanosystems. The samples were observed without any additional treatment.

2.3.8. Color

The CIE-$L^*a^*b^*$ color properties of the Mts were recorded from a Macbeth® colorimeter (Color-Eye 2445 model, illuminant D65 and 10° observer, New Windsor, New York, USAcity, state, country). Color difference (ΔE^*) values among the modified Mts, with and without added JFE, and the control (NMt), were determined utilizing Equation (4) [23]:

$$\Delta E = \sqrt{\Delta a^2 + \Delta b^2 + \Delta L^2} \quad (4)$$

where $\Delta L = L^*_{control} - L^*_{sample}$, $\Delta a = a^*_{control} - a^*_{sample}$ and $b^*_{control} - b^*_{sample}$.

The yellowness index (*YI*) of the Mts was determined according to ASTM D1925-70 [27], while the whiteness index (*WI*) was calculated as follows [28]:

$$WI = 100 - (100 - L)^2 + a^2 + b^2 \tag{5}$$

The chromaticity (*C**) and hue angle (°*h*) values were analyzed following García-Tejeda et al. [29]. Three readings were taken for each Mt nanosystem. The data were reported as mean values ± SD.

2.3.9. Response to pH Changes

With the purpose of determining the response of the Mt nanosystems to different pH stimuli, samples of each nanosystem (0.1 g of Mt) were placed in 4 mL solutions of pH = 1, 7 and 13, made from NaOH (0.1M) and HCl. (0.1M). Color changes in the Mts were then evaluated from images that were acquired with an 8.1-megapixel Cyber-shot Sony camera model DSC-H3 (Tokyo, Japan).

2.3.10. DPPH• Antioxidant Activity

The total antioxidant activity of each Mt nanosystem was obtained utilizing the 2,2-diphenyl-2-picrylhydrazyl (DPPH•) radical methodology employed by Molyneux [30], with a UV-visible spectrometer UV-1601 PC (Shimadzu Corporation, Kyoto, Japan) at 517 nm. The antioxidant activity of the Mts analyzed was reported as the DPPH• inhibition percentage, and determined following Equation (6):

$$\%Inhibition = (A_0 - A_{60}) * (A_0)^{-1} * 100 \tag{6}$$

where A_0 and A_{60} are the absorbance values of the blank sample and the sample-containing radical, respectively. Three determinations were made *per* nanosystem. The results were expressed as mean values ± SD.

2.3.11. Antimicrobial Activity

The antimicrobial activity of the Mts was tested by exactly following the agar diffusion methodology described by Gutiérrez et al. [23]. The zone of inhibition on solid media assay was utilized to determine the antimicrobial properties of the Mts against two typical pathogen microorganisms: *Escherichia coli* O157:H7 (32158, American Type Culture Collection—Gram-negative bacteria) and *Listeria monocytogenes innocua* (Gram-positive bacteria) provided by CERELA (Centro de Referencia de Lactobacilos, Tucumán, Argentina). Each assay was performed in triplicate on two separate experimental runs.

2.3.12. Cytotoxicity Assay

Cell viability was evaluated using the (3-(4,5-dimethylthiazol-2-yl)-2,5-diphenyl tetrazolium bromide (MTT) method, which is based on the ability of viable cells to metabolically reduce a yellow tetrazolium salt (MTT; Invitrogen) to purple crystals of formazan [31]. This reaction is given when the mitochondrial reductases are active. The cell line used for this study was normal human lung fibroblasts (MRC-5, ATCC CCL-171). The cells were grown in 96-well plates (3×10^4 cells/well) for 24 h at 37 °C. Approximately 10 mg of each Mt were weighed in Eppendorf tubes, and 1 mL of distilled water was then added to each tube, in order to obtain suspensions at 1% w/v. These suspensions were stored in the dark for 24 h at 5 °C, and then micro-filtered using a syringe and membrane filter. Solutions of each Mt nanosystem at two different concentrations (0.25% and 0.5%) were then prepared. For the mitochondrial viability bioassays, cell monolayers were incubated in one of the Mt solutions for 24 h at 37 °C. Thereafter, the monolayers were laved with 1 mL of fetal bovine serum (FBS, Internegocios S.A., Buenos Aires, Argentina) and fixed with methanol (99.8%, Biopak—product code: 1655.08, El Salvador, Argentina) at ambient temperature during 10 min. The cells were then dyed with a solution of methylene blue (Merck, Darmstadt, Germany) for 10 min. The dye solution was then discarded, and the plate was laved with water. After this, the absorbance was determined at 570 cm^{-1}, using a Thermo Scientific model 2200 (Waltham, MA, USA) microplate reader. According to Leon et al. [32], this colorimetric bioassay is strongly correlated to cell proliferation measured by cell counting in a Neubauer chamber under the conditions described above. Results were reported as percent cell viability relative to natural Mt (control nanosystem, NMt). Three trials were performed *per* Mt nanosystem, and the average values ± SD were reported.

2.3.13. Cell Morphology

In order to analyze any changes to cell morphology after the cytotoxicity assay, the samples were treated as outlined in Section 2.3.12, and then observed under an optical microscope (Olympus IX51, Tokyo, Japan) at 50×. At least three microphotographs of each sample were taken with an 8.0-megapixel video camera imaging system (Olympus IMAGE RS, Tokyo, Japan) attached to a personal computer.

2.4. Statistic Analysis

All of the resulting data were analyzed using OriginPro 8 (Version 8.5, Northampton, MA, USA) software. The data were first evaluated with an analysis of variance (ANOVA), and significant results were subsequently analyzed using Duncan's multiple range tests ($p \leq 0.05$) to compare mean values.

3. Results and Discussion

3.1. X-Ray Diffraction (XRD)

The NMt displayed a diffraction peak at $2\theta = 7.10°$ associated with a 001 basal spacing of 12.5 Å (Figure 1). Similar results were reported by Tunç and Duman [33,34] for pure Mt (12.7 Å). The interlayer spacings of the MtMB and MtMD (*d*-values ≅ 18.5 Å and 26.5 Å, respectively) were notably wider compared to that of the NMt, representing an increase of about 6.0 Å and 14.0 Å, respectively (Table 1). This suggests that the dimethyl dihydrogenated tallow ammonium organo-modifying agent increased interlayer spacing to a greater degree than the dimethyl benzylhydrogenated tallow ammonium organo-modifying agent, possibly because the former is a smaller compound than the latter, and can thus penetrate the interlayer spacing of the Mt more easily. de Azeredo [35] indicated that this behavior is related to cation exchange reactions between inorganic cations and organic cations in Mts.

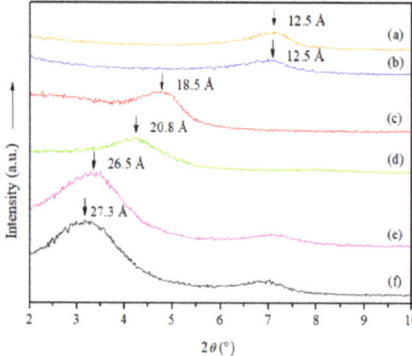

Figure 1. X-ray diffraction patterns of the different Mt nanosystems studied: (a) natural montmorillonite (NMt), (b) natural montmorillonite-containing Jamaica flower extract (NMt + JFE), (c) montmorillonite modified with dimethyl benzylhydrogenated tallow ammonium (MtMB), (d) montmorillonite modified with dimethyl benzylhydrogenated tallow ammonium containing Jamaica flower extract (MtMB + JFE), (e) montmorillonite modified with dimethyl dihydrogenated tallow ammonium (MtMD) and (f) montmorillonite modified with dimethyl dihydrogenated tallow ammonium containing Jamaica flower extract (MtMD + JFE).

Table 1. Differences in interplanar distances (Δ_{id}), mole fractions of the Jamaica flower extract (X_{JFE}) loaded within the Mt nanosystems, moisture content (MC), and color parameters of the different Mt nanosystems.

Parameter	NMt	NMt + JFE	MtMB	MtMB + JFE	MtMD	MtMD + JFE
Δ_{id} (Å)	-	0.0 ± 0.1 [a]	6.0 ± 0.1 [b]	8.3 ± 0.1 [c]	14.0 ± 0.1 [d]	14.8 ± 0.1 [e]
X_{JFE}	-	0.00 ± 0.01 [a]	-	0.10 ± 0.01 [c]	-	0.05 ± 0.01 [b]
MC (%)	5.6 ± 0.7 [d]	9.5 ± 0.4 [e]	1.8 ± 0.2 [b]	2.6 ± 0.2 [b,c]	1.4 ± 0.1 [a]	2.3 ± 0.3 [b]
L^*	83 ± 5 [d]	49 ± 2 [a]	65 ± 5 [b]	69 ± 2 [b,c]	60.4 ± 0.6 [b]	68 ± 5 [b,c]
a^*	1.9 ± 0.1 [c]	1.48 ± 0.07 [b]	1.20 ± 0.08 [a]	1.19 ± 0.01 [a]	1.43 ± 0.08 [b]	1.99 ± 0.01 [c]
b^*	10.6 ± 0.7 [d]	6.3 ± 0.1 [a]	11.7 ± 0.7 [d]	8.2 ± 0.1 [c]	12.0 ± 0.3 [d,e]	7.70 ± 0.09 [b]
Color difference (ΔE)	0.00 ± 0.00 [a]	34 ± 2 [d]	18 ± 5 [b]	14 ± 2 [b]	22.4 ± 0.6 [b,c]	15 ± 5 [b]
WhitenessIndex (WI)	80 ± 4 [d]	49 ± 2 [a]	63 ± 5 [b]	68 ± 2 [b,c]	58.6 ± 0.7 [b]	67 ± 4 [b,c]
C^*	10.8 ± 0.7 [d]	6.5 ± 0.1 [a]	11.7 ± 0.7 [d]	8.3 ± 0.1 [c]	12.1 ± 0.4 [d,e]	7.95 ± 0.09 [b]
h (°)	190.4 ± 0.2 [d]	193.1 ± 0.4 [e]	185.84 ± 0.01 [a]	188.22 ± 0.01 [c]	186.8 ± 0.2 [b]	194.5 ± 0.2 [f]

Equal letters in the same row indicate no statistically significant differences ($p \leq 0.05$). Mt nanosystems studied: natural montmorillonite (NMt), natural montmorillonite-containing Jamaica flower extract (NMt + JFE), montmorillonite modified with dimethyl benzylhydrogenated tallow ammonium (MtMB), montmorillonite modified with dimethyl benzylhydrogenated tallow ammonium containing Jamaica flower extract (MtMB + JFE), montmorillonite modified with dimethyl dihydrogenated tallow ammonium (MtMD) and montmorillonite modified with dimethyl dihydrogenated tallow ammonium containing Jamaica flower extract (MtMD + JFE).

The 001 basal reflections from MtMB and MtMD increased from the addition of Jamaica flower extract (JFE—100% polymeric anthocyanin) (MtMB and MtMD nanosystems compared to their respective analogous nanosystems containing JFE (MtMB + JFE and MtMD + JFE)) (Figure 1). This demonstrates that polymeric anthocyanins were intercalated (nanopacked/nanoencapsulated/loaded) in the silicate interlayer spaces [36]. It is worth noting that this behavior was not observed for the NMt nanosystem (NMt nanosystem compared to their respective analogous nanosystem containing JFE (NMt + JFE). The reason for this may be because the polymeric anthocyanins were larger than the basal spacing of the NMt (12.5 Å), thus preventing the JFE from entering and increasing the interlayer spacing.

3.2. Thermogravimetric Analysis (TGA)

The TGA curves (Figure 2) were analyzed in order to detect variations in the thermal stability of the nanosystems assessed; they were also used to determine the mole fraction of the JFE (X_{JFE}) that was loaded into the Mts. Three stages of thermal degradation in the Mt nanosystems were observed: the first, from ambient temperature (30 °C) up to around 150 °C, was attributed to the evaporation of physisorbed water molecules; the second, between 150 °C and 450 °C, related to the slow loss of water molecules that were initially occluded between the interlayers; and the third, from 650 °C, resulted from the dehydroxylation of the structural OH groups of the Mts [23].

In the temperature region between 30 °C and 120 °C, the TGA curves of the JFE-containing Mt nanosystems were slightly displaced towards lower temperatures with regard to those of the Mt nanosystems without JFE. This suggests that possibly the polar sites (Lewis sites) available on the Mts were sterically hindered by polymeric anthocyanins, thereby weakening and reducing Mt–water molecule interactions [37].

The NMt nanosystem exhibited thermal behavior similar to that reported by Merino et al. [38] for natural bentonite, and had a lower thermal resistance than the organo-modified Mt nanosystems. This is consistent with Zhang et al. [39], who indicated that organoammonium groups grafted on the silicates show high thermal stability, and only begin to degrade at 400 °C. This phenomenon was more pronounced in the organo-Mt nanosystems that were modified with dimethyl dihydrogenated tallow ammonium, compared to the nanosystems that were modified with dimethyl benzylhydrogenated tallow ammonium. This suggests that the dimethyl-dihydrogenated-tallow-ammonium-modified reagent can enter the interlayer spacing more easily, and interact more strongly with the Mt than the dimethyl benzylhydrogenated tallow ammonium, thereby increasing the thermal resistance of the former. The JFE-containing Mt nanosystems (NMt + JFE, MtMB + JFE and MtMD + JFE) showed a lower heat resistance compared to their analogous nanosystems without added JFE (NMt, MtMB and MtMD). This fact is possibly related to the thermal degradation of the interleaved organic matter that was contained in the interlayer spaces of the Mts.

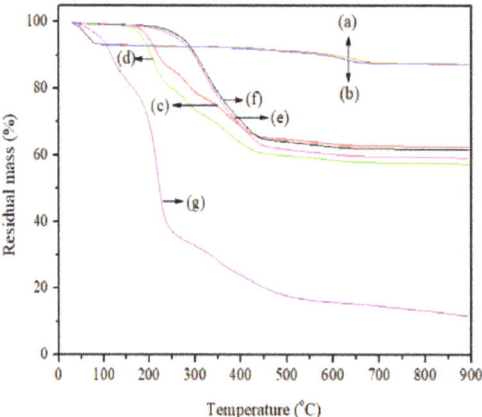

Figure 2. TGA curves of the Mt nanosystems and the extract studied: (a) natural montmorillonite (NMt), (b) natural montmorillonite containing Jamaica flower extract (NMt + JFE), (c) montmorillonite modified with dimethyl benzylhydrogenated tallow ammonium (MtMB), (d) montmorillonite modified with dimethyl benzylhydrogenated tallow ammonium containing Jamaica flower extract (MtMB + JFE), (e) montmorillonite modified with dimethyl dihydrogenated tallow ammonium (MtMD), (f) montmorillonite modified with dimethyl dihydrogenated tallow ammonium containing Jamaica flower extract (MtMD + JFE) and (g) Jamaica flower extract (JFE).

Interestingly, during the last stage of thermal degradation, MtMD showed a greater mass loss than MtMB, suggesting that there was more organic material packed within the interlayer spaces of the Mt organo-modified with dimethyl dihydrogenated tallow ammonium. This is in line with the results discussed so far.

The maximum thermal degradation temperature of the JFE (100% polymeric anthocyanin) was around 211 °C. According to the calculations made from the TGA curves, NMt + JFE did not nanopack the polymeric anthocyanins found in the JFE (Table 1). This agrees with the results obtained from the XRD diffractograms (see Section 3.1). The highest mole fraction of JFE (X_{JFE}) nanopacked into the Mts was obtained for MtMB + JFE (~0.10) (Table 1), whereas MtMD + JFE only nanopacked half of this amount (X_{JFE}~0.05). These results can be explained as follows: the organo-modification of Mt with dimethyl benzylhydrogenated tallow ammonium (MtMB) caused enough of an increase in the interlayer spacing to enable the entry of the polymeric anthocyanins. In addition, few modifying agent molecules were found within the interlayer spacing, thus leaving room for more JFE molecules to enter into it. In contrast, in MtMD + JFE, the molecules of the modifying agent (dimethyl dihydrogenated tallow ammonium; MtMD) easily entered the Mt structure. Therefore, the molar volume of polymeric anthocyanins nanopacked in MtMD + JFE was lower, despite having greater interlayer spacing.

3.3. Field Emission Scanning Electron Microscopy (FESEM)

The NMt nanoparticles were transformed slightly, from a spherical morphology (NMt, Figure 3a) to mainly irregular morphologies, either after organo-modification or when the JFE was added. The average size of the Mt particles also increased, from around 12 ± 2 μm (NMt and NMt + JFE) to 30 ± 6 μm, for the other nanosystems evaluated. Notwithstanding, no statistically significant variations among the particle sizes of the organo-modified Mts, with or without the addition of JFE, were observed, i.e., both organo-modifications led to an increase in particle size, regardless of the modifying agent used. These results fit well with the increase in interlayer spacing for these nanosystems, which was evidenced from the XRD patterns. Öztop and Shahwan [40] also reported that alkaline hydrothermal treatment-modified Mt has a similar behavior to that described above.

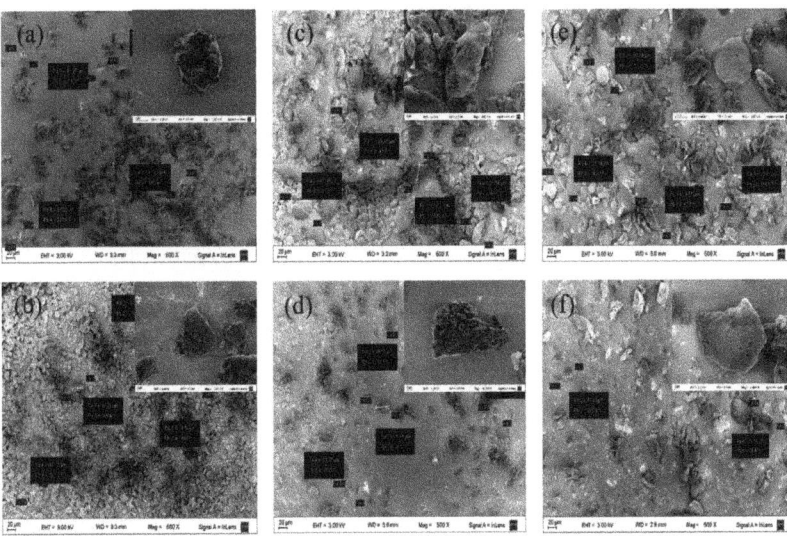

Figure 3. FESEM micrographs of the different Mt nanosystems studied: (**a**) natural montmorillonite (NMt), (**b**) natural montmorillonite containing Jamaica flower extract (NMt + JFE), (**c**) montmorillonite modified with dimethyl benzylhydrogenated tallow ammonium (MtMB), (**d**) montmorillonite modified with dimethyl benzylhydrogenated tallow ammonium containing Jamaica flower extract (MtMB + JFE), (**e**) montmorillonite modified with dimethyl dihydrogenated tallow ammonium (MtMD) and (**f**) montmorillonite modified with dimethyl dihydrogenated tallow ammonium containing Jamaica flower extract (MtMD + JFE). Some particle sizes obtained can be seen in the boxes from each image.

3.4. Moisture Content (MC)

The MC significantly differed among the treatments ($p \leq 0.05$) as follows: MtMD < MtMB < NMt (Table 1). This shows that both organo-modifying agents reduced the hydrophilicity of the Mt nanosystems, with this being more significant when dimethyl dihydrogenated tallow ammonium was used compared to dimethyl benzylhydrogenated tallow ammonium. Since the former (MtMD) is less polar than the latter (MtMB), it seems reasonable to suppose that it compensates polar sites (Lewis sites) within the Mt structure, thus decreasing the Mt nanosystems' hydrophilicity. This is in line with the TGA results (see Section 3.2). The MC of the JFE-containing Mts was statistically higher ($p \leq 0.05$) than their respective analogous Mt nanosystems without JFE. This is in line with results published by Gutiérrez et al. [23] for natural and modified nano-clays, with and without added blueberry extract (BE), and suggests that the addition of JFE increases the vulnerability of Mt nanosystems to water uptake from the ambient environment.

3.5. Attenuated Total Reflectance Fourier Transform Infrared (ATR/FTIR) Spectroscopy

The FTIR spectra for the manufactured Mt nanosystems over the entire absorption range (Figure 4A) showed absorption peaks at around 3624 cm^{-1}, which can be related to the structural OH groups in the Mts, and the bands centered at 3422 cm^{-1} associated with the stretching and bending vibrations of the hydroxyl (O-H) groups of the free water molecules physisorbed in the Mt nanosystems [38,41]. The peak intensity at 3422 cm^{-1} was greater in the JFE-containing Mt nanosystems than their analogous nanosystems without JFE (Figure 4B), suggesting that available cations were exchanged for protons from polar groups during JFE incorporation, thus raising the number of available OH groups [41]. This fits well with the MC results (see Section 3.4) that showed that the JFE-containing Mt nanosystems were more hydrophilic, i.e., higher MC values correlated with stronger absorption bands that were associated with free water and O-H groups. The bands situated at 2917 cm^{-1} and 2844 cm^{-1} corresponded to CH_2 asymmetric and CH_2 symmetric stretching vibrations within the modifying agents, and another band centered at 1468 cm^{-1} was associated with the deformation vibrations of CH_2/CH_3 that were also within the modifying agents [38]. The absorption bands positioned at 990 cm^{-1} were associated to Si-O groups in plane vibration, while the absorption bands at 916, 880 and 800 cm^{-1} were assigned to Al-Al-OH, Al-Fe-OH and Al-Mg-OH bending vibrations, respectively [42–44]. A decrease in the intensities of the bands located at 814 cm^{-1} (O-Si-O asymmetric stretching) and 990 cm^{-1} (Si-O stretching) was also evidenced, and was attributed to variations in the Si environment.

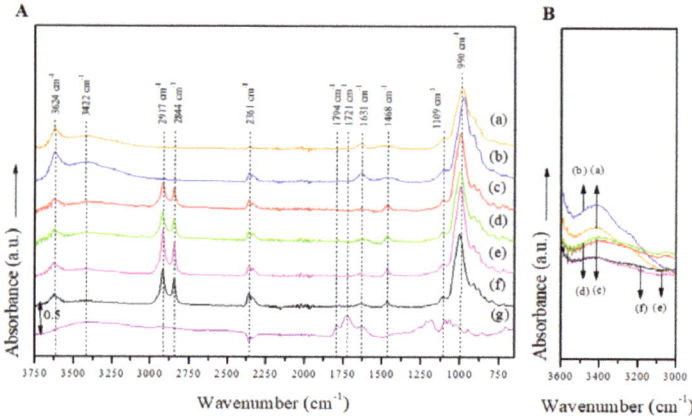

Figure 4. (A) FTIR spectra of the Mt nanosystems and the extract studied in the entire absorption range: (a) natural montmorillonite (NMt), (b) montmorillonite modified with dimethyl benzylhydrogenated tallow ammonium (MtMB), (c) montmorillonite modified with dimethyl dihydrogenated tallow ammonium (MtMD), (d) natural montmorillonite containing Jamaica flower extract (NMt + JFE), (e) montmorillonite modified with dimethyl benzylhydrogenated tallow ammonium containing Jamaica flower extract (MtMB + JFE), (f) montmorillonite modified with dimethyl dihydrogenated tallow ammonium containing Jamaica flower extract (MtMD + JFE) and (g) Jamaica flower extract (JFE). (B) FTIR spectra in the absorption range corresponding to C-O groups (OH stretching) of the Mt nanosystems and the extract evaluated: (a) NMt, (b) MtMB, (c) MtMD, (d) NMt + JFE, (e) MtMB + JFE and (f) MtMD + JFE.

The FTIR spectrum for the JFE (100% polymeric anthocyanin) showed all of the absorption peaks that corresponded to the active compound, all of which have been well characterized elsewhere in the literature [45].

3.6. Raman Spectroscopy and Confocal Laser Scanning Microscopy (CLSM)

In order to learn more about the structure of the manufactured Mts, Raman spectra were acquired (Figure 5). However, neither the organo-modified Mt nanosystems nor those with added JFE showed clear bands; thus, we were unable to discover more details about the structure of the Mt nanosystems. Gutiérrez et al. [23] indicated that the fluorescence phenomenon limits the acquisition of Raman spectra. To confirm this hypothesis, the Mt nanosystems were scanned using confocal laser scanning microscopy (CLSM) (Figure 6). The CLSM images confirmed the self-fluorescence of the Mt nanosystems developed. This is possibly related to multiple oxygen-containing functional groups. These functional groups can induce numerous localized energy levels within the n-π* gap, which decentralizes the excited electrons [46]. Notably, the fluorescence in the JFE-containing Mt nanosystems (NMt + JFE, MtMB + JFE and MtMD + JFE) was higher compared to their analogous nanosystems without added JFE (NMt, MtMB and MtMD). These fluorescence-inducing properties could make JFE (100% polymeric anthocyanin) promising for the development of theragnostic nanosystems, since, as will be discussed in the following sections, it could be used in the treatment of different diseases, and at the same time be localized by the fluorescence phenomenon. On a different note, and as expected, the morphology of the Mt nanosystems observed by CLSM (see Figure 6) was similar to that observed using FESEM (see Figure 3).

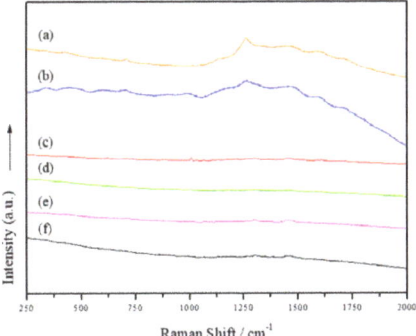

Figure 5. Raman spectra of the different Mt nanosystems studied: (**a**) natural montmorillonite (NMt), (**b**) natural montmorillonite containing Jamaica flower extract (NMt + JFE), (**c**) montmorillonite modified with dimethyl benzylhydrogenated tallow ammonium (MtMB), (**d**) montmorillonite modified with dimethyl benzylhydrogenated tallow ammonium containing Jamaica flower extract (MtMB + JFE), (**e**) montmorillonite modified with dimethyl dihydrogenated tallow ammonium (MtMD) and (**f**) montmorillonite modified with dimethyl dihydrogenated tallow ammonium containing Jamaica flower extract (MtMD + JFE).

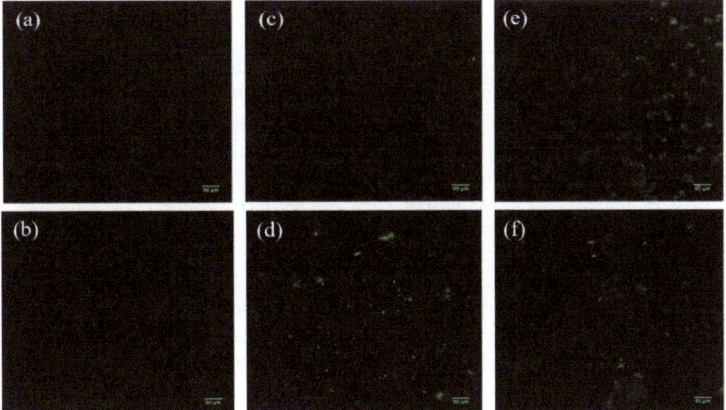

Figure 6. CLSM micrographs of the different Mt nanosystems studied: (**a**) natural montmorillonite (NMt), (**b**) natural montmorillonite containing Jamaica flower extract (NMt + JFE), (**c**) montmorillonite modified with dimethyl benzylhydrogenated tallow ammonium (MtMB), (**d**) montmorillonite modified with dimethyl benzylhydrogenated tallow ammonium containing Jamaica flower extract (MtMB + JFE), (**e**) montmorillonite modified with dimethyl dihydrogenated tallow ammonium (MtMD) and (**f**) montmorillonite modified with dimethyl dihydrogenated tallow ammonium containing Jamaica flower extract (MtMD + JFE).

3.7. Color

The results of the color analyses (Table 1) revealed that both organo-modification and/or the addition of JFE resulted in darker Mt nanosystems (see L^* values). The a^* values showed that all of the Mt nanosystems tended towards a red coloration; however, no clear tendency due to organo-modification or the incorporation of JFE was observed. The b^* values, notwithstanding, showed that although all of the Mt nanosystems tended towards yellow, this was notoriously less pronounced ($p \leq 0.05$) in those containing JFE, i.e., the addition of JFE produced less yellow nanosystems.

According to Obón et al. [47], obvious color differences identifiable by the human eye yield color difference (ΔE) values higher than 5. In this study, all of the treatments carried out on the NMt led to changes in coloration that were easily detectable by the naked eye.

The NMt had the highest whiteness index (WI) compared to the treated Mt nanosystems, which is consistent with the L^* values.

A decline in the chromaticity (C^*) values was appreciated by incorporating JFE into the Mt nanosystems. Similar results were obtained by Gutiérrez et al. [23] for nano-clays that contained polymeric anthocyanins extracted from blueberries. These C^* values were not drastically altered by the organo-modifications carried out on the Mt nanosystems.

The hue angles ($°h$) of the clay nanosystems were correctly located in the quadrants of the CIE $L^*a^*b^*$ color chart.

3.8. Response to pH Changes

The images of the color variations in the Mt nanosystems under different pH conditions (Figure 7) revealed that the JFE-containing Mts (NMt + JFE, MtMB + JFE and MtMD + JFE) showed a bathochromic effect, i.e., they changed color depending on the pH. These Mt nanosystems had a pink coloration at pH = 1, which was provoked by the formation of the flavylium cation (also called 2-phenyl-benzopyrylium, and consists of two aromatic groups: a benzopyrilium and a phenolic ring) or the oxonium ion, which are the most stable polymeric anthocyanin structures. At pH = 13, however, the JFE-containing Mt nanosystems showed a yellow coloration, possibly because of the quinoidal structure of the polymeric anthocyanins that formed at pH > 8 [48]. Our research group previously developed pH-sensitive nano-clays by incorporating polymeric anthocyanins extracted from blueberries. However, the color changes observed were different from those of the current study [23]. Color changes due to pH were produced by hydroxyl groups of the phenolic rings, and by the benzopirilium [49,50]. This suggests that the flavonoid structures of polymeric anthocyanins differ, depending on the plant source. It is worth noting that at least 20 different polymeric anthocyanin structures are known, and it is estimated that there may be as many as 150 [51]. The bathochromic or pH-sensitive effect observed in the JFE-containing Mt nanosystems is very interesting for the development of composite materials aimed at manufacturing pH-sensitive ("intelligent") food nanopackaging, or as nanosensors for theragnostic devices.

It should also be noted that the JFE at pH = 13 produced a slight emission of gases. This is because at alkaline pH, the loss of a proton from, and the addition of water to, the polymeric anthocyanin structure leads to a balance between the pseudobase hemicetal (carbinol) and chalcone (an open chain). Both hemicetal and chalcone are quite unstable forms and, at pH values higher than 7, are rapidly degraded by oxidation with air [49,52]. This means that at pH = 13, these JFE-containing Mt nanosystems undergo an irreversible color change.

Fascinatingly, not only did the color of the JFE-containing Mt nanosystems change, that of the solution they were in also changed, slightly. This verifies the diffusion of polymeric anthocyanin molecules towards the aqueous medium. Hence, polymeric anthocyanins can work as tracers of events taking place within the Mt nanosystems. This suggests that the polymeric anthocyanins that were nanopackaged within the interlayer spaces created van der Waals-type interactions between the Mts and the pigment. Otherwise, the chromophore groups of the polymeric anthocyanins would prevent pH-induced color variations in the Mt nanosystems.

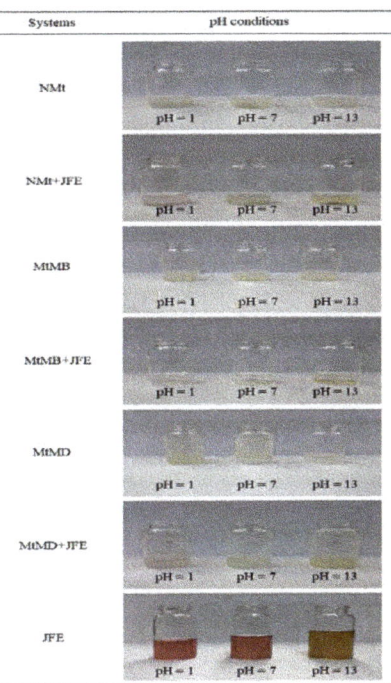

Figure 7. Response of the Mt nanosystems evaluated under different pH conditions: natural montmorillonite (NMt), natural montmorillonite containing Jamaica flower extract (NMt + JFE), montmorillonite modified with dimethyl benzylhydrogenated tallow ammonium (MtMB), montmorillonite modified with dimethyl benzylhydrogenated tallow ammonium containing Jamaica flower extract (MtMB + JFE), montmorillonite modified with dimethyl dihydrogenated tallow ammonium (MtMD), montmorillonite modified with dimethyl dihydrogenated tallow ammonium containing Jamaica flower extract (MtMD + JFE) and Jamaica flower extract (JFE).

3.9. DPPH• Antioxidant Activity and Antimicrobial Activity

The results of the antioxidant activity study showed that JFE is a highly active substance (~77% inhibition of the DPPH• radical—Figure 8). Similar results were reported by other authors elsewhere [3]. As expected, the Mt nanosystems without added JFE showed negligent oxidant activity (~1.0%). Gutiérrez et al. [23] reported similar values for this type of clay system. In contrast, all JFE-containing Mt nanosystems showed an antioxidant activity of around 50%. This confirms the stabilizing effect of the Mt nanosystems on the polymeric anthocyanins found in the JFE. Similar behavior was observed by Ribeiro et al. [20] for anthocyanins that were obtained from acerola juice, and stabilized with Mt. Importantly, all of the polymeric anthocyanin-stabilizing Mt nanosystems displayed similar antioxidant activities, suggesting that the polymeric anthocyanins were stabilized whether or not they intercalated within the interlayer spacing of the Mts (see Section 3.1). This meant that they could also be stabilized by chemical interactions that occurred between them and the surfaces of the Mt nanosystems.

The antioxidant activity of these polymeric anthocyanin-stabilizing Mt nanosystems represents a promising alternative for the treatment of different diseases, as has been discussed in the literature. This means that JFE could also be utilized for the production of medicinal and functional foods [1].

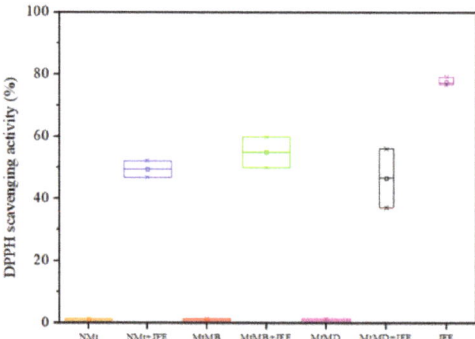

Figure 8. DPPH• cavenging activity of the Mt nanosystems and the extract evaluated: natural montmorillonite (NMt), natural montmorillonite containing Jamaica flower extract (NMt + JFE), montmorillonite modified with dimethyl benzylhydrogenated tallow ammonium (MtMB), montmorillonite modified with dimethyl benzylhydrogenated tallow ammonium containing Jamaica flower extract (MtMB + JFE), montmorillonite modified with dimethyl dihydrogenated tallow ammonium (MtMD), montmorillonite modified with dimethyl dihydrogenated tallow ammonium containing Jamaica flower extract (MtMD + JFE) and Jamaica flower extract (JFE).

In regards to the antimicrobial properties of the Mt nanosystems, other authors have demonstrated that both JFE and two commercial modified Mt nanosystems, Cloisite 20A and Cloisite 30B, show antimicrobial activity [35,53]. In this study, however, the polymeric anthocyanin-stabilizing Mt nanosystems that were developed had no antimicrobial activity against either *Escherichia coli* O157:H7 or *Listeria monocytogenes innocua*. According to Gutiérrez et al. [23], nano-clays that contained BE within their structure exhibited a similar trend. In spite of this, no microbiological growth was recorded for the polymeric anthocyanin-stabilizing Mt nanosystems.

3.10. Cytotoxicity Assay and Cell Morphology

The cell viability assays showed that natural Mts (NMt and NMt + JFE) are not cytotoxic to human cells, even at doses as high as 0.5% *w/v* (Figure 9). These results were validated through morphological observations of cells that were exposed to these two Mt nanosystems: no damage to cell ultrastructure was observed (Figure 10a,b). In fact, cell viability was significantly greater ($p \leq 0.05$) after exposure to the higher dose (0.5% *w/v*) of the polymeric anthocyanin-stabilizing Mt nanosystem (NMt + JFE), than it was after exposure to the lower dose (0.25% *w/v*). This suggests that a dose of JFE at 0.5% *w/v* could be beneficial for the growth of healthy human cells. In contrast, the Mt nanosystems modified with dimethyl benzylhydrogenated tallow ammonium (MtMB and MtMB + JFE) showed high cytotoxicity, even at the lower dose (0.25% *w/v*), i.e., cell viability was reduced after exposure to these systems. These results were confirmed by optical microscopy (Figure 10c,d), which revealed that the Mt nanosystems organo-modified with dimethyl benzylhydrogenated tallow ammonium destroyed the cell membrane, leaving the nucleus exposed, and leading to a collapse of the cell structure. These results are regrettable, since the MtMB + JFE polymeric anthocyanin-stabilizing nanosystem nanopacked the highest amounts of JFE within its structure (see Section 3.2). Similar behavior was reported in the literature for clays that were organo-modified by silylation [24]. In contrast, the Mt systems organo-modified with dimethyl dihydrogenated tallow ammonium (MtMD and MtMD + JFE) had a similar effect on cell viability as the NMt nanosystems (NMt and NMt + JFE), except that a slight decrease in viability was evidenced at the lower dose (0.25% *w/v*). This was also observed in samples examined under the optical microscope (Figure 10e,f).

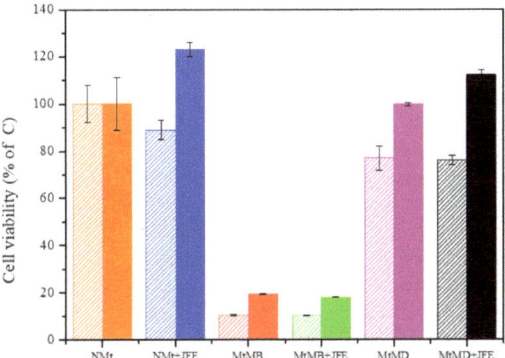

Figure 9. Cell viability of the different Mt nanosystems evaluated at two different doses (dose 0.25% *w/v*—segmented boxes; dose 0.5% *w/v*—solid boxes). Mt nanosystems: natural montmorillonite (NMt), natural montmorillonite containing Jamaica flower extract (NMt + JFE), montmorillonite modified with dimethyl benzylhydrogenated tallow ammonium (MtMB), montmorillonite modified with dimethyl benzylhydrogenated tallow ammonium containing Jamaica flower extract (MtMB + JFE), montmorillonite modified with dimethyl dihydrogenated tallow ammonium (MtMD) and montmorillonite modified with dimethyl dihydrogenated tallow ammonium containing Jamaica flower extract (MtMD + JFE).

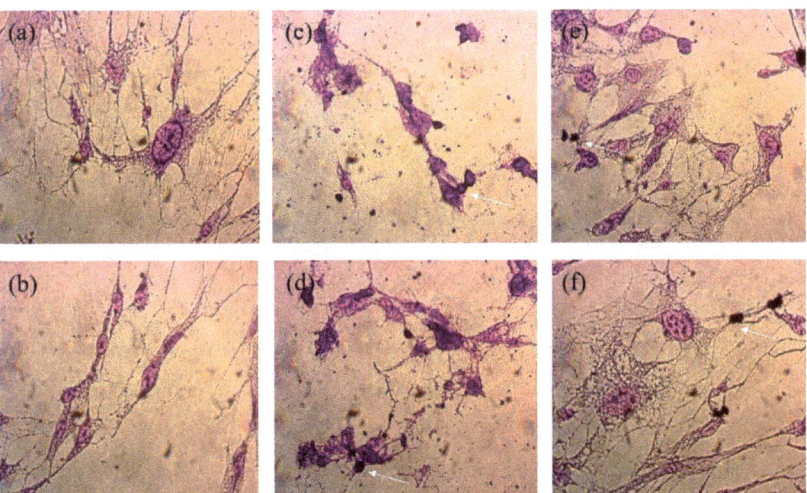

Figure 10. Optical micrographs (at 50× of magnification) of the morphology of cells exposed to the different Mt nanosystems studied: (**a**) natural montmorillonite (NMt), (**b**) natural montmorillonite containing Jamaica flower extract (NMt + JFE), (**c**) montmorillonite modified with dimethyl benzylhydrogenated tallow ammonium (MtMB), (**d**) montmorillonite modified with dimethyl benzylhydrogenated tallow ammonium containing Jamaica flower extract (MtMB + JFE), (**e**) montmorillonite modified with dimethyl dihydrogenated tallow ammonium (MtMD) and (**f**) montmorillonite modified with dimethyl dihydrogenated tallow ammonium containing Jamaica flower extract (MtMD + JFE). The arrows denote cell death of normal human lung fibroblast cells as an example. Many other similar dead cells can be seen from the same field of view in each image.

4. Conclusions

Montmorillonites (Mt) were shown to have the capacity to form polymeric anthocyanin-stabilizing nanosystems by two mechanisms: (1) intercalation of the bioactive compound within the interlayer spacing of the Mts, and (2) van der Waals-type chemical interactions between the polymeric anthocyanins and the outer surfaces of the Mts. Unfortunately, the Mts that were organo-modified with dimethyl benzylhydrogenated tallow ammonium (MtMB and MtMB + JFE) were cytotoxic for the cell line evaluated (normal human lung fibroblast, MRC-5, ATCC CCL-171), thus limiting their application as food additives. The other two polymeric anthocyanin-stabilizing nanopackaging systems that were developed, however, show promise for the development of food additives and nano-sensors, with applications in the food and biotechnology industries. This is due to their pH-sensitive behavior, self-fluorescence properties and antioxidant activity, although sadly, they did not show antimicrobial activity. Remarkably, the polymeric anthocyanin-stabilizing and nanopackaging systems were more hydrophilic compared to their analogous nanosystems without polymeric anthocyanins. Finally, the novelty of this study highlights the importance and the crucial point of the toxicological assessment of nanopackaging materials; as demonstrated here, extraordinary food packaging nanosystems able to stabilize and encapsulate active compounds can be obtained, but their food toxicity may limit their use as food contact material. To the best of the authors' knowledge, these results provide a significant advance in the understanding of this field, since very few studies have characterized and evaluated the toxicity of these kinds of materials.

Author Contributions: T.J.G. directed this research, fabricated the samples, carried out the evaluations of XRD, TGA, FESEM, moisture content, ATR/FTIR and Raman spectroscopy; CLSM and response to pH changes; depurated and analyzed the data, and drafted this paper. I.E.L. performed the cytotoxicity and cell morphology assays. A.G.P. tested the color parameters and the DPPH$^\bullet$ antioxidant and antimicrobial activities. V.A.A. contributed to the acquisition of the main funds for this research. All authors have read and agreed to the published version of the manuscript.

Funding: This research was funded by the Consejo Nacional de Investigaciones Científicas y Técnicas (CONICET) (grant PIP 2021-2023, PIBAA 2022-2023 (28720210100300CO) and PIP 2051 (2021-2024)), Universidad Nacional de Mar del Plata (UNMdP) (grant UNMdP 15/G618-ING622/21), Agencia Nacional de Promoción Científica y Tecnológica (ANPCyT) (grant PICT-2020-SERIEA-03137 and PICT 2019-2322) and Universidad Nacional de La Plata (UNLP X053).

Institutional Review Board Statement: Not applicable.

Data Availability Statement: Transparency data associated with this article can be found in the online version at https://doi.org/10.17632/hxjftyt7k8.1.

Acknowledgments: T. J. Gutiérrez would like to thank Mirian Carmona-Rodríguez for her valuable contribution. Many thanks also to Andrés Torres Nicolini for all his assistance in this investigation.

Conflicts of Interest: The authors declare no conflict of interest.

References

1. Castañeda-Ovando, A.; de Lourdes Pacheco-Hernández, M.; Páez-Hernández, M.E.; Rodríguez, J.A.; Galán-Vidal, C.A. Chemical Studies of Anthocyanins: A Review. *Food Chem.* **2009**, *113*, 859–871. [CrossRef]
2. Gonçalves, A.C.; Falcão, A.; Alves, G.; Lopes, J.A.; Silva, L.R. Employ of Anthocyanins in Nanocarriers for Nano Delivery: In Vitro and in Vivo Experimental Approaches for Chronic Diseases. *Pharmaceutics* **2022**, *14*, 2272. [CrossRef]
3. Jabeur, I.; Pereira, E.; Barros, L.; Calhelha, R.C.; Soković, M.; Oliveira, M.B.P.P.; Ferreira, I.C.F.R. *Hibiscus sabdariffa* L. as a Source of Nutrients, Bioactive Compounds and Colouring Agents. *Food Res. Int.* **2017**, *100*, 717–723. [CrossRef]
4. Ali, B.H.; Cahliková, L.; Opletal, L.; Karaca, T.; Manoj, P.; Ramkumar, A.; Al Suleimani, Y.M.; Al Za'abi, M.; Nemmar, A.; Chocholousova-Havlikova, L.; et al. Effect of Aqueous Extract and Anthocyanins of Calyces of *Hibiscus sabdariffa* (Malvaceae) in Rats with Adenine-Induced Chronic Kidney Disease. *J. Pharm. Pharmacol.* **2017**, *69*, 1219–1229. [CrossRef] [PubMed]
5. Mozaffari-Khosravi, H.; Jalali-Khanabadi, B.-A.; Afkhami-Ardekani, M.; Fatehi, F.; Noori-Shadkam, M. The Effects of Sour Tea (*Hibiscus sabdariffa*) on Hypertension in Patients with Type II Diabetes. *J. Hum. Hypertens.* **2008**, *23*, 48. [CrossRef]
6. Panaitescu, M.; Lengyel, E. Monitoring the Antibacterial Activity of *Hibiscus sabdariffa* Extracts. *Manag. Sustain. Dev.* **2017**, *9*, 31–34. [CrossRef]
7. Abedi-Firoozjah, R.; Yousefi, S.; Heydari, M.; Seyedfatehi, F.; Jafarzadeh, S.; Mohammadi, R.; Rouhi, M.; Garavand, F. Application of Red Cabbage Anthocyanins as pH-Sensitive Pigments in Smart Food Packaging and Sensors. *Polymers* **2022**, *14*, 1629. [CrossRef]
8. Gutiérrez, T.J. Active and Intelligent Films Made from Starchy Sources/Blackberry Pulp. *J. Polym. Environ.* **2018**, *26*, 2374–2391. [CrossRef]

9. Gutiérrez, T.J. Advanced Materials Made from Reactive and Functional Polymers: Editor's Insights. In *Reactive and Functional Polymers Volume Three*; Gutiérrez, T.J., Ed.; Springer International Publishing: Cham, Switzerland, 2020; pp. 1–4.
10. Bracone, M.; Merino, D.; González, J.; Alvarez, V.A.; Gutiérrez, T.J. Nanopackaging from Natural Fillers and Biopolymers for the Development of Active and Intelligent Films. In *Natural Polymers: Derivatives, Blends and Composites*; Ikram, S., Ahmed, S., Eds.; Nova Science Publishers: New York, NY, USA, 2016; pp. 119–155. ISBN 978-1-63485-831-1.
11. Gutiérrez, T.J.; Suniaga, J.; Monsalve, A.; García, N.L. Influence of Beet Flour on the Relationship Surface-Properties of Edible and Intelligent Films Made from Native and Modified Plantain Flour. *Food Hydrocoll.* **2016**, *54*, 234–244. [CrossRef]
12. Merz, B.; Capello, C.; Leandro, G.C.; Moritz, D.E.; Monteiro, A.R.; Valencia, G.A. A Novel Colorimetric Indicator Film Based on Chitosan, Polyvinyl Alcohol and Anthocyanins from Jambolan (*Syzygium cumini*) Fruit for Monitoring Shrimp Freshness. *Int. J. Biol. Macromol.* **2020**, *153*, 625–632. [CrossRef]
13. Gutiérrez, T.J.; Alvarez, V.A. Bionanocomposite Films Developed from Corn Starch and Natural and Modified Nano-Clays with or without Added Blueberry Extract. *Food Hydrocoll.* **2018**, *77*, 407–420. [CrossRef]
14. Gutiérrez, T.J.; Toro-Márquez, L.A.; Merino, D.; Mendieta, J.R. Hydrogen-Bonding Interactions and Compostability of Bionanocomposite Films Prepared from Corn Starch and Nano-Fillers with and without Added Jamaica Flower Extract. *Food Hydrocoll.* **2019**, *89*, 283–293. [CrossRef]
15. Gutiérrez, T.J.; Herniou-Julien, C.; Álvarez, K.; Alvarez, V.A. Structural Properties and in Vitro Digestibility of Edible and pH-Sensitive Films Made from Guinea Arrowroot Starch and Wastes from Wine Manufacture. *Carbohydr. Polym.* **2018**, *184*, 135–143. [CrossRef] [PubMed]
16. Gutiérrez, T.J. Surface and Nutraceutical Properties of Edible Films Made from Starchy Sources with and without Added Blackberry Pulp. *Carbohydr. Polym.* **2017**, *165*, 169–179. [CrossRef] [PubMed]
17. Toro-Márquez, L.A.; Merino, D.; Gutiérrez, T.J. Bionanocomposite Films Prepared from Corn Starch with and without Nanopackaged Jamaica (*Hibiscus sabdariffa*) Flower Extract. *Food Bioprocess Technol.* **2018**, *11*, 1955–1973. [CrossRef]
18. Gutiérrez, T.J. Are Modified Pumpkin Flour/Plum Flour Nanocomposite Films Biodegradable and Compostable? *Food Hydrocoll.* **2018**, *83*, 397–410. [CrossRef]
19. Sinela, A.; Rawat, N.; Mertz, C.; Achir, N.; Fulcrand, H.; Dornier, M. Anthocyanins Degradation during Storage of *Hibiscus sabdariffa* Extract and Evolution of Its Degradation Products. *Food Chem.* **2017**, *214*, 234–241. [CrossRef]
20. Ribeiro, H.L.; de Oliveira, A.V.; de Brito, E.S.; Ribeiro, P.R.V.; Souza Filho, M.d.s.M.; Azeredo, H.M.C. Stabilizing Effect of Montmorillonite on Acerola Juice Anthocyanins. *Food Chem.* **2018**, *245*, 966–973. [CrossRef]
21. de Moura, S.C.S.R.; Berling, C.L.; Germer, S.P.M.; Alvim, I.D.; Hubinger, M.D. Encapsulating Anthocyanins from *Hibiscus sabdariffa* L. Calyces by Ionic Gelation: Pigment Stability during Storage of Microparticles. *Food Chem.* **2018**, *241*, 317–327. [CrossRef]
22. Gutiérrez, T.J.; Álvarez, K. Biopolymers as Microencapsulation Materials in the Food Industry. In *Advances in Physicochemical Properties of Biopolymers*; Masuelli, M., Renard, D., Eds.; Bentham Science Publishers: Sharjah, United Arab Emirates, 2017; pp. 296–322.
23. Gutiérrez, T.J.; Ponce, A.G.; Alvarez, V.A. Nano-Clays from Natural and Modified Montmorillonite with and without Added Blueberry Extract for Active and Intelligent Food Nanopackaging Materials. *Mater. Chem. Phys.* **2017**, *194*, 283–292. [CrossRef]
24. Maisanaba, S.; Ortuño, N.; Jordá-Beneyto, M.; Aucejo, S.; Jos, Á. Development, Characterization and Cytotoxicity of Novel Silane-Modified Clay Minerals and Nanocomposites Intended for Food Packaging. *Appl. Clay Sci.* **2017**, *138*, 40–47. [CrossRef]
25. Wang, M.K.; Wang, S.L.; Wang, W.M. Rapid Estimation of Cation-Exchange Capacities of Soils and Clays with Methylene Blue Exchange. *Soil Sci. Soc. Am. J.* **1996**, *60*, 138–141. [CrossRef]
26. Dai, J.; Gupte, A.; Gates, L.; Mumper, R.J. A Comprehensive Study of Anthocyanin-Containing Extracts from Selected Blackberry Cultivars: Extraction Methods, Stability, Anticancer Properties and Mechanisms. *Food Chem. Toxicol.* **2009**, *47*, 837–847. [CrossRef] [PubMed]
27. *ASTM D1925-70*; Test Method for Yellowness Index of Plastics. ASTM International: West Conshohocken, PA, USA, 2021. Available online: https://www.astm.org/d1925-70r88e01.html (accessed on 26 October 2022).
28. Hsu, C.-L.; Chen, W.; Weng, Y.-M.; Tseng, C.-Y. Chemical Composition, Physical Properties, and Antioxidant Activities of Yam Flours as Affected by Different Drying Methods. *Food Chem.* **2003**, *83*, 85–92. [CrossRef]
29. García-Tejeda, Y.V.; López-González, C.; Pérez-Orozco, J.P.; Rendón-Villalobos, R.; Jiménez-Pérez, A.; Flores-Huicochea, E.; Solorza-Feria, J.; Bastida, C.A. Physicochemical and Mechanical Properties of Extruded Laminates from Native and Oxidized Banana Starch during Storage. *LWT Food Sci. Technol.* **2013**, *54*, 447–455. [CrossRef]
30. Molyneux, P. The Use of the Stable Free Radical Diphenylpicrylhydrazyl (DPPH) for Estimating Antioxidant Activity. *Songklanakarin J. Sci. Technol.* **2004**, *26*, 211–219.
31. Mosmann, T. Rapid Colorimetric Assay for Cellular Growth and Survival: Application to Proliferation and Cytotoxicity Assays. *J. Immunol. Methods* **1983**, *65*, 55–63. [CrossRef]
32. Leon, I.E.; Porro, V.; Di Virgilio, A.L.; Naso, L.G.; Williams, P.A.M.; Bollati-Fogolín, M.; Etcheverry, S.B. Antiproliferative and Apoptosis-Inducing Activity of an Oxidovanadium(IV) Complex with the Flavonoid Silibinin against Osteosarcoma Cells. *JBIC J. Biol. Inorg. Chem.* **2014**, *19*, 59–74. [CrossRef]
33. Tunç, S.; Duman, O. Preparation and Characterization of Biodegradable Methyl Cellulose/Montmorillonite Nanocomposite Films. *Appl. Clay Sci.* **2010**, *48*, 414–424. [CrossRef]

34. Tunç, S.; Duman, O. Preparation of Active Antimicrobial Methyl Cellulose/Carvacrol/Montmorillonite Nanocomposite Films and Investigation of Carvacrol Release. *LWT Food Sci. Technol.* **2011**, *44*, 465–472. [CrossRef]
35. de Azeredo, H.M.C. Antimicrobial Nanostructures in Food Packaging. *Trends Food Sci. Technol.* **2013**, *30*, 56–69. [CrossRef]
36. Ouellet-Plamondon, C.M.; Stasiak, J.; Al-Tabbaa, A. The Effect of Cationic, Non-Ionic and Amphiphilic Surfactants on the Intercalation of Bentonite. *Colloids Surf. A Physicochem. Eng. Asp.* **2014**, *444*, 330–337. [CrossRef]
37. Zhu, J.; Qing, Y.; Wang, T.; Zhu, R.; Wei, J.; Tao, Q.; Yuan, P.; He, H. Preparation and Characterization of Zwitterionic Surfactant-Modified Montmorillonites. *J. Colloid Interface Sci.* **2011**, *360*, 386–392. [CrossRef] [PubMed]
38. Merino, D.; Ollier, R.; Lanfranconi, M.; Alvarez, V. Preparation and Characterization of Soy Lecithin-Modified Bentonites. *Appl. Clay Sci.* **2016**, *127*, 17–22. [CrossRef]
39. Zhang, J.; Gupta, R.K.; Wilkie, C.A. Controlled Silylation of Montmorillonite and Its Polyethylene Nanocomposites. *Polymer* **2006**, *47*, 4537–4543. [CrossRef]
40. Öztop, B.; Shahwan, T. Modification of a Montmorillonite–Illite Clay Using Alkaline Hydrothermal Treatment and Its Application for the Removal of Aqueous Cs^+ Ions. *J. Colloid Interface Sci.* **2006**, *295*, 303–309. [CrossRef]
41. D'Amico, D.A.; Ollier, R.P.; Alvarez, V.A.; Schroeder, W.F.; Cyras, V.P. Modification of Bentonite by Combination of Reactions of Acid-Activation, Silylation and Ionic Exchange. *Appl. Clay Sci.* **2014**, *99*, 254–260. [CrossRef]
42. Madejová, J.; Komadel, P. Baseline Studies of the Clay Minerals Society Source Clays: Infrared Methods. *Clays Clay Miner.* **2001**, *49*, 410–432. [CrossRef]
43. Farmer, V.C.; Russell, J.D. The Infra-Red Spectra of Layer Silicates. *Spectrochim. Acta* **1964**, *20*, 1149–1173. [CrossRef]
44. Nayak, P.S.; Singh, B.K. Instrumental Characterization of Clay by XRF, XRD and FTIR. *Bull. Mater. Sci.* **2007**, *30*, 235–238. [CrossRef]
45. Pereira, V.A.; de Arruda, I.N.Q.; Stefani, R. Active Chitosan/PVA Films with Anthocyanins from *Brassica oleraceae* (Red Cabbage) as Time–Temperature Indicators for Application in Intelligent Food Packaging. *Food Hydrocoll.* **2015**, *43*, 180–188. [CrossRef]
46. Zhang, Y.; Hu, Y.; Lin, J.; Fan, Y.; Li, Y.; Lv, Y.; Liu, X. Excitation Wavelength Independence: Toward Low-Threshold Amplified Spontaneous Emission from Carbon Nanodots. *ACS Appl. Mater. Interfaces* **2016**, *8*, 25454–25460. [CrossRef]
47. Obón, J.M.; Castellar, M.R.; Alacid, M.; Fernández-López, J.A. Production of a Red–Purple Food Colorant from *Opuntia Stricta* Fruits by Spray Drying and Its Application in Food Model Systems. *J. Food Eng.* **2009**, *90*, 471–479. [CrossRef]
48. Steyn, W.J. Prevalence and Functions of Anthocyanins in Fruits. In *Anthocyanins: Biosynthesis, Functions, and Applications*; Winefield, C., Davies, K., Gould, K., Eds.; Springer: New York, NY, USA, 2009; pp. 86–105. ISBN 978-0-387-77335-3.
49. Garzón, G.A. Anthocyanins as Natural Colorants and Bioactive Compounds: A Review. *Acta Biol. Colomb.* **2008**, *13*, 27–36.
50. Wong, D.W.S. *Química de Los Alimentos: Mecanismos y Teoría*; Acribia: Zaragoza, Spain, 1994; ISBN 8420007757.
51. Badui, S. *Química de Los Alimentos*, 5th ed.; Longman de México Editores: Mexico City, Mexico, 1999.
52. Hutchings, J.B. *Food Color and Appearance*, 2nd ed.; Aspen Publishers, Inc.: Boston, MA, USA, 1999.
53. Abreu, A.S.; Oliveira, M.; de Sá, A.; Rodrigues, R.M.; Cerqueira, M.A.; Vicente, A.A.; Machado, A.V. Antimicrobial Nanostructured Starch Based Films for Packaging. *Carbohydr. Polym.* **2015**, *129*, 127–134. [CrossRef] [PubMed]

Review

Methodologies to Assess the Biodegradability of Bio-Based Polymers—Current Knowledge and Existing Gaps

João Ricardo Afonso Pires [1], Victor Gomes Lauriano Souza [1,2,*], Pablo Fuciños [2], Lorenzo Pastrana [2] and Ana Luísa Fernando [1]

[1] MEtRiCS, Departamento de Ciências e Tecnologia da Biomassa, NOVA School of Science and Technology, FCT NOVA, Campus de Caparica, Universidade NOVA de Lisboa, 2829-516 Caparica, Portugal; jr.pires@campus.fct.unl.pt (J.R.A.P.); ala@fct.unl.pt (A.L.F.)

[2] International Iberian Nanotechnology Laboratory (INL), Av. Mestre José Veiga s/n, 4715-330 Braga, Portugal; pablo.fucinos@inl.int (P.F.); lorenzo.pastrana@inl.int (L.P.)

* Correspondence: victor.souza@inl.int

Abstract: Our society lives in a time of transition where traditional petroleum-based polymers/plastics are being replaced by more sustainable alternative materials. To consider these bioproducts as more viable options than the actual ones, it is demanded to ensure that they are fully biodegradable or compostable and that there is no release of hazardous compounds to the environment with their degradation. It is then essential to adapt the legislation to support novel specific guidelines to test the biodegradability of each biopolymer in varied environments, and consequently, establish consistent data to design a coherent labeling system. This review work aims to point out the current standards that can serve as a basis for the characterization of biopolymers' biodegradation profile in different environments (soil, compost, and aquatic systems) and identify other laboratory methodologies that have been adopted for the same purpose. With the information gathered in this work, it was possible to identify remaining gaps in existing national and international standards to help establish new validation criteria to be introduced in future research and policies related to bioplastics to boost the sustainable progress of this rising industry.

Keywords: biopolymers; compost; soil; aquatic systems; biodegradation assay; food packaging

1. Introduction

The emergence of different plastics in daily products is directly related to the growth of life's quality since the 1950s, presenting new opportunities and more social benefits for all mankind [1]. Considering its exceptional intrinsic properties [2], versatility [3], reduced cost [4], and ease of processing [5], it was effortless to achieve the status of most handled material by distinct industries. Among the industries that introduce plastic in their products, the packaging sector stands out globally, with many millions of tons being produced annually, accounting for almost 50% of the total weight. The numbers for the packaging sector are far superior to the other industries, mainly due to single-use plastics, sometimes disposed of after brief periods of working life [6,7]. In spite of all the fascinating aspects undoubtedly linked to conventional plastic, their non-renewable character, unsustainable use, and short lifetime related with their tolerance to degradation cause a serious and realistic environmental problem, bringing out the responsibility to discover innovative and better ways to upgrade the disposal systems [8,9]. Especially in some countries, the lack of viable waste disposal alternatives to the clogged landfills boosts the accumulation of this non-degradable resource in water bodies, leaving a trail of pollution that can be seen from the shoreline to the ocean, causing potential health problems for the population worldwide [1]. In fact, the extensive production of plastics and their non-selective disposal have increased the number of microplastics (MPs) which have been detected in different environments, to the point of being considered already ubiquitous in nature. MPs assimilate the surrounding pollutants (chemicals, heavy metals, microorganisms, etc.) through different types of interactions, increasing the probability of becoming loaded with potential toxic agents. Accordingly, the inhalation and ingestion of these microparticles constitute a relevant exposure route to the health of all living animals who ingest this type of nourishment [10,11].

Regardless of some of the European Union (EU) members having banned landfill applications, there are not yet sufficient alternatives to arouse a paradigm transition, wherein incineration is not an ecological option due to the emission of pollutant gases, in which about 2.8 kg of CO_2 is evolved in the combustion of 1 kg of plastic. Furthermore, recycling still faces difficulties in separating certain plastics from contaminants, remaining a low-yielding process that needs to be properly adjusted [12–14]. Of all the plastic used in food packaging, about 9–10% is recycled, and only half of that percentage survives more than one recycling cycle. These remarkably low rates are associated with factors such as packaging composed by distinct layers of plastic material grades, contaminants present in the packaging (e.g., pigments, inks, and metals), unavailability of suitable waste sorting systems, and deficit of informative sorting and recycling labels [15]. Following the recommendations published in the document "Use of Recycled Plastics in Food Packaging (Chemistry Considerations): Guidance for Industry" by the Food and Drug Administration (FDA), the recommended maximum level of a chemical contaminant present in recycled material, if there is a migration to food, should not exceed 1.5 micrograms/person/day (0.5 parts per billion (ppb) dietary concentration) [16].

Composting is equally an alternative option to overcome the environmental impacts associated with plastics, but the resistance to microbial decomposition severely limits this approach [17,18]. For example, common plastics found in food and beverage packaging, such as polyethylene terephthalate (PET) and high-density polyethylene (HDPE), have an expected life span of 450 and 600 years, respectively [19]. However, undoubtedly, novel insights on waste management and recovery have increased among the industrial sector, the civil and scientific community, formalizing the stimulus for the establishment of a sustainable and bio-based society [20]. The problems exposed above, coupled with consumers' awareness for healthier and more beneficial products, have guided the replacement of traditional plastics for modern eco-friendly materials. Currently, the use of bioplastics in packaging merely represents values around 1–2% of global plastic packaging sales; however, the United Nations (UN) already considered these biomaterials fundamental to achieving the sustainable development goals [6]. Hereupon, it is essential to continue investing in scientific research and development of biodegradable and bio-based alternatives and adopt contemporary greener policies based on sustainable growth to reduce or even extinguish the negative footprint that petrochemical polymers induce in the environment [21,22].

Bio-based plastics/polymers, recurrently denominated in the literature as bioplastics or biopolymers, have lately been assigned as the logical candidates to substitute conventional plastics due to their renewability, high abundance, accessibility, low cost, reduced toxicity, and biodegradable character [23]. They consist of a chain-like molecule of covalently bonded monomers, in which the difference for polymers resides solely in the prefix "bio", indicating their biological or renewable origin [12]. Biopolymers can be divided into two vast groups, natural and synthetic, based on their origin. The natural ones are extracted directly from renewable sources, such as proteins (e.g., collagen or gelatin), polysaccharides (e.g., starch, cellulose, or chitin/chitosan), or lipids (e.g., waxes and free fatty acids) or could be produced by microbiological processes such as polyhydroxyalkanoate (PHA) or poly(3-hydroxybutyrate) (PHB). Poly(lactic acid) (PLA) and poly(vinyl alcohol) (PVA) represent some examples of synthetic-produced biopolymers [12,24,25]. Biodegradability is, along with the renewable character, the key point of interest attributed when compared with traditional plastics. It is expected that bio-based plastics can be easily degraded when exposed to sunlight, through photo-oxidation, or fragmentation by the action of living organisms that secrete extracellular enzymes in the presence (aerobic degradation) or absence (anaerobic degradation) of oxygen. The organic chemical substances are then converted into harmless small molecular byproducts, such as carbon dioxide, water, inorganic compounds, methane, and biomass (Figure 1) [23]. Furthermore, some of these elements can also be compostable, at a rate consistent with familiar materials [8].

Nevertheless, not all bio-based polymers are biodegradable, and, in opposition, not every fossil-based polymer is non-biodegradable. Indeed, the biodegradation rate differs for each bioplastic, depending on the external environment factors, intrinsic physico-chemical properties of the biopolymer, or the characteristics of the filler in the case of blends/composites [2,14,26]. For this reason, it is indispensable to mark an explicit distinction between bio-based plastics and biodegradable plastics. Following the instructions set by the American Society for Testing and Materials (ASTM), bio-based plastic is described as "a plastic containing organic carbon of renewable origin such as agricultural, plant, animal, fungi, microorganisms, marine, or forestry materials living in a natural environment in equilibrium with the atmosphere" [27], while biodegradable plastic is " a degradable plastic in which the degradation results from the action of naturally-occurring microorganisms such as bacteria, fungi, and algae" [28]. In favor of establishing a generic pattern, the European Bioplastics Association identifies bioplastics as a diverse family of materials, which can be bio-based, biodegradable, or both [29]. It is equally worthy to remark that biopolymer degradation can be hostage from other agents. It was identified that during biological degradation, enzymes produced by micro-organisms are responsible for the degradation. In contrast, in chemical degradation, the agents are oxygen, water, alkaline substances, acids, and solvents. A unique variation of chemical degradation is hydrolytic degradation, and other types of degradation are physical and atmospheric [30].

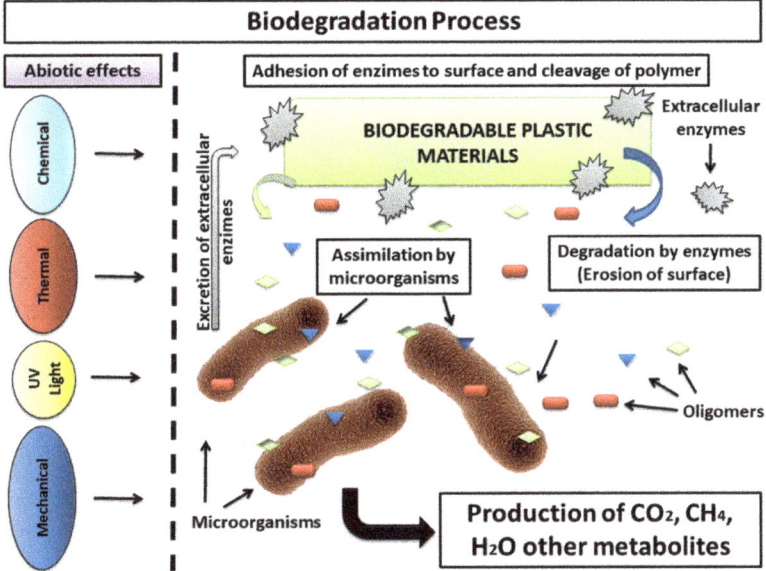

Figure 1. Biodegradation process scheme (Adapted from Souza 2015 [31]).

Thereby, to successfully perform the task of attenuating ecological impact and cut the waste connected with the end of life of single-use plastics, each novel bio-based plastic must be intensively deliberated and optimized earlier than being marketed, ensuring that it is fully biodegradable and that any detrimental ecotoxicological effects emerge from such degradation [21]. This is particularly important when new and innovative solutions with bioplastics are being tested to turn them more commercially available to reduce the dependence of some industries on fossil-based-plastic, such as the food packaging industry. Additionally, to make it feasible for a biopolymer to match the functionality of traditional plastics in food packaging, and as a result go from conception to market, it is pivotal to reinforce their matrix, modifying and upgrading their characteristics. Therefore, another undergoing solution is the incorporation of organic or inorganic nanoparticles to reinforce the film, and the addition of active elements, such as extracts and essential oils, to increase the antimicrobial and antioxidant properties, nonetheless, these new added elements could slower the biopolymer degradation [9].

As follows, the aim of this work was to gather the knowledge and discuss: (i) the current standard methodologies that can be applied to test and monitor the biodegradation of the bioplastics; (ii) the effect of environmental conditions, chemical composition, and additives on biodegradability; (iii) the verification of research works that test alternative approaches to analyze the biodegradability of bioplastics in various environments. The information gathered supports the identification of bottlenecks in existing national and international standards used to characterize plastics degradation so that concepts related to bioplastics can be correctly established and categorize the existing limits to bioplastics degradation to identify potential opportunities to boost the sustainable progress of this rising industry.

2. Current Biodegradation Assays Methodologies

The methodologies currently used to assess the biodegradability of biopolymers can be organized according to the place where the plastics are being disposed, namely: (i) in soils; (ii) in compost; (iii) in aquatic systems (Figure 2).

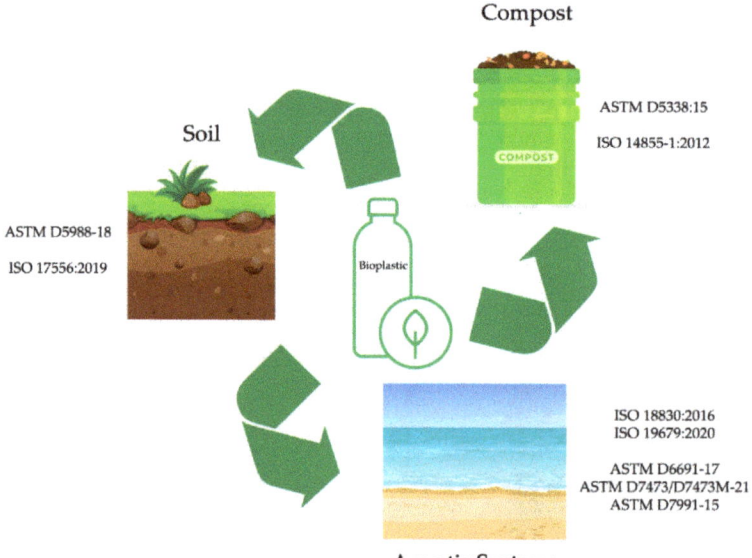

Figure 2. Main standards adopted for assessing the biodegradability of bioplastics.

2.1. Biodegradation Assays in Soil

Over time, several biological, geological, and hydrologic phenomena shaped the earth's crust creating new porous land surfaces that usually are defined as soil. The uncertainty generated by the complex number of natural variables means that on our planet it is fairly easy to find soils with uncommon temperatures, pH, moisture, and amounts of oxygen that consequently affect local microbiota activity and composition [32,33]. Under these circumstances, it is tough to find an ecosystem as rich in biodiversity as soils. In just one gram of it, a population of about 10 billion microorganisms that live directly in communion with the existing fauna, containing thousands of bacteria and fungi unique species is established [33]. Biodegradation, as the name implies, is directly related to the degradation of macromolecular chains through the action of microorganisms. Accordingly, the correlation between this process and the vastness of microorganisms in soils promotes a more susceptible and desirable bioplastic degradability study than in other locations [13,34].

Examining the literature, it is noted that miscellaneous techniques to evaluate the biodegradation of biopolymers in the soil are being adopted, since currently there are no European or International specifications or criteria settled for this purpose. This is a little bittersweet since some European markets already sell packaging that claims to biodegrade in soil [32]. It is not imperative to follow standards to interpret that a product based on an organic material shows potential to be biodegradable; however, the problem resides in the determination of the exact mass loss and time required by each bioplastic to decompose before receiving that classification [13,34]. For instance, it is perceived that the complete degradation of poly(lactic acid) in natural conditions may take up to several years [35] and it takes some time to start degrading, as demonstrated in an eleven-month test performed in the Mediterranean soil environment by Rudnik and Briassoulis [36]. Yet, a bio-based film made from starch and poly(vinyl alcohol) achieved 90% biodegradation within 28 days using a similar soil burial method [37]. In view of this, it has to be taken into account that each biopolymer is biodegraded in a specific way, reckoning variable intrinsic factors such as crystallinity, chemical structure, molecular weight, surface area, and crosslinks, and likewise is limited by some environmental soil factors such as temperature, moisture, pH, and microbial composition [13,18]. Since the bioplastic definition is so broad, it is required to define proper legislation that indicates whether the "biodegradable" categorization is ordinary to all or differs individually.

It is imperative to interrogate how the biodegradability of these novel materials can be measured and labeled [38]. Essentially, while any specific guidelines have been defined, the scientific community is majorly following two soil conditions for the assembly of biodegradation processes. If the aim is only to determine the intrinsic biodegradability instead of studying the secondary toxicity effects of it in soil, the laboratory assays seem to be the conventional choice due to a question of time. Indeed, following ISO or ASTM standards previously defined to test the biodegradation of traditional plastics, the samples are prepared under controlled parameters, which optimizes the process [38–41]. On the other hand, if it is intended to perform a realistic simulation, the biopolymer should be directly buried in outdoor soil respecting the local environmental conditions [18,42]. However, it is equally meaningful to bear in mind that these two methodologies must be complementary. Despite ensuring repeatability, standardized tests are designed to trial under optimal conditions of biodegradation, and these contrast with the sub-optimal conditions (scarce nutrients, insufficient water activity, low temperature, and reduced gas exchange) found in real soils that would constrain the adequate growth of microbes, hence slowing down the biodegradation response [32,38].

The most actual standard methodologies for biodegradation of plastics in soil are ASTM D5988-18 "standard test method for determining aerobic biodegradation of plastic materials in soil" [43] and ISO 17556:2019 "plastics—determination of the ultimate aerobic biodegradability of plastic materials in soil by monitoring the oxygen demand in a respirometer or the amount of carbon dioxide evolved" [44]. Briefly, ASTM D5988-18 measures the carbon dioxide developed by microorganisms as a function of time of exposure, thus measuring the degree of biodegradability relative to a reference material. Similarly, the principle behind the ISO 17556:2019 yields the optimum rate of biodegradation of plastic material in a test soil by controlling the oxygen consumption or the carbon dioxide production [41]. For both cases, the two standards research settings are reasonably simple and similar as stated in Table 1. Basically, the bioplastic sample is buried in closed vessels containing previously prepared soil. Subsequently, the vessels are subjected to a temperature range favorable to the growth of mesophilic microorganisms jointly with optimal conditions of moisture and oxygen. Although both indicate the same temperature range, ISO 17556:2019 establishes an ideal temperature of 25 °C.

Table 1. Some soil technical specifications for each standard.

Specifications	ASTM D5988-18	ISO 17556:2019
Type	The soil should be natural and fertile, preferably "sandy loam" collected from fields and forests not exposed to pollutants. The soil used should be a mixture of natural and fertile soils collected from the surface layers of at least 3 diverse locations.	Natural soil from fields and/or forests may be used as an inoculum to simulate biodegradation in a specific natural environment.
Particle size	Less than 2 mm	Less than 5 mm
Room Temperature	$20 - 28 \pm 2\,°C$	$20 - 28 \pm 2\,°C$ but in preference $25\,°C$
pH	6–8	6–8
Moisture holding Capacity (MHC)	80–100%	40–60%
Running Time	Should not exceed 6 months	Should not exceed 6 months but may be extended up to 24 months
Reference Material	Cellulose or starch	Microcrystalline cellulose powder, ashless cellulose filters or poly(3-hydroxybutyrate)
Validation Criteria	The test is considered valid if after 6 months more than 70% biodegradation is achieved for the reference material and if the amount of CO_2 evolved from the control reactors is within 20% of the mean at the plateau phase or at the end of the test.	The test is considered valid if the degree of biodegradation of the reference material is more than 60% at the plateau phase or at the end of the test and if the BOD (biological oxygen demand) values of or the amount of CO_2 evolved from the controls are within 20% of the mean at the plateau phase or at the end of the test.

These standards should not merely serve as a guide but also as base methods for possible new approaches. Based on this premise, Šerá et al. [41] used ISO 17556:2019 as the basic methodology to test an innovative method that allows the acceleration of biodegradation of slowly degradable polyesters in the soil, raising the control temperature from 25 °C to 37 °C to decrease the analysis time. A moderate increase in temperature aimed to increase the number of extremophilic species but without a drastic change in microorganisms existing at 25 °C, and therefore, be liable for assessing ultimate biodegradation under representative conditions of ambient temperature. In another study, Pischedda et al. [38] observed the effect of temperature in the biodegradation of bioplastics in soil following ASTM D5988-18. They used pellets of a commercial biodegradable plastic material to be tested at 28, 20, and 15 °C. On balance, it was possible to draw a parallel between the temperature and the speed of biodegradation. These two recent studies marked an initial phase towards deploying alternative methodologies to simulate field dissipation kinetics contemplating the effects of soil temperature.

A parameter that still generates many doubts is the running time, raising some valid questions before setting up the experiment. Both ASTM D5988-18 and ISO 17556:2019 state that the test should not exceed six months, but what should represent the amount of biodegradation that the sample must show to consider the test finished? Should there be complete biodegradation or is it acceptable to merely go beyond the validation values of the materials used as reference? In addition, in which days the bioplastic degradation evolution should be assessed? In two different works, in which was used poly(vinyl alcohol)/starch blended, different experiment designs were tested. It should be registered that the tests were carried out under the laboratory simulated conditions without ensuing any of the standards previously presented, and both evaluated the biodegradation through weight loss percentage over time. One decided that the samples should be collected every 7 days for up to 35 days [37], while the other determined that the samples should be unburied every 10 days for up to 50 days [45]. Sen and Das [37] noted almost total film biodegradation within a period of 35 days, while Kaur et al. [45] registered 96.6% weight loss after 50 days. Adopting these two works as examples, this slightly illustrates the complexity of knowledge that presently can be found in the literature. The biodegradability of similar biomaterials is being tested differently and, therefore, sustaining a broad range of values. This lack of uniformity constitutes a giant challenge of how the authentic biodegradation values for each biomaterial should be established. It has already been mentioned that numerous variables condition affects the polymer biodegradation, but the present standards are limited to encompass all of them. A synergic approach employing various methodologies should be followed when dealing with the biodegradation of bioplastics [46]. Novelty studies are performing extensive assessments and thus presenting more consistent values of biodegradability in soil (Table 2).

Table 2. Recent methodologies to monitor biodegradation in soil.

Biopolymer *	Soil Conditions	Methodology	Ref.
Chitosan/Corn cob	Laboratory-controlled conditions	ASTM D5988	[39]
PHBV/Olive Pomace	Laboratory-controlled conditions	ASTM D5988	[40]
Cellulose based	Laboratory-controlled conditions	ASTM D5988	[51]
PBAT/Nanocellulose	Laboratory-controlled conditions	ASTM D5988	[52]
Polyurethane (PU)/Starch	Laboratory-controlled conditions	ASTM D5988, morphology and chemical characterization, visual analysis	[48]
Starch/Nanocellulose	Laboratory-controlled conditions	ISO 17556	[53]
Poly(lactic acid) (PLA)/Glycerol	Laboratory-controlled conditions	ISO 17556	[54]
Mater-Bi (Mixture of PBAT, starch, and additives)	Laboratory-controlled conditions	ISO 17556, ecotoxicological analysis	[55]
PHA, PBS, and PBAT/PLA	Laboratory-controlled conditions	ISO 17556, microrganismo characterization	[41]
PHB mixed with natural fillers	Laboratory-controlled conditions	Mass Loss	[47]
Starch/Nanocellulose	Outdoor soil conditions	Mass Loss	[42]
Poly(vinyl alcohol) (PVA)/Starch	Outdoor soil conditions	Mass Loss	[45]
PVA/Starch	Laboratory-controlled conditions	Mass loss, biofilm area, soil characterization, and visual analysis	[37]
PCL, PHB, PLA, and PBS	Laboratory-controlled conditions and outdoor soil conditions	Mass loss, mechanical properties, and microorganism characterization	[49]
Starch based	Laboratory-controlled conditions	Mass loss, morphology analysis, and mechanical properties	[56]
PBS/Sugarcane Fiber	Laboratory-controlled conditions	Mass loss, morphology analysis, and thermal characterization	[57]
PHA	Outdoor soil conditions	Mass loss and morphology and chemical analysis	[50]
PLA and PLA/Starch	Outdoor soil conditions	Mass loss, morphology and chemical analysis, and thermal characterization	[18]
PVA/Starch and PVA/Starch mixed with different natural fillers	Laboratory-controlled conditions	Mass loss, morphology and chemical analysis, and soil characterization	[58]

* Poly(3-hydroxybutyrate-co-3-hydroxyvalerate) (PHBV), Poly(butylene adipate-co-terephthalate) (PBAT), Polyhydroxyalkanoate (PHA), Poly(1,4-butylene succinate) (PBS), Poly(3-hydroxybutyrate) (PHB), Polycaprolactone (PCL), Poly(1,4 butylene) succinate (PBS), and Poly(vinyl alcohol) (PVA).

Mass loss is the most basic and widely used biodegradation index. Typically, the analysis of this specification is conducted through experimental measurement that consists of collecting samples at different times, drying until achieving a constant weight [42,45,47]. Complementary, another way to quantify the mass loss is through molecular weight measurement using gel permeation chromatography [18]. The morphology analysis with scanning electron microscopy (SEM) is a good complement assay that allows to state the effect of the ingress of water in the biopolymer, which is called surface erosion [48]. Moreover, with this observation, it is possible to analyze and identify the microorganisms that grow superficially during the degradation process. Šerá et al. [41] identified a massive mesh of fungal filaments and fungal spores coating the entire surface of fast degrading materials (PHA and PBS). In opposition, the authors discovered that slowly degradable polymers (PBAT/PLA) and experimental polyester network (denominated ICL-PN) were uncovered with microorganisms. Al Hosni et al. [49] also observed that a range of microorganisms has grown in distinct polymers, taking advantage of them as nutrients, particularly when under starvation and lacking essential nutrition, therefore representing a crucial role in the polymer biodegradation. Subsequently, to improve the characterization, rRNA gene sequencing was performed to identify the fungal isolates. Besides the morphology, the biopolymer surfaces' chemical composition should also be evaluated. Spectroscopy was used to estimate the biodegradation development through the changes in specific functional groups of the bioplastics [46]. Tai et al. [48] and Sabapathy et al. [50] used Fourier transform infrared spectroscopy (FTIR) to measure the level of degradation, observing the shift and also intensity of specific infrared peaks.

In another intriguing study, Ibrahim et al. [56] compared the variations in the mechanical properties with the biodegradation behavior of starch-based composites after being reinforced with different lignocellulosic fibers. This is helpful to determine the influence of the filler type in biopolymer's degradation process. The team discovered that biocomposites' tensile strength and modulus of elasticity decreased more than 50% during the first week, and then further gradual deterioration inevitably took place until the end of the experiment.

Before ending this prerogative of ideas, it is worth recalling something mentioned in the introduction. To consider bioplastics as a viable option to one day replace traditional plastics it is necessary to ensure that they are fully biodegradable and that no harmful ecotoxicological effects result from such degradation. Once, during degradation, the biopolymer repeatedly breaks down and drops many different elements to the soil, with the released molecules fulfilling a significant role in plant metabolism, either inhibiting or stimulating flora growth; consequently, it is fundamental to consider ecotoxicological assessments. Seed germination and plant growth bioassay are the most frequent techniques used to complete this evaluation [59]. The *modus operandi* is simple, the biopolymer is buried in soil in an incredibly high dose and suffers biodegradation. From time to time, a sample of the soil is tested with bioassays for ecotoxicity herein with a control sample not exposed to the biopolymer and a control sample exposed to a GRAS (generally recognized as safe) substance, such as cellulose. To rule out negative ecotoxicity effects, no significant difference must be perceived between the test samples and control samples [56]. Still, regarding soil quality, another fascinating topic to be studied in the future is the effect that the degradation of biopolymers can have on the remediation of soils contaminated with heavy metals. It is known that, for example, chitosan demonstrates excellent metal-binding properties [60]. Why not combine the useful with the pleasant and enhance the application of bio packages at the end of life by depositing them in contaminated soil?

Briefly, it is mandatory to establish a European or international standard that helps to complement the already existing ISO and ASTM standards so that it is possible to characterize with more certainty the biodegradability of each biopolymer that is entering the market. Only then will it be possible to state with certainty which are the best techniques for depositing these in soils to guarantee total and safe biodegradability.

2.2. Biodegradation Assays in Compost

Composting is an aerobic method of solid waste management, wherein the presence of microorganisms, under controlled conditions, biodegradable materials are biologically decomposed into humus, which is a good nutrition source for strengthening soil productivity and agricultural yield. This process may be incredibly helpful in the disposal problem and to diminish greenhouse gases emissions since the compost is a beneficial organic amendment and substrate that can be reintroduced into the economic system [17,61–63].

Analyzing the most recent scientific data is possible to verify that a vast majority of the assays on biopolymers degradation under aerobic composting conditions have been following more standardizations than in soils once these tests are mainly being carried out at laboratory level. As perceived in Table 3, the most adopted standards are the ISO 14855-1:2012 "determination of the ultimate aerobic biodegradability of plastic materials under controlled composting conditions—method by analysis of evolved carbon dioxide—Part 1: General method" [64] and, analogously, the American Society for Testing and Materials has the ASTM D5338-15 "standard test method for determining aerobic biodegradation of plastic materials under controlled composting conditions, incorporating thermophilic temperatures" [65]. Both determine the ultimate aerobic biodegradability (means by which microorganisms entirely consume a chemical or organic substance in the presence of oxygen) of plastics based on organic compounds under controlled composting conditions by measuring the percentage conversion of the carbon into carbon dioxide and the degree of disintegration of the plastic at the end of the test.

The standards mentioned above were designed with the intention of simulating typical aerobic composting conditions, by exposing the test material to an inoculum derived from the organic fraction of solid mixed municipal waste (Table 4). The composting takes place in an environment wherein temperature, aeration, and humidity are accurately monitored and controlled. The percentage of CO_2 involved is quantified through acid-base titration or by employing a direct measurement such as infrared or gas chromatography. Additionally, the ISO 14855 has a Part 2, the ISO 14855-2:2018 [66], differing from the first part by implementing a gravimetric method to measure the biopolymer mineralization. The material sample disintegration is estimated comparing the total dry solids of the initial test amount with the retrieved fractions of material superior to 2 mm.

Equally, as seen formerly for soils, the existent standards are still limited and the conditions disparities between home and industrial composting may lead to a significant difference in the biopolymer degradation character [13]. The fragmentation emerges quicker in an industrial composter because thermophilic temperatures are most easily achieved on a large scale and constitute less risk of microplastics exhibition to the environment as the degradation occurs within a sealed and controlled system. In opposition, it is much more difficult to achieve these temperatures in small-scale residential composting units, often referred to as "backyard" or "home" composting, and the toxicological effects on the environment are by far distinct [3]. Until now, no international standards have been presented concerning specifications for domestic compostability. However, some national grades, mostly based on EN 13432 "requirements for packaging recoverable through composting and biodegradation—test scheme and evaluation criteria for the final acceptance of packaging" [67], have been implemented with the aim of certifying the home compostability of bioplastics. Moreover, TÜV AUSTRIA BELGIUM and DIN CERTCO provide certification for home compostability following the Australian standard AS 5810 "biodegradable plastics—biodegradable plastics suitable for home composting" [68]. To characterize the biopolymer in a less idealized and more realistic composting environment, the ISO 16929:2021 "determination of the degree of disintegration of plastic materials under defined composting conditions in a pilot-scale test" [69] could represent the most adequate alternative. This one determines the degree of disintegration of materials in a pilot-scale test under defined composting conditions, being much more representative of microbial populations than laboratory-scale setups.

Table 3. Recent methodologies to monitor biodegradation in compost.

Biopolymer *	Compost Origin	Methodologies to Monitor Biodegradation	Ref.
PLA, PLA/Chitosan, PLA/Nanocellulose, and PLA/Gum	Food waste	ASTM D5338, mass loss, morphology and chemical analysis, thermal characterization, contact angle, and microorganism characterization	[63]
PVA PVA/Nanocellulose	MSW	ASTM D5338, visual analysis, mass loss, morphology and chemical analysis, thermal characterization, and compost quality (including physico-chemical parameters and ecotoxicological analysis)	[17]
PHA-based	MSW	ASTM D5338	[70]
Starch/Carboxymethyl Chitosan	Comercial	ASTM D5338	[71]
Chitosan	Not Specified	ASTM D-5338 and ISO 14855-1, visual analysis, morphology and chemical analysis, thermal characterization, and microorganism characterization	[61]
PVA/Starch	MSW	ISO 14855-1, ISO 20200, visual analysis, morphology analysis, thermal characterization	[72]
Nano-reinforced PLA	MSW	ISO 14855-1, ISO 20200 with adaptations of ISO 16929, visual analysis, and compost quality (including physico-chemical parameters and ecotoxicological analysis)	[73]
Gelatin, Chitosan, and/or Sodium Caseinate based	Mature compost (10%) mixed with synthetic biowaste (contained sawdust (40%), rabbit food (30%), starch (10%), sugar (5%), oil (4%), and urea (1%))	ISO 20200, visual analysis, mass loss, and chemical analysis	[62]
PHA based	Mature compost (10%) mixed with synthetic biowaste (contained sawdust (40%), rabbit food (30%), starch (10%), sugar (5%), oil (4%), and urea (1%))	ISO 20200	[70]
PLA and PLA/Silica	Not speficied compost fermented with biomass (2:1)	Modified ISO 17556	[74]
PLA, PLA/TAC, and PLA/PHB/TAC	MSW	CO_2 concentration, mass loss, chemical analysis, and thermal characterization	[75]
PCL, PHB, PLA, and PBS	Comercial	Mass loss, mechanical properties, and microorganism characterization	[49]
Starch/Montmorillonite	Not Specified	Mass loss and ecotoxicological analysis	[76]
Nano-reinforced PLA	MSW	Mass loss and microorganism characterization	[77]

* Poly(lactic acid) (PLA), Poly(vinyl alcohol) (PVA), Polyhydroxyalkanoate (PHA), Plasticizer triacetine (TAC), Poly(3-hydroxybutyrate) (PHB), Polycaprolactone (PCL), Poly(1,4 butylene) succinate (PBS), and municipal solid waste (MSW).

Two other standards were developed for more embracing purposes, the ASTM D6400-21 "standard specification for labeling of plastics designed to be aerobically composted in municipal or industrial facilities" [78] and ISO 17088:2021 "plastics—organic recycling—specifications for compostable plastics" [79]. Besides the evaluation of biodegradability through CO_2 measurement and disintegration, the ASTM D6400 additionally includes elemental analysis, plant germination (phytotoxicity), and mesh filtration of the resulting particles. In fact, if it is only intended to perform a biodegradation test without the extensive range of compost analysis, ASTM D5338 is enough. In ISO 17088:2021, apart from the intention of study biodegradability and disintegration such as in ISO 14855-1:2012, this normative equally has as purpose the evaluation of negative consequences on the composting process and facility and negative effects on the quality of the resulting compost, including the presence of high levels of regulated metals and other harmful components. Both agree on the identification parameters for material to be considered a compostable material. The polymer should pass in the toxicity tests, and after 84 days less than 10% of the initial weight of the material should disintegrate into fragments that can go across a 2 mm sieve, a further 90% of the organic carbon (absolute or relative) shall be converted to CO_2 after 180 days. A summary of the presented methodologies can be found in Table 5.

Table 4. Inoculum technical specifications for ASTM D5338:15 and ISO 14855:2012.

Specifications	ASTM D5338:15	ISO 14855-1:2012
Inoculum	Industrial compost soil that has a maturity level of 2–4 months	Stabilized, mature compost derived (four-month old), if possible, from composting the organic fraction of solid municipal waste
Particle size	-	0.5 cm to 1 cm
Room temperature	Incubated in the dark at 58 °C ± 2 °C	Incubated in the dark at 58 °C ± 2 °C
pH	7.0–8.2	7.0–9.0
Total dry solids content	between 50% and 55% of the wet solids	between 50% and 55% of the wet solids
Volatile solids content	To produce more than 50 mg but less than 150 mg of CO_2 per gram of volatile solids after 10 days of incubation	No more than about 15% of the wet or 30% of the dry solids
Moisture holding capacity (MHC)	50%	50–60%
Running time	Minimum of 90 days and should not exceed 6 months	Minimum of 90 days and should not exceed 6 months
Reference material	Celullose	TLC (thin-layer chromatography)-grade cellulose with a particle size of less than 20 μm
Validation criteria	If: (a) more than 2 g of volatile fatty acids per kilogram of dry matter in the composting vessel is formed, the test must be regarded as invalid; (b) a minimum of 70% for cellulose within 45 days is not observed with the positive reference, the test must be regarded as invalid and should be repeated, using new inoculum; (c) the deviation of the percentage of biodegradation of the positive reference is greater than or equal to 20% at the end of the test, then the test shall be regarded as invalid.	The test is considered as valid if: (a) the degree of biodegradation of the reference material is more than 70% after 45 days; (b) the difference between the percentage biodegradation of the reference material in the different vessels is less than 20% at the end of the test; (c) the inoculum in the blank has produced more than 50 mg but less than 150 mg of carbon dioxide per gram of volatile solids after 10 days of incubation.

The biopolymer biodegradation assessment is an expansion study field still pursuing uniformity, and as such, the addition of these tests in scientific investigations represents a heroic attempt to prove the outstanding features presented by the prefix "bio" when compared with traditional plastics. Nonetheless, it is equally visible that only a few numbers of studies manage to achieve a complete approach to the subject. Biodegradability tests, in both soil and compost, are complex, extensive, and expensive, and few research teams have laboratory conditions to execute them, most often limiting themselves to following existing standards with the addition of one or another characterization. ASTM D6400 and ISO 17088 are the most comprehensive standards, but they still lack some fundamental characterizations. As follows, it is essential to study the degradation of the biopolymer at various levels and under a multiplicity of environmental conditions to be able to accurately establish complete guidelines. Acknowledging the actual barriers in the following paragraphs, some interesting results of a set of recent articles that assess different biopolymer blends biodegradation from unique perspectives are presented, confirming the importance of complementing the standardization with the study of other parameters.

Table 5. Summary of ASTM and ISO standards for plastic degradation in compost.

Standard	Name	Scope
ASTM D5338:15	Standard test method for determining aerobic biodegradation of plastic materials under controlled composting conditions, incorporating thermophilic temperatures	Determination of the ultimate aerobic biodegradability
ISO 14855-1:2012	Determination of the ultimate aerobic biodegradability of plastic materials under controlled composting conditions—method by analysis of evolved carbon dioxide—Part 1: general method	
ISO 14855-2:2018	determination of the ultimate aerobic biodegradability of plastic materials under controlled composting conditions—method by analysis of evolved carbon dioxide—Part 2: gravimetric measurement of carbon dioxide evolved in a laboratory-scale test	
ISO 16929:2021	Determination of the degree of disintegration of plastic materials under defined composting conditions in a pilot-scale test	Determination of the disintegration degree of plastic materials in composting
ISO 20200:2015	Plastics—determination of the degree of disintegration of plastic materials under simulated composting conditions in a laboratory-scale test	
ASTM D6400-21	Standard specification for labeling of plastics designed to be aerobically composted in municipal or industrial facilities	Establishes the requirements for identifying plastics and products made from plastics that compost satisfactorily in industrial and municipal aerobic composting facilities. Includes biodegradation, disintegration, and environmental safety testing requirements and criteria
ISO 17088:2021	Plastics—organic recycling—specifications for compostable plastics	

A biodegradable material is not necessarily compostable. As seen in the compostability requirements stated in ASTM D6400 and ISO 17088, a material to be considered compostable must not only respect certain biodegradability and disintegration rates but also demand to pass in the environmental safety tests. The methods for evaluating the ecotoxicity of compostable polymer materials are typically based on the use of plants, soil fauna (earthworms), aquatic fauna (*Daphnia*), algae (green algae), and microbes (luminescent bacteria). To confirm that biodegradable materials are not automatically compostable, Gutiérrez et al. [76] analyzed the ecotoxicity of starch/montmorillonite films by examining the growth of the primary root of lettuce (*Lactuca sativa*) seedlings exposed to three different concentrations. They observed that the produced films, besides being marked as biodegradable, revealed to be non-compostable material at high concentrations (100 µg/mL) as measured by their effect on lettuce seedlings.

Salehpour et al. (2018) [17] investigated the effect that the incorporation of cellulose nanofibers (CNF) had on the biodegradation of poly(vinyl alcohol). They conjugate the ASTM D5338 to measure the amount of mineralized carbon, with more extensive analyses of the decomposition process resorting to mass loss, visual, morphology and chemical analysis, and thermal characterization. Moreover, they monitored the impact of the biodegradation in the compost, performing physicochemical and ecotoxicological evaluation by growing cress and spinach. At the end of the assay, it was observed that the PVA/nanocellulose had lower biodegradation rates than the neat PVA. This can be attributed to the difficulty that water molecules suffer to enter the polymer matrix, consequently reducing the microbial growth and activity and to the good interfacial bonding between CNF and PVA that makes it very difficult to break the strong bonds of the film. Another remarkable observation during the thermal analysis was the increase in the biopolymer crystallinity over the composting time. Such can be explained by the evidence that microbes easily degrade the random amorphous component of the polymer, rather than the crystalline phases. Positively, the ecotoxicological assay concluded that no negative chemical effects on the germination and growing of the studied plants were verified during the biodegradation process. In the following work, Kalita et al. [63] aimed to study the aerobic biodegradation behavior of modified PLA-based biocomposites under composting conditions by adding to ASTM D5338 other analytical techniques such as molecular weight, differential scanning calorimetry (DSC), contact angle analysis, and microbial colony count. According to the authors, these techniques allow to know how the composting temperature and the change in molecular activity affects the test material wettability. Globally, a decreased trend was observed in contact angles values, suggesting that sample surfaces were becoming more hydrophilic. Meanwhile, as seen in the Salehpour et al. (2018) [17] study, crystallinity percentage increases in biodegraded samples due to loss of amorphous phase at a faster rate, which interestingly counters the previous results because generally contact angles for polymeric materials demonstrate an increasing tendency when the crystallinity percentage is higher. This investigation also offers information about the microorganism counts in compost for each blend. Nutrient colonies agar and plate count agar were the chosen plates to determine colony growth. It was discovered that colony formation was much higher in PLA/Chitosan than in PLA/Nanocellulose and PLA/Gum Arabic. The analyzed plates, using optical microscopy, also demonstrate that gram-negative bacteria were predominant.

It is crucial to recognize the diversity of microorganisms present in compost ecosystems to know which taxonomic class produces the most adequate degradation enzymes to each biopolymer, making the composting more effective. This was the aim that supported the research paper "Tailoring the microbial community for improving the biodegradation of chitosan films in composting environment" by Altun et al. [61]. The work was divided into two parts. In the first place, the authors investigated the biodegradation of chitosan films in controlled composting reactors and analyzed microbial diversity via PCR-denaturing gradient gel electrophoresis (DGGE). Regarding this, it was observed that the dominant taxonomic groups at the phylum level were Ascomycota and Proteobacteria. In the second part, the amount of carbon dioxide emitted by two specific groups of microorganisms added to the inoculum was assessed. The ensemble is composed of Bacillus (*Bacillus circulans*, *Bacillus licheniformis*), Actinobacteria (*Streptomyces roseolus*, *Streptomyces zaomyceticus*), and fungi (*Penicillium islandicum*, *Penicillium chrysogenum*) and were capable of producing chitosanase that increased the chitosan degradation rate. In conclusion, the design and optimization of the microbial community used are important to enhance biodegradation efficiency. In a similar research, Castro-Aguirre et al. [77] isolated and identified the PLA-degrading microbial strains present in compost (*Geobacillus thermoleovorans*) and introduced such strains with bioaugmentation purposes. This is a promising technique used to accelerate the biodegradation of compostable plastics.

The lack of result reproducibility due to the excessive variability of sources and the different factors that affect biodegradability make it very arduous to evaluate it correctly, whatever the type of environment. However, studies show that performing a complete biodegradation analysis, with controlled and constant monitoring, helps to reduce the error margin. The amount of CO_2 produced during the process is probably the best indicator of the material's biodegradability, but this is not enough to evaluate its composting. It is essential to combine this test with others to determine the mechanisms of biodegradation as well as to evaluate its impacts on the environment. Currently, following the guidelines proposed in the ASTM D6400 and ISO 17088, it is already possible to have a good idea of how biopolymers behave under controlled composting situations, but these standards still fail to simulate real conditions.

2.3. Biodegradation Assays in Aquatic Systems

Plastic litter is the main responsible for the catastrophic scenario of pollution observed nowadays in aquatic systems. This pollutant naturally embarks on its journey in effluents from the wastewater treatment process, reaching inland freshwaters such as rivers and lakes, and then keeps on treading the path to the oceans where it settles and continues disintegrating into micro- and nano-plastics (less than 5 mm). The formed debris may affect the growth, development, behavior, reproduction, and mortality of marine and freshwater fauna. Moreover, the microorganisms existent in the varied waters are suitable hosts for covering the microplastics surface and when ingested by the fauna, they can turn into a public health problem if that food chain reaches humans [80,81]. Few studies evince concern in knowing the mechanisms and conditions of degradation of bioplastics in aquatic systems, most probably because there is still a preconceived idea that they are easily degraded in any context [81,82]. In this circumstance, the scientific community, with research focused on the subject of bio-based plastic, must have as a social responsibility not to transmit the inaccurate idea that the future use of these is seen as an ultimate solution to the problem of marine pollution without causing any harmful effect in the environment.

The aquatic ecosystem englobes a huge variability of habitats with markedly distinct environmental conditions. Each one influences the biopolymer biodegradability in a particular way, therefore, predicting this feature in aquatic systems is an even more challenging task when compared with terrestrial locations [83]. The water surface, for example, holds more moderate temperatures, receiving UV light from the sun and oxygen from the air, factors that speed up the abiotic degradation. In opposition, the lack of these in deep waters reduces photo- and thermo-oxidative degradation. The hydrolysis rate is equally affected by the type of water, wherever freshwater contains more microorganism concentration than in seawater rich in salt. Some material properties such as density affect the rate of degradation; if the plastic is floating, the degradation is completely different from that of a plastic concentrated in deep waters [84,85]. Within the marine habitat, it is also necessary to consider the geographical location, for instance, the disintegration of a biopolymer in sediments next to where the waves break is not the same as in the open ocean with calm waters [86,87]. Concerning this, biodegradation should be assessed in three significant marine habitats: (a) the pelagic zone, where plastic is floating neutrally buoyantly in the water; (b) the beach sediment zone, which is periodically covered by water due to waves or tide, defined as the eulittoral; (c) the sublittoral benthic zone, where plastic is sunken to the seafloor.

Currently, the first conclusive studies on the biopolymers biodegradation in aquatic systems are appearing, largely due to the thematic novelty and long duration of the experiment. There is an interconnection between marine and freshwater ecosystems since water derived from waste treatment systems, rivers, and lakes conducts much pollution to the oceans; consequently, the rate of biodegradation of a polymer could be dependent on two different aquatic media [81]. Although, the published results are mainly focused on coastal marine habitats where the residence of plastics is more consistent. It can also be perceived that those studies rarely follow an existing evaluation standard because it is difficult to adapt them to real aquatic conditions. At the present moment, there are no European or international standards focused on studying biodegradation in freshwater systems, and to our knowledge, there is no development for creating one related to inland water bodies. However, regarding marine habitats, some ISO and ASTM standards are available.

The International Organization for Standardization indicates two norms for the assessment of aerobic plastic biodegradability, which are ISO 18830:2016 "plastics—determination of aerobic biodegradation of non-floating plastic materials in a seawater/sandy sediment interface—method by measuring the oxygen demand in closed respirometer" [88] and ISO 19679:2020 "plastics—determination of aerobic biodegradation of non-floating plastic materials in a seawater/sediment interface—method by analysis of evolved carbon dioxide" [89]. As the name implies, the two methodologies are very similar, centering on experimental laboratory simulations under controlled conditions using seawater and sediments found in different sublittoral benthic zones. The difference between them resides in the biodegradation evaluation, where one assay measures the oxygen demand and the other keeps up the CO_2 evolution. Furthermore, a standard that determinates ultimate biodegradation in anaerobic conditions can be found under the denomination of ISO 14853:2016 "plastics—determination of the ultimate anaerobic biodegradation of plastic materials in an aqueous system—method by measurement of biogas production" [90]. Two more ISO standards were developed with the intention of simulating a water column scenario, ISO 23977-1:2020 "plastics—determination of the aerobic biodegradation of plastic materials exposed to seawater—Part 1: method by analysis of evolved carbon dioxide" [91] and ISO 23977-2:2020 "plastics—determination of the aerobic biodegradation of plastic materials exposed to seawater—Part 2: method by measuring the oxygen demand in closed respirometer" [92]. The five exposed ISOs are intended for laboratory testing, but the urgent necessity to start methodic testing plastic biodegradation in situ conditions boosted the publication of more standards. To test it in the sea surface ISO 15314:2018 was created, "plastics—methods for marine exposure" [93] and for seafloor and beach scenario ISO 22766:2020 exists, "plastics—determination of the degree of disintegration of plastic materials in marine habitats under real field conditions" [94]. ASTM also proposes three standards about this subject, ASTM D6691-17 "standard test method for determining aerobic biodegradation of plastic materials in the marine environment by a defined microbial consortium or natural sea water inoculum" [95], ASTM D7473/D7473M-21 "standard test method for weight attrition of non-floating plastic materials by open system aquarium incubations" [96], and ASTM D7991-15 "standard test method for determining aerobic biodegradation of plastics buried in sandy marine sediment under controlled laboratory conditions" [97]. The ASTM D6691-17 scope resides in the employment of the measurements of CO_2 evolution under controlled conditions to determine the degree and rate of aerobic biodegradation of plastic materials to a pre-grown population of at least ten aerobic marine microorganisms living in natural seawater. Complementary to this test, ASTM D7473/D7473M-21 has the objective to predict real-world experiences based on the dimension and tax of biodegradation data for the same test material, bringing into play visual proof of biodegradation and determination of the weight loss as a function of time during the exposure to seawater in a flow-through system. Finally, ASTM D7991-15 [97] simulates the environmental conditions found in the tidal zone, exposing the plastic to collected sediment and seawater. A resume about this extensive information can be revised in the following Table 6.

Table 6. Summary of ASTM and ISO standards for plastic degradation in aquatic systems.

Standard	Name	Scope
ISO 18830:2016	Plastics—determination of aerobic biodegradation of non-floating plastic materials in a seawater/sandy sediment interface—method by measuring the oxygen demand in closed respirometer	Determination of the degree and rate of aerobic biodegradation of plastic materials when settled on marine sandy sediment at the interface between seawater and the seafloor by measuring the oxygen demand in a closed respirometer.
ISO 19679:2016	Plastics—determination of aerobic biodegradation of non-floating plastic materials in a seawater/sediment interface—method by analysis of evolved carbon dioxide	Determination of the degree and rate of aerobic biodegradation of plastic materials when settled on marine sandy sediment at the interface between seawater and the seafloor by measuring the evolved carbon dioxide.
ISO 14853:2016	Plastics—determination of the ultimate anaerobic biodegradation of plastic materials in an aqueous system—method by measurement of biogas production	Determination of the ultimate anaerobic biodegradability of plastics by anaerobic microorganisms by exposing the test material to sludge for a period of up to 90 days, which is longer than the normal sludge retention time (25 to 30 days) in anaerobic digesters.

Table 6. *Cont.*

Standard	Name	Scope
ISO 23977-1:2020	Plastics—determination of the aerobic biodegradation of plastic materials exposed to seawater—Part 1: method by analysis of evolved carbon dioxide	Determination of the degree and rate of the aerobic biodegradation level of plastic materials. Biodegradation is determined by measuring the CO_2 evolved from plastic materials when exposed to seawater sampled from coastal areas under laboratory conditions.
ISO 23977-2:2020	Plastics—determination of the aerobic biodegradation of plastic materials exposed to seawater—Part 2: method by measuring the oxygen demand in closed respirometer	Determination of the degree and rate of the aerobic biodegradation level of plastic materials. Biodegradation of plastic materials is determined by measuring the oxygen demand in a closed respirometer when exposed to seawater sampled from coastal areas under laboratory conditions.
ISO 15314:2018	Plastics—methods for marine exposure	Description of three methods for the exposure of plastics in a marine environment. Method A covers exposures where specimens float on the surface, method B covers exposures where specimens are partially immersed method C covers exposures where specimens are completely immersed.
ISO 22766:2020	Plastics—determination of the degree of disintegration of plastic materials in marine habitats under real field conditions	Determination of the degree of disintegration of plastic materials exposed to marine habitats under real field conditions. The marine areas under investigation are the sandy sublittoral and the sandy eulittoral zone where plastic materials can either be placed intentionally.
ISO 62:2008	Plastics—determination of water absorption	Determination of the moisture absorption properties in the "through-the-thickness" direction of flat or curved-form solid plastics. Determination of the amount of water absorbed by plastic specimens of defined dimensions when immersed in water or when subjected to humid air under controlled conditions.
ASTM D6691-17	Standard test method for determining aerobic biodegradation of plastic materials in the marine environment by a defined microbial consortium or natural sea water inoculum	Determination of the degree and rate of aerobic biodegradation of plastic materials (including formulation additives) exposed to pre-grown population of at least ten aerobic marine microorganisms of known genera or the indigenous population existing in natural seawater.
ASTM D7473/D7473M-21	Standard test method for weight attrition of non-floating plastic materials by open system aquarium incubations	Determination of the weight loss as a function of time of non-floating plastic materials (including formulation additives) when incubated under changing, open marine aquarium conditions, which is representative of aquatic environments near the coasts and near the bottom of a body of water in the absence of sunlight, particularly UV and visible portions of the spectrum.
ASTM D7991-15	Standard test method for determining aerobic biodegradation of plastics buried in sandy marine sediment under controlled laboratory conditions	Determination of the biodegradation level of plastic materials exposed to laboratory conditions that simulate the environment found in the sandy tidal zone. The tidal zone, that is, the part of the coast affected by the tides and movement of the waves, is the borderline between sea and land, frequently a sandy area that is kept constantly damp by the lapping of the waves.
ASTM D570-98(2018)	Standard test method for water absorption of plastics	Determination of the relative rate of absorption of water by plastics when immersed. This test method is intended to apply to the testing of all types of plastics, including cast, hot-molded, and cold-molded resinous products, and both homogeneous and laminated plastics in rod and tube form and in sheets 0.13 mm (0.005 in.) or greater in thickness.

Table 6. *Cont.*

Standard	Name	Scope
ASTM D5229/D5229M-20	Standard test method for moisture absorption properties and equilibrium conditioning of polymer matrix composite materials	Determination of moisture absorption or desorption properties in the "through-the-thickness" direction for single-phase Fickian solid materials in flat or curved panel form. Procedures for conditioning test coupons prior to use in other test methods are also covered, either to an essentially moisture-free state to equilibrium in a standard laboratory atmosphere environment, or to equilibrium in a non-laboratory environment. Procedures for determining the moisture loss during elevated temperature testing are also included, as well as moisture loss resulting from thermal exposure after removal from the conditioning environment, such as during strain gauge bonding.

The aforementioned standards demonstrate limitations that call into question their reliability when it is intended to be used as a basis for in situ experiments. These limitations were reflected in the excellent review "Biodegradability standards for carrier bags and plastic films in aquatic environments: a critical review" by Harrison et al. (2018) [81], of which we recommend a careful reading. Concisely, these limitations can be cataloged in five different origins: (a) approaches to inoculum preparation and test conditions; (b) absence of specific guidelines for employing different test materials; (c) insufficient statistical replication; (d) a lack of adequate procedures for unmanaged aquatic environments; (f) shortcoming toxicity testing and the impacts of plastic litter on aquatic ecosystems. Furthermore, the arduous climatic conditions often experienced in marine environments, such as storms, strong currents, and waves can compromise the experiment. This experience may also encounter other adversities such as coming into conflict with fishing activities or even hitting recreational boats, which in the last case can lead to sample sabotage and theft. Nevertheless, this type of experimental setup requires exorbitant costs matched only with private financing or grants directed to large-scale projects. This set of factors drives many research groups away from in situ testing and deciding to focus on more simplified laboratory results. This does not mean that it is not possible to make an interesting characterization about the biodegradability of a certain biopolymer in freshwater or seawater, it just becomes more difficult to develop more complex models. This in itself is a significant contribution, orienting future research and proving that more targeted studies are mandatory to directly expose the influence of different biodegradation factors [87,98].

The biodegradation assessment tests that complement the standards, already presented for soil and compost, can equally be applied to aquatic systems. In addition, water absorption, a method to test hydrolytic degradation, is another laboratory test widely used in parallel and which can be useful to predict degradation in systems involving water. The soaked water impregnates the polymer matrix and changes the water gradient in the space between the surface and the inner part of the material [30]. This parameter is crucial to study biopolymers reinforced with fillers, since the reinforcement introduces structural changes in the biopolymer that can affect hydrophilicity. Additionally, the more water the biopolymer is able to absorb, the more likely it is that colonies of microorganisms form. This test also has defined guidelines that can be followed through the ASTM D570-98(2018) "standard test method for water absorption of plastics" [4], ASTM D5229/D5229M-20 "standard test method for moisture absorption properties and equilibrium conditioning of polymer matrix composite materials" [99], or ISO 62:2008 "plastics—determination of water absorption" [100]. Complementarily, solubility analysis can be coupled into the study. The test procedure for the water absorption test is simple, the samples are dried in an oven for an assigned time and temperature and then set in a desiccator to cool. Forthwith, upon cooling the specimens are weighed. The material subsequently emerges in the water at controlled conditions, often 23 °C for 24 h or until equilibrium. Specimens are removed, dried, and weighed. The water intake percentage is the difference in the weight of the specimen before and after the tests. However, 24 h is too short to properly help in the development of a complex biopolymer degradation profile, thus a new experimental design should be created where the standard is adapted to longer times. For instance, Hassan et al. [101] and Jiménez-Rosado et al. [102] observed the hydrolysis degradation during hours, while Kasmuri et al. [103] and Kumar Thiagamani et al. [104] decided to measure it in a question of days. Furthermore, the experiments could use various types of waters, not being limited to distilled water.

Following the predicate that for the aquatic systems, trusty field test methods and standards for assessing and authenticating biodegradation are lacking; it is crucial to expose two very recent and interesting works. Briassoulis et al. [86] evaluated the standard ISO 19679:2020 [89] and proposed modifications and adjustments to improve the validity and confidence of the methodology in some aspects. Foremost, the authors included agitation in the water surface by a floating magnetic stirrer to simulate wave motion. This modification has the intention of transporting more oxygen from the water surface to the sediments and with that opens the door to fast biodegradation. Generally speaking, they proposed the method should have bigger bioreactors, larger test material quantities mixed with the essential nutrients, and required continuous surface stirring together with a higher threshold. To validate all these alterations, they also recreated field experiments marking the importance of the proposed test to the natural sublittoral conditions. Additionally, Lott et al. [87] presented novel field tests to assess the performance of biodegradable plastics under natural marine conditions. To validate this methodology, it was successfully applied in three coastal habitats (eulittoral, benthic, and pelagic) and in two climate zones (Mediterranean Sea and tropical Southeast Asia). Likewise, they developed a stand-alone mesocosm (or tank) system independent of the direct access to seawater. Mesocosm tests occupy a leading role as a methodological link between field and lab tests because these impersonate a medium that better approximates the real environment than in small-scale laboratory tests. This work is relevant because the outcome has supported the development of the new ISO 22766:2020.

Two other fresh and remarkable works deserve appreciation in this review. Bagheri et al. [82] designed a one-year comparative degradation study of six different polymers (five taken from the so-called biodegradable polyesters, including poly(lactic-co-glycolic acid) (PLGA), PCL, PLA, PHB, Ecoflex, and one well-known non-degradable polymer poly(ethylene terephthalate) (PET). The polymers were immersed in artificial seawater and freshwater under controlled conditions in a thermostatic chamber at 25 °C and under fluorescence light (16 h light and 8 h dark). Analyzing this study it was possible to conclude that under similar conditions only PLGA presented 100% bulk degradation in both mediums, while PHB, for example, simply showed 8% of degradation after 365 days. The amorphous nature of the polymer could be a possible explanation of the faster hydrolysis and complete degradation of PLGA, making diffusion of water easy all throughout the bulk. Relative to the comparison of the different types of aquatic environments, a similar degradation it was observed but with little tendency to find fast degradation results majorly in freshwater. In the second study, a Japanese team investigated the growth of the bacterial consortium on the bioplastic surface in freshwaters [105]. The authors also isolated and identified the bacteria responsible for that degradation. Briefly, freshwater from 5 different Japanese locations was used to test the biodegradability of 6 distinct bioplastics. The bioplastics were soaked in freshwater inside vial bottles and were incubated at 30 °C with a 150 rpm slow shaking. The formation of significant growth of microorganisms in the bioplastic surface was observed after two weeks. The authors concluded that *Acidovorax* and *Undibacterium* were the predominant genera in most of the samples.

As a conclusion of this chapter, it is worth recalling that very little is known about the potential of each biopolymer in different aquatic environments, but it is known that degradation is affected by a large number of variables that need to be explored in order to be able to implement novel reliable and more robust research methodologies.

3. Final Remarks

Our society is currently experiencing a period of extreme change due to the urgent focus on combating climate change. Consumers are constantly confronted with new products labeled as natural, biodegradable, or compostable. However, how is this classification achieved? It is essential to create social strategies to educate consumers and companies on how to manage and classify bio-based products in order to minimize the dumping of these residues in inappropriate places where the biodegradation processes would take a longer time or not occur at all and to defy an increasingly established problem in our society, so-called "green-washing". Thus, the ideas presented in this review summarized the current knowledge about the methodologies that have been adopted to access the biopolymer degradation in different environments (soil, compost, and aquatic systems). Ultimately, the work highlights the uncertainties, difficulties, and existing gaps that still constrain the accurate assessment of the biodegradability of a bio-based plastic entering the market. The information gathered here is valuable to help industrial companies categorize the existing limits to bioplastic degradation and identify potential opportunities to boost the sustainable progress of the food packaging industry towards the production of cleaner and environmentally friendlier packaging, meeting consumers' and market's expectations for the future of this important sector of industrial production.

Currently, the global bioplastics sector has presented fast growth because of the continuously emerging range of bio-based and biodegradable polymers production and rising interest in investing in this sector. In a recent report, it was foreseen that its production will expand from 2.11 million tons in 2018 to 2.62 million tons in 2023, with Europe leading the rank of research and development of bioplastics, while Asia stands as a major hub for bioplastic production and consumption [106]. Bio-based industry consortiums in partnership with the European Union (EU) are investing about EUR 3.7 billion on large-scale flagship projects to encourage new technologies in this field.

Due to the differences in the properties of biodegradable and non-biodegradable polymers, a lot of research is still needed to develop biodegradable polymers or polymer blends/composites that have the necessary properties to replace most of the current non-biodegradable ones. Moreover, when moving toward the goal of widely spreading the production and use of bio-based biodegradable polymers, another constraint arises as a political challenge may occur to educate people to properly dispose of these biodegradable plastics in such a way that they can be transferred to the correct dedicated composting sites for effective biodegradation [107].

When compared with traditional plastics based on non-renewable sources, it is undeniable that the environment benefits from the shorter degradation rates presented by the most diverse bio-based polymers. However, the biodegradation rate is dependent not only on the biopolymer chemical structure but also on the surrounding environmental conditions. Moreover, if other components are incorporated in the bioplastic, such as reinforcement fillers or active agents, it is probable that the biodegradation profile will change.

Currently, there are no official guidelines to characterize the biodegradation profile of a bio-based polymer; consequently, it is difficult to guarantee that products made from them are labeled correctly. The research teams and certifying companies are compelled to base their results on individual standards used to evaluate conventional plastics. In addition, in many works presented in this review, the results were acquired through the application of ISO and ASTM standards and complemented with other data supported by extra laboratory methodologies. The discrepancy of approaches applied in each work is reflected on the different biodegradability rates identified for the same biopolymer. Therefore, it is urgent to create a policy framework that establishes specifications and validation criteria that fit as a foundation for new biodegradability standards adapted for each specific biopolymer and that can be applied in different environments. Indeed, several procedures can be followed to evaluate the biodegradability of bio-based polymers, two of which are: (a) control of the oxygen consumption or the carbon dioxide production; (b) via determination of mass loss (to evaluate disintegration of the plastic). Yet, it is essential to study the degradation of the biopolymer at various levels. Analyzing the different compounds produced during the decomposition process chemically, and/or analyzing mass loss, and the visual, morphological and thermal characterization, could provide hints on biodegradation mechanisms. Therefore, it is suggested that future standardized protocols to measure bio-base polymer degradation should also be rooted in complementing the standardization with the study of other parameters, namely through chemical analysis of the compounds liberated.

A bioplastic solely represents a sustainable solution if it is fully biodegradable and if any adverse ecotoxicological effects arise from its degradation. To ensure this, it is fundamental characterization tests are executed not only at the laboratory scale but also in situ through ecotoxicological assessments. Additionally, the microbiota identification present in the ecosystems is also a key aspect to take into consideration in order to determine which taxonomic class produces the most adequate degradation enzymes to each biopolymer. This complementarity requires significant financial effort and is extremely difficult to implement for the majority of the research teams.

It is therefore clear that much remains to be done, and legislators have a key role in establishing standards that help harmonize the biodegradability assays applicable to a biopolymer in order to secure the safety of these new products when distributed in the market.

Author Contributions: Conceptualization, J.R.A.P., V.G.L.S. and A.L.F.; methodology, J.R.A.P.; resources, V.G.L.S. and A.L.F.; writing—original draft preparation, J.R.A.P.; writing—review and editing, V.G.L.S., A.L.F., P.F. and L.P.; supervision, V.G.L.S. and A.L.F.; funding acquisition, A.L.F., P.F. and L.P. All authors have read and agreed to the published version of the manuscript.

Funding: This research was funded by national funding from the FCT, the Foundation for Science and Technology, and through the individual research grants of J.R.A.P. (SFRH/BD/144346/2019). This work was supported by the Mechanical Engineering and Resource Sustainability Center—MEtRICs, which is financed by national funds from FCT/MCTES (UIDB/04077/2020 and UIDP/04077/2020). This work was also funded by FlexFunction2Sustain | Innovation for nano-functionalized flexible plastic surfaces (Horizon2020, IA—Innovation Action, Grant agreement ID 862156).

Institutional Review Board Statement: Not applicable.

Informed Consent Statement: Not applicable.

Data Availability Statement: Data is contained within the article.

Conflicts of Interest: The authors declare no conflict of interest.

References

1. Pires, J.R.A.; Souza, V.G.L.; Fernando, A.L. Production of Nanocellulose from Lignocellulosic Biomass Wastes: Prospects and Limitations. In *Innovation, Engineering and Entrepreneurship*; Machado, J., Soares, F., Veiga, G., Eds.; Lecture Notes in Electrical Engineering; Springer International Publishing: Cham, Switzerland, 2019; Volume 505, pp. 719–725. ISBN 9783319913339.
2. Kawashima, N.; Yagi, T.; Kojima, K. How do bioplastics and fossil-based plastics play in a circular economy? *Macromol. Mater. Eng.* **2019**, *304*, 1900383. [CrossRef]
3. Kubowicz, S.; Booth, A.M. Biodegradability of plastics: Challenges and misconceptions. *Environ. Sci. Technol.* **2017**, *51*, 12058–12060. [CrossRef] [PubMed]
4. D570-98(2018); Standard Test Method for Water Absorption of Plastics. ASTM International: West Conshohocken, PA, USA, 2018.
5. Mangaraj, S.; Yadav, A.; Dash, S.K.; Mahanti, N.K. Application of biodegradable polymers in food packaging industry: A comprehensive review. *J. Packag. Technol. Res.* **2018**, *3*, 77–96. [CrossRef]
6. Rosenboom, J.; Langer, R.; Traverso, G. Bioplastics for a circular economy. *Nat. Rev. Mater.* **2022**, *7*, 117–137. [CrossRef] [PubMed]
7. Ncube, L.K.; Ude, A.U.; Ogunmuyiwa, E.N.; Zulkifli, R.; Beas, I.N. An overview of plastic waste generation and management in food packaging industries. *Recycling* **2021**, *6*, 12. [CrossRef]
8. Fortunati, E.; Luzi, F.; Yang, W.; Kenny, J.M.; Torre, L.; Puglia, D. Bio-based nanocomposites in food packaging. In *Nanomaterials for Food Packaging*; Elsevier: Amsterdam, The Netherlands, 2018; pp. 71–110. ISBN 9780323512718.

9. Pires, J.; De Paula, C.D.; Gomes, V.; Souza, L.; Fernando, A.L. Understanding the barrier and mechanical behavior of different nanofillers in chitosan films for food packaging. *Polymers* **2021**, *13*, 721. [CrossRef] [PubMed]
10. Luo, H.; Liu, C.; He, D.; Xu, J.; Sun, J.; Li, J.; Pan, X. Environmental behaviors of microplastics in aquatic systems: A systematic review on degradation, adsorption, toxicity and biofilm under aging conditions. *J. Hazard. Mater.* **2022**, *423*, 126915. [CrossRef]
11. Wieland, S.; Balmes, A.; Bender, J.; Kitzinger, J.; Meyer, F.; Frm, A.; Roeder, F.; Tengelmann, C.; Wimmer, B.H.; Laforsch, C.; et al. From properties to toxicity: Comparing microplastics to other airborne microparticles. *J. Hazard. Mater.* **2022**, *428*, 128151. [CrossRef]
12. Shankar, S.; Rhim, J. Bionanocomposite films for food packaging applications. *Ref. Modul. Food Sci.* **2018**, 1–10. [CrossRef]
13. Emadian, S.M.; Onay, T.T.; Demirel, B. Biodegradation of bioplastics in natural environments. *Waste Manag.* **2017**, *59*, 526–536. [CrossRef]
14. Thakur, S.; Chaudhary, J.; Sharma, B.; Verma, A.; Tamulevicius, S.; Thakur, V.K. Sustainability of bioplastics: Opportunities and challenges. *Curr. Opin. Green Sustain. Chem.* **2018**, *13*, 68–75. [CrossRef]
15. Sundqvist-Andberg, H.; Åkerman, M. Sustainability governance and contested plastic food packaging—An integrative review. *J. Clean. Prod.* **2021**, *306*, 127111. [CrossRef]
16. *Use of Recycled Plastics in Food Packaging (Chemistry Considerations): Guidance for Industry*; U.S. Food & Drug Administration: Rockville, MD, USA, 2021.
17. Salehpour, S.; Jonoobi, M.; Ahmadzadeh, M.; Siracusa, V.; Rafieian, F.; Oksman, K. Biodegradation and ecotoxicological impact of cellulose nanocomposites in municipal solid waste composting. *Int. J. Biol. Macromol.* **2018**, *111*, 264–270. [CrossRef] [PubMed]
18. Lv, S.; Zhang, Y.; Gu, J.; Tan, H. Soil burial-induced chemical and thermal changes in starch/poly (lactic acid) composites. *Int. J. Biol. Macromol.* **2018**, *113*, 338–344. [CrossRef]
19. Qin, Z.-H.; Mou, J.; Chao, C.Y.H.; Chopra, S.S.; Daoud, W.; Leu, S.; Ning, Z.; Tso, C.Y.; Chan, C.K.; Tang, S.; et al. Biotechnology of plastic waste degradation, recycling, and valorization: Current advances and future perspectives. *ChemSusChem* **2021**, *14*, 4103–4114. [CrossRef]
20. Fernando, A.L.; Duarte, M.P.; Vatsanidou, A.; Alexopoulou, E. Environmental aspects of fiber crops cultivation and use. *Ind. Crop. Prod.* **2015**, *68*, 105–115. [CrossRef]
21. Lambert, S.; Wagner, M.; Wagner, M. Environmental performance of bio-based and biodegradable plastics: The road ahead. *Chem. Soc. Rev.* **2017**, *46*, 6855–6871. [CrossRef]
22. García, A.; Gandini, A.; Labidi, J.; Belgacem, N.; Bras, J. Industrial and crop wastes: A new source for nanocellulose biorefinery. *Ind. Crops Prod.* **2016**, *93*, 26–38. [CrossRef]
23. Souza, V.G.L.; Fernando, A.L. Nanoparticles in food packaging: Biodegradability and potential migration to food—A review. *Food Packag. Shelf Life* **2016**, *8*, 63–70. [CrossRef]
24. Brigham, C. Biopolymers: Biodegradable alternatives to traditional plastics. In *Green Chemistry: An Inclusive Approach*; Elsevier Inc.: Amsterdam, The Netherlands, 2018; pp. 753–770. ISBN 9780128095492.
25. Youssef, A.M.; El-Sayed, S.M. Bionanocomposites materials for food packaging applications: Concepts and future outlook. *Carbohydr. Polym.* **2018**, *193*, 19–27. [CrossRef]
26. Garrison, T.F.; Murawski, A.; Quirino, R.L. Bio-Based Polymers with Potential for Biodegradability. *Polymers* **2016**, *8*, 262. [CrossRef] [PubMed]
27. D6866-21; Standard Test Methods for Determining the Biobased Content of Solid, Liquid, and Gaseous Samples Using Radiocarbon Analysis. ASTM International: West Conshohocken, PA, USA, 2021.
28. D883-20b; Standard Terminology Relating to Plastics. ASTM International: West Conshohocken, PA, USA, 2020.
29. European Bioplastics. What are Bioplastics? Available online: https://www.european-bioplastics.org/bioplastics/ (accessed on 17 February 2020).
30. Andrzejewska, A.; Wirwicki, M.; Andryszczyk, M. Procedure for determining aqueous medium absorption in biopolymers. *AIP Conf. Proc.* **2017**, *1902*, 020060. [CrossRef]
31. Souza, V.G.L. Thesis Plan Proposal: Development of a Novel Bionanocomposite Based on Chitosan/MMT with Antioxidant Activity for Food Appliance. Ph.D. Thesis, Universidade Nova de Lisboa, Lisbon, Portugal, 2015.
32. Briassoulis, D.; Innocenti, F.D. Standards for soil biodegradable plastics. In *Soil Degradable Bioplastics for a Sustainable Modern Agriculture*; Malinconico, M., Ed.; Springer: Cham, Switzerland, 2017; pp. 139–168. ISBN 9783662541302.
33. Rudnik, E. Biodegradation of compostable polymers in various environments. In *Compostable Polymer Materials*; Elsevier: Amsterdam, The Netherlands, 2019; pp. 255–287.
34. Agarwal, S. Biodegradable polymers: Present opportunities and challenges in providing a microplastic-free environment. *Macromol. Chem. Phys.* **2020**, *221*, 2000017. [CrossRef]
35. Pattanasuttichonlakul, W.; Sombatsompop, N.; Prapagdee, B. Accelerating biodegradation of PLA using microbial consortium from dairy wastewater sludge combined with PLA-degrading bacterium. *Int. Biodeterior. Biodegrad.* **2018**, *132*, 74–83. [CrossRef]
36. Rudnik, E.; Briassoulis, D. Degradation behaviour of poly (lactic acid) films and fibres in soil under Mediterranean field conditions and laboratory simulations testing. *Ind. Crop. Prod.* **2011**, *33*, 648–658. [CrossRef]
37. Sen, C.; Das, M. Biodegradability of starch based self-supporting antimicrobial film and its effect on soil quality. *J. Polym. Environ.* **2018**, *26*, 4331–4337. [CrossRef]

38. Pischedda, A.; Tosin, M.; Degli-Innocenti, F. Biodegradation of plastics in soil: The effect of temperature. *Polym. Degrad. Stab.* **2019**, *170*, 109017. [CrossRef]
39. Chan, M.Y.; Koay, S.C. Biodegradation and thermal properties of crosslinked chitosan/corn cob biocomposite films by electron beam irradiation. *Polym. Eng. Sci.* **2019**, *59*, E59–E68. [CrossRef]
40. Lammi, S.; Gastaldi, E.; Gaubiac, F. How olive pomace can be valorized as fillers to tune the biodegradation of PHBV based composites. *Polym. Degrad. Stab.* **2019**, *166*, 325–333. [CrossRef]
41. Šerá, J.; Serbruyns, L.; De Wilde, B.; Koutný, M. Accelerated biodegradation testing of slowly degradable polyesters in soil. *Polym. Degrad. Stab.* **2020**, *171*, 109031. [CrossRef]
42. Balakrishnan, P.; Geethamma, V.G.; Gopi, S.; George, M.; Huski, M.; Kalarikkal, N.; Volova, T.; Rouxel, D.; Thomas, S. Thermal, biodegradation and theoretical perspectives on nanoscale confinement in starch/cellulose nanocomposite modified via green crosslinker. *Int. J. Biol. Macromol.* **2019**, *134*, 781–790. [CrossRef] [PubMed]
43. D5988-18; Standard Test Method for Determining Aerobic Biodegradation of Plastic Materials in Soil. ASTM International: West Conshohocken, PA, USA, 2018.
44. 17556:19; Plastics—Determination of the Ultimate Aerobic Biodegradability of Plastic Materials in Soil by Measuring the Oxygen Demand in a Respirometer or the Amount of Carbon Dioxide Evolved. ISO: Geneva, Switzerland, 2019.
45. Kaur, K.; Jindal, R.; Maiti, M.; Mahajan, S. Studies on the properties and biodegradability of PVA/Trapanatans starch(N-st) composite films and PVA/N-st-g-poly(EMA) composite films. *Int. J. Biol. Macromol.* **2018**, *123*, 826–836. [CrossRef] [PubMed]
46. Ruggero, F.; Gori, R.; Lubello, C. Methodologies to assess biodegradation of bioplastics during aerobic composting and anaerobic digestion: A review. *Waste Manag. Res.* **2019**, *37*, 959–975. [CrossRef] [PubMed]
47. Thomas, S.; Shumilova, A.A.; Kiselev, E.G.; Baranovsky, S.V.; Vasiliev, A.D.; Nemtsev, I.V.; Kuzmin, A.P.; Sukovatyi, A.G.; Pai, R.; Avinash; et al. Thermal, mechanical and biodegradation studies of biofiller based poly-3-hydroxybutyrate biocomposites. *Int. J. Biol. Macromol.* **2019**, in press. [CrossRef]
48. Tai, N.L.; Adhikari, R.; Shanks, R.; Adhikari, B. Aerobic biodegradation of starch—Polyurethane flexible films under soil burial condition: Changes in physical structure and chemical composition. *Int. Biodeterior. Biodegrad.* **2019**, *145*, 104793. [CrossRef]
49. Al Hosni, A.S.; Pittman, J.K.; Robson, G.D. Microbial degradation of four biodegradable polymers in soil and compost demonstrating polycaprolactone as an ideal compostable plastic. *Waste Manag.* **2019**, *97*, 105–114. [CrossRef] [PubMed]
50. Sabapathy, P.C.; Devaraj, S.; Devaraj, S.; Kathirvel, P. Polyhydroxyalkanoate production from statistically optimized media using rice mill effluent as sustainable substrate with an analysis on the biopolymer´s degradation potential. *Int. J. Biol. Macromol.* **2019**, *126*, 977–986. [CrossRef]
51. Otoni, C.G.; Lodi, B.D.; Lorevice, M.V.; Leitão, R.C.; Ferreira, M.D.; De Moura, M.R.; Mattoso, L.H.C. Optimized and scaled-up production of cellulose-reinforced biodegradable composite films made up of carrot processing waste. *Ind. Crop. Prod.* **2018**, *121*, 66–72. [CrossRef]
52. Pinheiro, I.F.; Ferreira, F.V.; Souza, D.H.S.; Gouveia, R.F.; Lona, L.M.F.; Morales, A.R.; Mei, L.H.I. Mechanical, rheological and degradation properties of PBAT nanocomposites reinforced by functionalized cellulose nanocrystals. *Eur. Polym. J.* **2017**, *97*, 356–365. [CrossRef]
53. Bagde, P.; Nadanathangam, V. Mechanical, antibacterial and biodegradable properties of starch film containing bacteriocin immobilized crystalline nanocellulose. *Carbohydr. Polym.* **2019**, *222*, 115021. [CrossRef]
54. Borowicz, M.; Paciorek-sadowska, J.; Isbrandt, M.; Grzybowski, Ł. Glycerolysis of poly (lactic acid) as a way to extend the "life cycle" of this material. *Polymers* **2019**, *11*, 1963. [CrossRef]
55. Sforzini, S.; Oliveri, L.; Chinaglia, S.; Viarengo, A. Application of biotests for the determination of soil ecotoxicity after exposure to biodegradable plastics. *Front. Environ. Sci.* **2016**, *4*, 68. [CrossRef]
56. Ibrahim, H.; Mehanny, S.; Darwish, L.; Farag, M. A comparative study on the mechanical and biodegradation characteristics of starch-based composites reinforced with different lignocellulosic fibers. *J. Polym. Environ.* **2018**, *26*, 2434–2447. [CrossRef]
57. Huang, Z.; Qian, L.; Yin, Q.; Yu, N.; Liu, T.; Tian, D. Biodegradability studies of poly (butylene succinate) composites filled with sugarcane rind fiber. *Polym. Test.* **2018**, *66*, 319–326. [CrossRef]
58. Aleixo Moreira, A.; Mali, S.; Yamashita, F.; Bilck, A.P.; de Paula, M.T.; Merci, A.; Oliveira, A.L.M. de Biodegradable plastic designed to improve the soil quality and microbiological activity. *Polym. Degrad. Stab.* **2018**, *158*, 52–63. [CrossRef]
59. Miteluț, A.C.; Popa, E.E.; Popescu, P.A.; Rapa, M.; Popa, M.E. Soil ecotoxicity assessment after biodegradation of some polymeric materials. *Agron. Ser. Sci. Res. Stiinț. Ser. Agron.* **2019**, *61*, 538–543.
60. Pires, J.R.A.; Souza, V.G.L.; Fernando, A.L. Valorization of energy crops as a source for nanocellulose production—Current knowledge and future prospects. *Ind. Crop. Prod.* **2019**, *140*, 111642. [CrossRef]
61. Altun, E.; Çelik, E.; Ersan, H.Y. Tailoring the microbial community for improving the biodegradation of chitosan films in composting environment. *J. Polym. Environ.* **2020**, *28*, 1548–1559. [CrossRef]
62. Bonilla, J.; Sobral, P.J.A. Disintegrability under composting conditions of films based on gelatin, chitosan and/or sodium caseinate containing boldo-of-Chile leafs extract. *Int. J. Biol. Macromol.* **2020**, *151*, 178–185. [CrossRef]
63. Kalita, N.K.; Kumar, M.; Mudenur, C.; Kalamdhad, A. Biodegradation of modified poly (lactic acid) based biocomposite films under thermophilic composting conditions. *Polym. Test.* **2019**, *76*, 522–536. [CrossRef]
64. 14855-1:2012; Determination of the Ultimate Aerobic Biodegradability of Plastic Materials under Controlled Composting Conditions—Method by Analysis of Evolved Carbon Dioxide—Part 1: General Method. ISO: Geneva, Switzerland, 2012.

65. D5338-15; Standard Test Method for Determining Aerobic Biodegradation of Plastic Materials under Controlled Composting Conditions, Incorporating Thermophilic Temperatures. ASTM International: West Conshohocken, PA, USA, 2015.
66. 14855-2:2018; Determination of the Ultimate Aerobic Biodegradability of Plastic Materials under Controlled Composting Conditions—Method by Analysis of Evolved Carbon Dioxide—Part 2: Gravimetric Measurement of Carbon Dioxide Evolved in a Laboratory. ISO: Geneva, Switzerland, 2018.
67. DIN EN 13432; Requirements for Packaging Recoverable Through Composting and Biodegradation—Test Scheme and Evaluation Criteria for the Final Acceptance of Packaging. Australian Standards: Sydney, NSW, Australia, 2000.
68. AS 5810-2010; Biodegradable Plastics—Biodegradable Plastics Suitable for Home Composting. Australian Standards: Sydney, NSW, Australia, 2010.
69. 16929:2021; Plastics—Determination of the Degree of Disintegration of Plastic Materials under Defined Composting Conditions in a Pilot-Scale Test. ISO: Geneva, Switzerland, 2021.
70. Cinelli, P.; Seggiani, M.; Mallegni, N.; Gigante, V.; Lazzeri, A. Processability and degradability of PHA-based composites in terrestrial environments. *Int. J. Mol. Sci.* **2019**, *20*, 284. [CrossRef]
71. Suriyatem, R.; Auras, R.A.; Rachtanapun, P. Improvement of mechanical properties and thermal stability of biodegradable rice starch—Based films blended with carboxymethyl chitosan. *Ind. Crop. Prod.* **2018**, *122*, 37–48. [CrossRef]
72. Cano, A.I.; Cháfer, M.; Chiralt, A.; González-Martínez, C. Biodegradation behavior of starch-PVA films as affected by the incorporation of different antimicrobials. *Polym. Degrad. Stab.* **2016**, *132*, 11–20. [CrossRef]
73. Balaguer, M.P.; Aliaga, C.; Fito, C.; Hortal, M. Compostability assessment of nano-reinforced poly (lactic acid) films. *Waste Manag.* **2016**, *48*, 143–155. [CrossRef] [PubMed]
74. Prapruddivongs, C.; Apichartsitporn, M.; Wongpreedee, T. Effect of silica resources on the biodegradation behavior of poly (lactic acid) and chemical crosslinked poly (lactic acid) composites. *Polym. Test.* **2018**, *71*, 87–94. [CrossRef]
75. Sedničková, M.; Pekařová, S.; Kucharczyk, P.; Janigová, I.; Kleinová, A.; Jochec, D.; Omaníková, L.; Perďochová, D.; Sedlařík, V.; Alexy, P.; et al. Changes of physical properties of PLA-based blends during early stage of biodegradation in compost. *Int. J. Biol. Macromol.* **2018**, *113*, 434–442. [CrossRef] [PubMed]
76. Gutiérrez, T.J.; Toro-Márquez, L.A.; Merino, D.; Mendieta, J.R. Hydrogen-bonding interactions and compostability of bionanocomposite films prepared from corn starch and nano-fillers with and without added Jamaica flower extract. *Food Hydrocoll.* **2019**, *89*, 283–293. [CrossRef]
77. Castro-Aguirre, E.; Auras, R.; Selke, S.; Rubino, M.; Marsh, T. Enhancing the biodegradation rate of poly(lactic acid) films and PLA bionanocomposites in simulated composting through bioaugmentation. *Polym. Degrad. Stab.* **2018**, *154*, 46–54. [CrossRef]
78. D6400-21; Standard Specification for Labeling of Plastics Designed to be Aerobically Composted in Municipal or Industrial Facilities. ASTM International: West Conshohocken, PA, USA, 2021.
79. 17088:2021; Plastics–Organic Recycling–Specifications for Compostable Plastics. ISO: Geneva, Switzerland, 2021.
80. Zambrano, M.C.; Pawlak, J.J.; Daystar, J.; Ankeny, M.; Goller, C.C.; Venditti, R.A. Aerobic biodegradation in freshwater and marine environments of textile micro fibers generated in clothes laundering: Effects of cellulose and polyester-based microfibers on the microbiome. *Mar. Pollut. Bull.* **2020**, *151*, 110826. [CrossRef]
81. Harrison, J.P.; Boardman, C.; Callaghan, O.; Delort, A.; Song, J.; Harrison, J.P. Biodegradability standards for carrier bags and plastic films in aquatic environments: A critical review. *R. Soc. Open Sci.* **2018**, *5*, 171792. [CrossRef]
82. Bagheri, A.R.; Laforsch, C.; Greiner, A.; Agarwal, S. Fate of so-called biodegradable polymers in seawater and freshwater. *Glob. Chall.* **2017**, *1*, 1700048. [CrossRef]
83. Beltrán-Sanahuja, A.; Casado-Coy, N.; Simó-Cabrera, L.; Sanz-Lázaro, C. Monitoring polymer degradation under different conditions in the marine environment. *Environ. Pollut.* **2020**, *259*, 113836. [CrossRef] [PubMed]
84. Weber, M.; Unger, B.; Mortier, N. Assessing marine biodegradability of plastic—Towards an environmentally relevant international standard test scheme. *Proc. Int. Conf. Microplastic Pollut. Mediterr. Sea* **2018**, 189–193. [CrossRef]
85. Kjeldsen, A.; Price, M.; Lilley, C.; Guzniczak, E.; Archer, I. A Review of Standards for Biodegradable Plastics. 2019. Available online: https://assets.publishing.service.gov.uk/government/uploads/system/uploads/attachment_data/file/817684/review-standards-for-biodegradable-plastics-IBioIC.pdf (accessed on 12 February 2022).
86. Briassoulis, D.; Pikasi, A.; Papardaki, N.G.; Mistriotis, A. Aerobic biodegradation of bio-based plastics in the seawater/sediment interface (sublittoral) marine environment of the coastal zone—Test method under controlled laboratory conditions. *Sci. Total Environ.* **2020**, *722*, 137748. [CrossRef] [PubMed]
87. Lott, C.; Eich, A.; Unger, B.; Makarow, D.; Battagliarin, G.; Schlegel, K.; Lasut, M.T.; Weber, M. Field and mesocosm test methods to assess the performance of biodegradable plastic under marine conditions. *bioRxiv* **2020**. [CrossRef]
88. 18830:2016; Plastics—Determination of Aerobic Biodegradation of Non-Floating Plastic Materials in a Seawater/Sandy Sediment Interface—Method by Measuring the Oxygen Demand in Closed Respirometer. ISO: Geneva, Switzerland, 2016.
89. 19679:2016; Plastics—Determination of Aerobic Biodegradation of Non-Floating Plastic Materials in a Seawater/Sediment Interface—Method by Analysis of Evolved Carbon Dioxide. ISO: Geneva, Switzerland, 2016.
90. 14853:2016; Plastics—Determination of the Ultimate Anaerobic Biodegradation of Plastic Materials in an Aqueous System—Method by Measurement of Biogas Production. ISO: Geneva, Switzerland, 2016.
91. 23977-1:2020; Plastics—Determination of the Aerobic Biodegradation of Plastic Materials Exposed to Seawater—Part 1: Method by Analysis of Evolved Carbon Dioxide. ISO: Geneva, Switzerland, 2020.

92. *23977-2:2020*; Plastics—Determination of the Aerobic Biodegradation of Plastic Materials Exposed to Seawater—Part 2: Method by Measuring the Oxygen Demand in Closed Respirometer. ISO: Geneva, Switzerland, 2020.
93. *15314:2018*; Plastics—Methods for Marine Exposure. ISO: Geneva, Switzerland, 2018.
94. *22766:2020*; Plastics—Determination of the Degree of Disintegration of Plastic Materials in Marine Habitats under Real Field Conditions. ISO: Geneva, Switzerland, 2020.
95. *D6691-17*; Standard Test Method for Determining Aerobic Biodegradation of Plastic Materials in the Marine Environment by a Defined Microbial Consortium or Natural Sea Water Inoculum. ASTM International: West Conshohocken, PA, USA, 2017.
96. *D7473/D7473M-21*; Standard Test Method for Weight Attrition of Non-Floating Plastic Materials by Open System Aquarium Incubations. ASTM International: West Conshohocken, PA, USA, 2021.
97. *D7991-15*; Standard Test Method for Determining Aerobic Biodegradation of Plastics Buried in Sandy Marine Sediment under Controlled Laboratory Conditions. ASTM International: West Conshohocken, PA, USA, 2015.
98. Dilkes-Hoffman, L.S.; Lane, J.L.; Grant, T.; Pratt, S.; Lant, P.A.; Laycock, B. Environmental impact of biodegradable food packaging when considering food waste. *J. Clean. Prod.* **2018**, *180*, 325–334. [CrossRef]
99. *D5229/D5229M-20*; Standard Test Method for Moisture Absorption Properties and Equilibrium Conditioning of Polymer Matrix Composite Materials. ASTM International: West Conshohocken, PA, USA, 2020.
100. *62:2008*; Plastics—Determination of Water Absorption. ISO: Geneva, Switzerland, 2008.
101. Hassan, M.M.; Le Guen, M.J.; Tucker, N.; Parker, K. Thermo-mechanical, morphological and water absorption properties of thermoplastic starch/cellulose composite foams reinforced with PLA. *Cellulose* **2019**, *26*, 4463–4478. [CrossRef]
102. Jiménez-Rosado, M.; Zarate-Ramírez, L.S.; Romero, A.; Bengoechea, C.; Partal, P.; Guerrero, A. Bioplastics based on wheat gluten processed by extrusion. *J. Clean. Prod.* **2019**, *239*, 117994. [CrossRef]
103. Kasmuri, N.; Safwan, M.; Zait, A. Enhancement of bio-plastic using eggshells and chitosan on potato starch based. *Int. J. Eng. Technol.* **2018**, *7*, 110–115. [CrossRef]
104. Kumar Thiagamani, S.M.; Krishnasamy, S.; Muthukumar, C.; Tengsuthiwat, J.; Nagarajan, R.; Siengchin, S.; Ismail, S.O. Investigation into mechanical, absorption and swelling behaviour of hemp/sisal fibre reinforced bioepoxy hybrid composites: Effects of stacking sequences. *Int. J. Biol. Macromol.* **2019**, *140*, 637–646. [CrossRef]
105. Morohoshi, T.; Oi, T.; Aiso, H.; Suzuki, T.; Okura, T.; Sato, S. Biofilm formation and degradation of commercially available biodegradable plastic films by bacterial consortiums in freshwater environments. *Microbes Environ.* **2018**. [CrossRef]
106. Rameshkumar, S.; Shaiju, P.; Connor, K.E.O. Bio-based and biodegradable polymers—State-of-the- art, challenges and emerging trends. *Curr. Opin. Green Sustain. Chem.* **2020**, *21*, 75–81. [CrossRef]
107. Luyta, A.S.; Malik, S.S. Can biodegradable plastics solve plastic solid waste accumulation. In *Plastics to Energy Fuel, Chemicals, and Sustainability Implications*; Al-Salem, S.M., Ed.; Elsevier: Amsterdam, The Netherlands, 2019; pp. 403–423.

MDPI
St. Alban-Anlage 66
4052 Basel
Switzerland
www.mdpi.com

Polymers Editorial Office
E-mail: polymers@mdpi.com
www.mdpi.com/journal/polymers

Disclaimer/Publisher's Note: The statements, opinions and data contained in all publications are solely those of the individual author(s) and contributor(s) and not of MDPI and/or the editor(s). MDPI and/or the editor(s) disclaim responsibility for any injury to people or property resulting from any ideas, methods, instructions or products referred to in the content.

www.ingramcontent.com/pod-product-compliance
Lightning Source LLC
LaVergne TN
LVHW070409100526
838202LV00014B/1422